ULTIMATE AMERICANS

Also by Tom Lowenstein

Some Aspects of Sea Ice Subsistence Hunting in Point Hope, Alaska, 1980

The Things That Were Said of Them: Shaman Stories and Oral Histories of the Tikiġaq People, 1990

Ancient Land:Sacred Whale: The Inuit Hunt and Its Rituals, 1993

Ancestors and Species: New and Selected Ethnographic Poetry, 2005

ULTIMATE AMERICANS

Point Hope, Alaska: 1826–1909

Tom Lowenstein

University of Alaska Press
Fairbanks

P.O. Box 756240
Fairbanks, AK 99775-6240
www.uaf.edu/uapress

Printed in the United States

This publication was printed on paper that meets the minimum requirements for ANSI/NISO x39.48–1992 (Permanence of Paper).

Library of Congress Cataloging-in-Publication Data

Lowenstein, Tom.
Ultimate Americans : Point Hope, Alaska, 1826–1909 / Tom Lowenstein.
p. cm.
Includes bibliographical references.
ISBN 978-1-60223-027-9 (cloth : alk. paper) — ISBN 978-1-60223-038-5 (pbk. : alk. paper)
1. Inupiat—Alaska—Point Hope—History. 2. Inupiat—First contact with Europeans.
3. Russians—Alaska—Point Hope. 4. Whaling—Alaska—Point Hope. 5. Assimilation (Sociology)—Alaska—Point Hope. 6. Point Hope (Alaska)—History—19th century. I. Title.

E99.E7L694 2008
979.8'7—dc22 20

Cover design by Dixon Jones

CONTENTS

LIST OF ILLUSTRATIONS

"ESKIMO! WHITE MAN!"

Sometime in the 1920s, Asatchaq composed a song about Eskimos and white men that was one of about six hundred songs that he made during his long life in the village of Tikiġaq, Alaska. "Eskimo! White Man" is a humorous composition of the sort that accompanied a dance witnessed by the Englishman Frederick W. Beechey when he landed near Point Hope in August 1826 and, for the first time in European annals, identified the people of the north Alaskan coast as Eskimos (Bockstoce 1977:16).

Even by the standards of the condensed and allusive songs of north Alaska, Asatchaq's composition is remarkable for its brevity. In addition to the *aya-ya* refrain syncopated against a regular drum beat, it contains just two words:

Eskimo! Naluaġmiu!

The first word, *Eskimo*, comes originally from the eastern Canadian Montagnais Indian language, but so far as Alaskan Eskimos were ever concerned, it is an English term. The second word, *naluaġmiu*, is what nineteenth-century Point Hope people called the white man. Broken into its two parts it is *naluaq* (bleached sealskin) + *miu* (person of place or type). White men were therefore "People of the Bleached Sealskin": scraped seal's hide that has been pegged on the ground and exposed to the sun preparatory to boot making. The witty character of the song, however, lies in the absurd fact that while the Native American is invoked with a white man's term ("Eskimo!"), the white man is proclaimed through a local coinage ("Naluaġmiu!").

One word is, perhaps intentionally, missing from this exclamatory call, which accompanied a dramatic and humorous dance. The missing word is the indigenous name for the Eskimos themselves. The Eskimos of north Alaska call themselves Iñupiaq (plural Iñupiat), which comes from *iñuk* (person) and the suffix *-piaq* (real, genuine). "Genuine people" regarded themselves as real human beings as opposed to animals, spirits, Alaskan Natives outside Iñupiaq territory, and, more recently, Euro-Americans, whom Point Hope came to know in the nineteenth century.

The song thus encapsulates a historical situation. Born in 1891, Asatchaq grew up with knowledge of traders, commercial whalers, and missionaries. Ships came and went, traders settled on Point Hope's south shore, the medical missionary John Driggs opened a school, which Asatchaq attended. Threatened with extinction through famines and the diseases brought by the white man, the Iñupiat were forced into partial dependence on imported goods and services. This was the paradoxical nature of Point Hope life during Asatchaq's childhood. The ironies were many, but Asatchaq's song expresses one of the most pungent, and his song playfully identifies the "old self" of the Iñupiaq through the use of the white man's term, while conversely invoking, with an Iñupiaq coinage, the new and "not quite real" people. The convergence and mutual assimilation of two very different worlds whose paradoxes are suggested so succinctly in Asatchaq's song is the subject of this study, which I respectfully dedicate to the composer, his ancestors, and today's Point Hope community.

PREFACE

> The Europeans introduced among the savages of North America firearms, ardent spirits, and iron; they taught them to exchange for manufactured stuffs the rough garments that had previously satisfied their untutored simplicity. Having acquired new tastes, without the arts by which they could be gratified, the Indians were obliged to have recourse to the workmanship of the whites; but in return for their productions the savage had nothing to offer except the rich furs that still abounded in his woods. Hence the chase became necessary, not merely to provide for his subsistence, but to satisfy the frivolous desires of Europeans. He no longer hunted merely to obtain food, but to procure the only objects of barter which he could offer. While the wants of the Natives were thus increasing, their resources continued to diminish.
>
> —Alexis de Tocqueville, *Democracy in America*

This is the third volume in a series of books about Tikiġaq (Point Hope), an Iñupiaq village on the Northwest Coast of Alaska where I worked with village elders on nine field trips between 1973 and 1989. The first volume in the series, *The Things That Were Said of Them*, was an edition of shaman stories translated from recordings made by the storyteller Asatchaq (1891–1980). That book was followed by *Ancient Land:Sacred Whale*, an evocation of ceremonials surrounding Point Hope's annual whale hunt. The present volume, in complement to the two earlier books about traditional life, explores what happened when Point Hope people met Euro-Americans.

The main action of this book takes place in the nineteenth century at a time when large areas of mainland North America had already been occupied by white settlers, but when wars, epidemics, and territorial appropriations continued to reduce Native populations. The impact of culture contact in Alaska was similar in many respects. But whereas it was usually the purpose of settlers on the U.S. mainland to acquire Native territory, few Alaskan Natives were deliberately displaced from their homes. Aleut people, who were the first to suffer the impact of Russian traders (1741–1867), represented a major exception. Even here, it was not the Aleutian Islands themselves but labor and sexual favors from women that the traders wanted; land itself, throughout Alaska, remained marginal to the white man's interest. Even in the gold-mining and administrative centers of Russian southeast Alaska, Tlingit Indians continued to occupy their territories. Similarly in arctic Alaska, land, as opposed to animal resources, was of minimal interest. And despite late-nineteenth-century Iñupiat migrations during a period of game depletion, the majority of the Eskimo populations of north Alaska similarly continued to live in their traditional territories.

Fig. 1. *Asatchaq (Jimmie Killigivuk) and Pauyuuraq (Jerry) on board the* Karluk, *summer 1913. Asatchaq and his family were major informants in this study. For Pauyuuraq's story, see Chapter 9. Photo gifted to the author by William Laird McKinlay, meteorologist to Stefansson's Arctic Expedition of 1913–18.*

While the Russians had traded in the southern area of Alaska since the 1740s, major contact between Iñupiat and Euro-Americans began only when commercial hunters pushed north through the Bering Strait in pursuit of the bowhead whale for oil and baleen ("whalebone") after the 1850s. By the time kerosene replaced whale oil as a lighting fuel, and baleen was displaced by spring steel in 1905, the bowhead population had been seriously diminished.

The decimation of the whale was of a piece with the destruction of the American buffalo. And just as the buffalo hunt represented an assault on Plains Indian culture, so the commercial whale hunt deprived Iñupiat hunters of subsistence and, in the course of this, undermined their cultural integrity. It was into this destabilized environment that the first missionaries to north Alaska arrived in the summer of 1890. By this date, epidemic diseases transmitted by commercial whalers had already disastrously reduced many Native populations, and it was the sincerely held purpose of the missionary societies who sent their representatives to Point Hope, Barrow, and Cape Prince of Wales to secure for the survivors a more stable, and of course Christianized, future.

But destructive to Native integrity as these diverse processes were, the Iñupiat seldom reacted to the white man in anger or hostility. There were no wars, and apart from the Gilley massacre of 1877 and some short-term conflict with missionary teachers at Cape Prince of Wales, there was almost no violence. Native people were, of course, largely impotent in the face of the big ships and superior firepower. With their numbers reduced, they were forced into accepting European goods in return for local resources for whose depletion the traders were themselves responsible. Commerce in tobacco and liquor also debilitated Native health, yet it was these very commodities that often most powerfully drew people to the traders. While the white man could come and go, Iñupiat were trapped in relationships on which they came to depend. Some of these contacts were overtly, sometimes violently, abusive; others were more friendly and useful to Natives. The purpose of this study is not, however, to launch an attack on nineteenth-century Euro-Americans. The men engaged in the commercial whale hunt were representatives of an age in which hunting for profit was an acceptable part of economic activity. What did ultimately outrage the American conscience was the largely inexplicable epidemiological side effects of contact and the abuse of Native women. It was these issues that the missionary initiative, in part, set out to address.

Uneasy feelings about depletion of Native resources had existed for some decades among a significant minority. Congress tried to legislate for the control of the buffalo hunt in 1870, but the attempt failed and was not repeated. This failure opened the way to unregulated assaults on both the northern and the southern herds, so it was scarcely surprising that no quota was applied to the whale hunt during the same period. Both buffalo and whale, along with the sea otter in the northern Pacific and the beaver in the Rocky Mountains, were thus threatened with extermination. As James Fenimore Cooper showed in his famous account of the passenger pigeon hunt in *The Pioneers*, there were too many reckless men with guns, while scrupulous woodsman like the fictional Natty Bumpo were few. In historical fact, the result by 1900 was the annihilation of America's most numerous bird species, estimated originally to have stood at five billion (J. Fenimore Cooper 1823: Chapter 22).

Knowledge of the whale hunt, as conducted by Americans, focused on the sperm whale hunt in the southern hemisphere, as both dramatized and condemned by Herman Melville. But Melville also knew about the more northerly species, and he included a warning about this whale's fate in one of the visionary meditations in *Moby-Dick* (see Chapter 5). But just as the crew of the *Pequod* are characterized in terms of heroic, if doomed, splendor, so the roughneck whalers who plied the Arctic were, on the whole, perceived by the American public either as unchurched and irredeemable scamps or as gallant adventurers engaged in patriotic commerce.

This commercial heroism was celebrated by Commander Charles H. Stockton, whose U.S. Navy frigate *Thetis* cruised to north Alaska in the summer of 1889, and who described the whaling masters he encountered as "a fine body of American seamen" (Stockton 1890; see Chapter 6). Early missionary observers, on the other hand, saw whaling crews as drunkards, jailbirds, and thieves, coming from a class of "roustabouts" who were "the

Fig. 2. *Anaqulutuq (second left), his wife Mammaŋniiña, their children, and an anonymous Iñupiaq on board the USS* Thetis, *summer 1889. This is the first surviving photograph of named Point Hope individuals. Anaqulutuq, wearing Siberian reindeer skin and bowler hat, and his family are dressed in new and costly materials. Photograph most likely by Commander Stockton, who appointed Anaqulutuq as Point Hope's "head man" that summer. Archives of the Episcopal Church, Austin, Texas*

lowest of the low" (Hare 1873). There is rather more truth in the second of these views, although both are somewhat exaggerated.

To his credit, it was the same Commander Stockton who first alerted church authorities to the dereliction that had, since the 1850s, been wreaked on Native Alaskans by commercial whalers. Since the only organizations with the interest and the will to accomplish anything in these regions were the churches, and because Stockton himself was a well-known and influential Episcopalian, his message—coinciding with the energetic plans of the Presbyterian Sheldon Jackson to evangelize Alaska—had quick effect in Point Hope, Barrow, and Cape Prince of Wales.

The ambiguities inherent in such a project were by no means obvious even to the most intelligent and well-meaning nineteenth-century churchmen. The complicated and fascinating career of Point Hope's first missionary teacher, John Driggs, is told here in sufficient detail to bring light, if not resolution, to the difficult issue as to how far Euro-American religious traditions could be reconciled with Iñupiaq worldviews and lifeways. The manner in which Point Hope responded to both goods and ideas brought by traders and churchmen constitutes a major part of this story.

The heart of this book thus lies in the threefold interaction among Native people, commercial whalemen, and the missionaries. In addition to an account of John Driggs's career, two long studies accompany this history. Chapters 9–12 contain a description of how Point Hope's "trading chief," Ataŋauraq, a shaman who began by mediating commerce between Point Hope and the whalers, abused his power, tyrannized his own community in the 1880s, and was eventually murdered. The second study (Chapters 23 and 24) examines the millenarian or revivalist movement (*uivvaqsaat*) that entered north Alaska in the 1880s with apocalyptic visions of salvation and which promised the return of ancestors and victims of disease. *Uivvaqsaat* is explored in some detail since it represented a direct response to the epidemic crisis. In presenting *uivvaqsaat,* I have also given an account of the Kobuk River visionary Maniiḷaq, whose teachings paralleled those of *uivvaqsaat* and whose influence reached both Point Hope and Barrow. Stories about Maniiḷaq's youth, which express the most subtle dimensions of Iñupiaq spirituality, are also given. Two larger-than-life shamanistic figures thus bestride this study. Maniiḷaq, who dissented from traditional shamanism, and Ataŋauraq, who reinvented himself as an American trader, were, in their different ways, both influential in preparing the ground for twentieth-century change and creating a milieu in which the missionaries would find success.

While alcohol and contact diseases destabilized populations in the nineteenth century, the period covered by this study had its happier conclusion. Point Hope and other Iñupiaq societies endured the afflictions experienced by almost all Native Americans. But depleted as Point Hope became, the community survived, and this book ends at the moment in 1909 when both commercial whaling and epidemic visitations had more or less run their course. This date also coincided with the deposition of John Driggs and the arrival of the missionary Augustus Hoare. The book concludes with Hoare's destruction of Point Hope's ancestral burial grounds, when human relics of many centuries were removed to a newly consecrated area. But if this was one kind of ending, it was by no means the endpoint to either Point Hope's population or its traditions. The community survived the upheavals of the nineteenth century, and today's population of almost eight hundred people, which continues to build with confidence on its long inheritance, is testimony to the endurance of this ancient people. It is this long endurance that has made the writing of this history possible.

ACKNOWLEDGMENTS

This book emerges from my own contact with the people of Point Hope, Alaska, and it is to this community, both past and present, that I owe my greatest debt. The hospitality and good humor of the Point Hope people have sustained me for more than thirty years, and I owe special thanks both to the present community and to those who have passed away since I lived and worked with them between 1973 and 1988. Among the elders to whom I continue to be grateful are Asatchaq Jimmie Killigivuk, Piquk Solomon Killigivuk, Dinah Aviq Frankson, Umigluk David Frankson, Sunshine and Qimiuraq Bob Tuckfield, Roy and Maligian Beatrice Vincent, Tukummiq Carol Omnik, George Omnik, Patrick Attungana, Uyaġaaġluk Laurie Kingik, Ayagiiyaq Lucy and Saġana Charlie Jensen, Aġniiñ Lily and Qiliġniq Donald Oktollik, John and Mollie Oktollik, Uqpik Augustus Kowunna, Daisey Oomittak, Walter Kowunna, and Tigluk Kirk Oviok. Of my friends and contemporaries in Point Hope, I would also like to thank Rex and Piquk Tuzroyluk, Elijah and Dorcus Rock, Rex Rock, Konrad Killigvivuk, and Peter Lisbourne.

To friends and colleagues I also owe thanks for help, advice, and information. First, since our meeting in Point Hope in 1975, to John R. Bockstoce, historian of all things Arctic. The work of Ernest S. "Tiger" Burch has also been indispensable. To Patricia Partnow I also owe special thanks. Her friendship and inspiration over more than thirty years has been constant. I have similarly profited from the friendship and linguistic knowledge of Lawrence D. Kaplan, of the Alaska Native Language Center at the University of Alaska. I owe equal thanks to Richard Bready, who has provided me with constant encouragement. His editorial input into many chapters made him a true collaborator in this project. I am also most indebted to Karin Rosenberg for identifying local historical societies of Delaware that led to contact with the Lamb and Turner families of Smyrna, Delaware, John Driggs's collateral descendants who provided me access to their great-uncle's invaluable papers. To these new friends in Delaware, I owe special thanks.

My work would have been impossible without the support of the staff of the Elmer E. Rasmuson Library, in particular Mary A. Larson and Rose Speranza; the archivists of the Episcopal Church, Austin, Texas; and to the late Terence Armstrong and Piers Vitebsky, currently director of anthropology at Scott Polar Research Institute. To Dr. Mary Larson I owe particular thanks for help in final revisions.

Others who have aided my work have been Dr. Robert Fortuine, Dr. Molly Lee, and Jonathan King. Thanks to Brigid MacCarthy, Nigel Wheale, John Welch, and to my daughters Elizabeth and Anna, who can't remember when I wasn't "writing my book."

I am also most grateful to the following organizations for funding research since 1975: the Alaska Humanities Forum, the Alaska Historical Commission, the British Academy, the Nuffield Foundation, the Society of Authors, the Arctic Institute of North America, the American Philosophical Society, and the John Simon Guggenheim Memorial Foundation. The British Museum's Department of Africa, Oceania, and the Americas has also greatly encouraged this work.

GLOSSARY

Alaska	Aleut word signifying "the Great Land." Territory purchased from the Russians in 1867. Statehood achieved in 1959.
Bear	The second U.S. Marine Revenue Cutter to cruise arctic Alaska
beluga	*Delphinapterus leucas*; white, dolphinlike whale
Bering Strait	The narrowest stretch of the Bering Sea dividing Alaska from Siberia
bowhead	*Balaena mysticetus*; large toothless baleen or "whalebone" whale
Chukchi	Native of Siberian coast opposite Alaska
Coast Guard	U.S. Revenue Cutter Service, an arm of the U.S. Treasury
contact	English word used to describe encounter between Native Americans and Euro-Americans
Corwin	The first U.S. Revenue Cutter to cruise arctic Alaska
Euro-American	Caucasian American, as opposed to the first Europeans in Alaska
iñua	Spirit owner, as in *tatqim iñua*, "moon spirit"
Iñupiaq (pl. Iñupiat)	"Real person"; North Alaskan Eskimo person and language. Because of the number of Iñupiaq terms in this book, "Eskimo" is also used for the sake of clarity.
Jabbertown	Trading station on Point Hope's south shore, 1887 to ca. 1914
maktak	Whale skin with blubber attached
***-miu* (pl. *-miut*)**	Suffix indicating affiliation to a social group
nuna	Land
orca	*Orcinus orca*; previously known as killer whale
the Point	The tip of the Point Hope peninsula
Point Barrow	Or Barrow; most northerly Iñupiaq village in Alaska
Revenue Cutter	Ship in U.S. territorial waters to enforce law and protect U.S. commercial interests and Alaskan Natives
shaman	Tribal mystic, healer, "medicine person"
Tikiġaq	Point Hope, Alaska; "Tigara" was a nineteenth-century spelling
Tikiġaġmiu (pl. *-miut*)	Point Hope resident
***umiaq* (pl. *umiat*)**	Eight-person open skinboat
***umialik* (pl. *umialit*)**	Whale-boat owner; man or woman of wealth

BRIEF CHRONOLOGY

c. 0–500 AD	Pre-Iñupiaq Ipiutak people colonize the south shore of the Point Hope peninsula. Ipiutak site "moved" as result of erosion to north side of peninsula.
600–800	Whale hunting made possible by the introduction of harpoon and drag-float technology
1741	Vitus Bering reaches Alaska
1743	Russian hunters arrive to exploit sea otter in southwest Alaska
1778	Captain James Cook reaches Icy Cape on Alaskan northwest coast
1780	Russian-American Company established
1784	First Russian settlement on Kodiak Island
1800	Tikiġaq (Point Hope) suffers defeats in regional warfare
1816–21	Russian naval expeditions to arctic Alaska
1826	Captain F. W. Beechey visits Point Hope peninsula
1848	Thomas Roys discovers bowhead whale herds north of Bering Strait
1867	U.S. purchase of Alaska
1877	U.S. Army abandons Alaska
1880s	Ataŋauraq's period of authority in Point Hope
1884	Alaska Organic Act brings civil government to Alaska
1884	Charles Brower in Point Hope
1887	Jabbertown whaling station established on Point Hope's south shore
1888	First arctic cruise of the Revenue Cutter Service
1889	Protestant Comity Agreement to send missionaries to Alaska
1889	Death of Ataŋauraq
1890	Arrival of missionaries at Point Hope, Point Barrow, and Cape Prince of Wales
1899	Gold discovered on Nome beaches
1900	The Great Sickness
1905	Disintegration of Jabbertown and the decline in arctic whaling
1908	John Driggs deposed. Arrival in Point Hope of Augustus Hoare.
1914	John Driggs dies

HISTORICAL CHARACTERS

Anaqulutuq (born ca. 1860): Point Hope "headman," friend of John Driggs
Aniqsuaq (died 1889–90): One of Ataŋauraq's assassins
Asatchaq (Jimmie Killigivuk) (1891–1980): Point Hope storyteller
Ataŋauraq (died 1889): Point Hope shaman and "chief"
Baranov, Alexander: Manager of the Russian-American Company in Alaska
Bayne, Peter: Born New Brunswick, Canada, whaler-trader at Jabbertown, 1887–1910
Beechey, Frederick William: Commander of HMS *Blossom*, in north Alaska, 1826
Belcher, Lieutenant Edward: Officer of HMS *Blossom*
Bodfish, Hartson: Whaler based in New Bedford and San Francisco
Brower, Charles: Whaler-trader, Point Hope and Point Barrow
Call, Dr. Samuel J.: Physician on U.S. Revenue Cutter *Bear*
Chapman, Reverend John: Established Anvik mission, Yukon, 1887
Cook, Captain James: Commander of *Resolution* and *Discovery* in north Alaska, 1778
Davis, Major General Jefferson: Commander of U.S. Army garrison, Sitka, 1867–77
Driggs, John Beach M.D.: Missionary to Point Hope, 1890–1908
Edson, Reverend Elijah: Visiting missionary to Point Hope, 1894–96
Franklin, Sir John: Arctic explorer of Northwest Passage, 1825–27
Hachmann, Johannes: Born Westphalia, Jabbertown whaler-trader, 1887–1910
Haverside: Captain of the supply steamer *Oscar and Hattie*
Healy, Michael: Commander of the U.S. Revenue Cutter *Bear*
Herman, Father: Russian Orthodox missionary on Kodiak Island, 1794
Hooper, Captain C.L.: Commander of U.S. Revenue Cutter *Corwin*
Jackson, Reverend Dr. Sheldon: General Agent of Education in Alaska
Kayuktuq: One of Ataŋauraq's assassins
Kelly, John W.: Jabbertown whaler-trader, author of booklet on arctic Alaska
Kiḷigvak (born ca. 1870): Whale-boat owner, husband of Niġuvana
Kimber, Reverend Joshua: Associate Secretary of the Episcopal Board of Missions
Klengenberg, Charlie: Jabbertown whaler-trader
Knapp, E. J.: New York lay preacher, visiting missionary to Point Hope
Knowles, Josiah: Director of the Pacific Steam Whaling Company, San Francisco
Koenig, Heinrich: Born Westphalia, Jabbertown whaler-trader, 1887–1910
Kotzebue, Otto von: Naval officer employed by Russian government, 1816
Langford, Dr. William S.: Secretary to the Episcopal Board of Missions, Church House, New York City, to whom Driggs's correspondence was addressed
Lopp, Thomas: Schoolteacher at Cape Prince of Wales, 1890

MacGregor, Captain: Of the whaler *Narwhal*
Maguire, Captain Rochfort: Commander of HMS *Plover* at Barrow, 1852–54
Maniiḷaq: Visionary/shaman/proto-Christian Kobuk Eskimo, active 1875
Marlin, Charlie: Jabbertown whaler-trader
McFarland, Amanda: Philanthropist and teacher at Wrangell, Alaska, 1877
McKenna, James: San Francisco–based whaler-trader
Muir, John: Naturalist and writer, traveled on *Corwin*, 1881
Nelson, Edward W.: Ethnographer, traveled on *Corwin*, 1881
Newth, Edwin W.: San Francisco–based whaler-trader
Niġuvana (born ca. 1875): Kiḷigvak's wife, Asatchaq's mother
Peard, Lieutenant George: Officer on HMS *Blossom*, 1826
Rainey, Froelich: Archaeologist and ethnographer at Point Hope, 1939–40
Rosse, Irving: Physician on *Corwin*
Rowe, Peter Trimble: Episcopalian Bishop of Alaska, appointed 1895
Samaruna (born ca. 1870): Niġuvana's brother, Asatchaq's uncle
Simpson, Dr. John: Physician on HMS *Plover*, Barrow, 1852–54
Stockton, Captain Charles H.: Of the U.S. Navy ship *Thetis*
Stuck, Hudson: Archdeacon of the Yukon and the Arctic Coast, 1904–20
Suuyuk the elder (born ca. 1850): Point Hope shaman
Suuyuk the younger (born ca. 1875): brother of Kiḷigvak, Ataŋauraq's victim
Thornton, Harrison: Missionary teacher, Cape Prince of Wales, 1890; killed 1893
Tuckfield, Joseph: Born Swansea, Wales; Jabbertown whaler-trader, 1889–1910
Veniaminov, Father Ivan: Missionary to Aleuts, 1824; Orthodox archbishop, Sitka
Woolfe, James: Trader and whaling agent at Point Hope and Point Barrow, 1884–1900

MISSIONARIES AT POINT HOPE 1890–1923

Dr. John Beach Driggs	1890–1908 (ordained as deacon of the church, 1903)
Reverend Elijah H. Edson	1894–95 (during Driggs's furlough)
Edward J. Knapp	1904–5
Reverend Augustus R. Hoare	1908–20
Reverend Frederick W. Goodman	1911–12 (during Hoare's furlough)
Reverend William A. Thomas	1918–19 (during Hoare's furlough)
Reverend William A. Thomas	1920–25
Reverend Frederick W. Goodman	1922–23

SPELLING AND TERMINOLOGY

Standardized north Alaskan orthography is used for Iñupiaq names and terms. Those transcribed by writers who predated standard twentieth-century spelling have mostly been converted. Some earlier spellings are retained where that older orthography contributes to a sense of the original writer's character. Given the large number of Iñupiaq terms in this study, modern place-names are used where possible in the interest of readability. And while the term "Iñupiaq" is used when necessary, "Eskimo" frequently replaces this.

PRONUNCIATION GUIDE

Iñupiaq names and terms are given in the standard orthography that is in general use throughout the Alaskan Iñupiaq region. For a detailed discussion of this see MacLean 1980.

Vowels

a is pronounced *uh* as in English "cut"
i is pronounced *ee* as in English "see"
u is pronounced *oo* as in English "boot"

The double vowels *aa, ii, uu* are roughly similar to their short counterparts, but the sound is lengthened.

In the Point Hope dialect, most vowel clusters are pronounced like a combination of the short vowels they contain. But *ai* is pronounced like the name of the letter *a,* as in English "say"; *au* is pronounced like a long *o* as in "so."

Consonants

Most Iñupiaq consonants correspond closely to English equivalents. The major exceptions are:

q	a uvular or back *k*, pronounced in the upper throat
ġ	a uvular or back *g*, like a French or German *r*
ḷ	a palatal *l*, similar to the *ll* in "million"
ñ	a sound like the Spanish ñ, or the first *n* in "onion"
ŋ	a sound like the *ng* in "singer"

Consonants also occur long, and these are written double, as in *Uivvaq,* Cape Lisburne; *qulliq,* "seal oil lamp"; *taaŋŋaq,* "tonic; liquor."

Stress

Iñupiaq words with only short vowels are stressed on the initial syllable: *nu*-na, "land"; *Ti*-kiġaq, "Point Hope"; *qal*-gi, "ceremonial house." Syllables containing a long vowel or a vowel cluster receive stress: u-nip-*kaaq,* "legend"; Ataŋ-*au*-raq; um-*ia*-lik, "skinboat owner."

Quick Pronunciation Guide to Selected Names and Terms

aŋatkuq	*ung*-ut-kok
Asatchaq	*Uh*-suh-chuk
Ataŋauraq	Uh-tung-*oh*-ruk
iñua	i-ny*oo*-uh
Iñupiaq	I-ny*oo*-pyuk
qalgi	*kul*-gi
Tikiġaq	*Ti*-ke-ruk
umialik	oo-*may*-lik

Map 1. Northwest Alaska

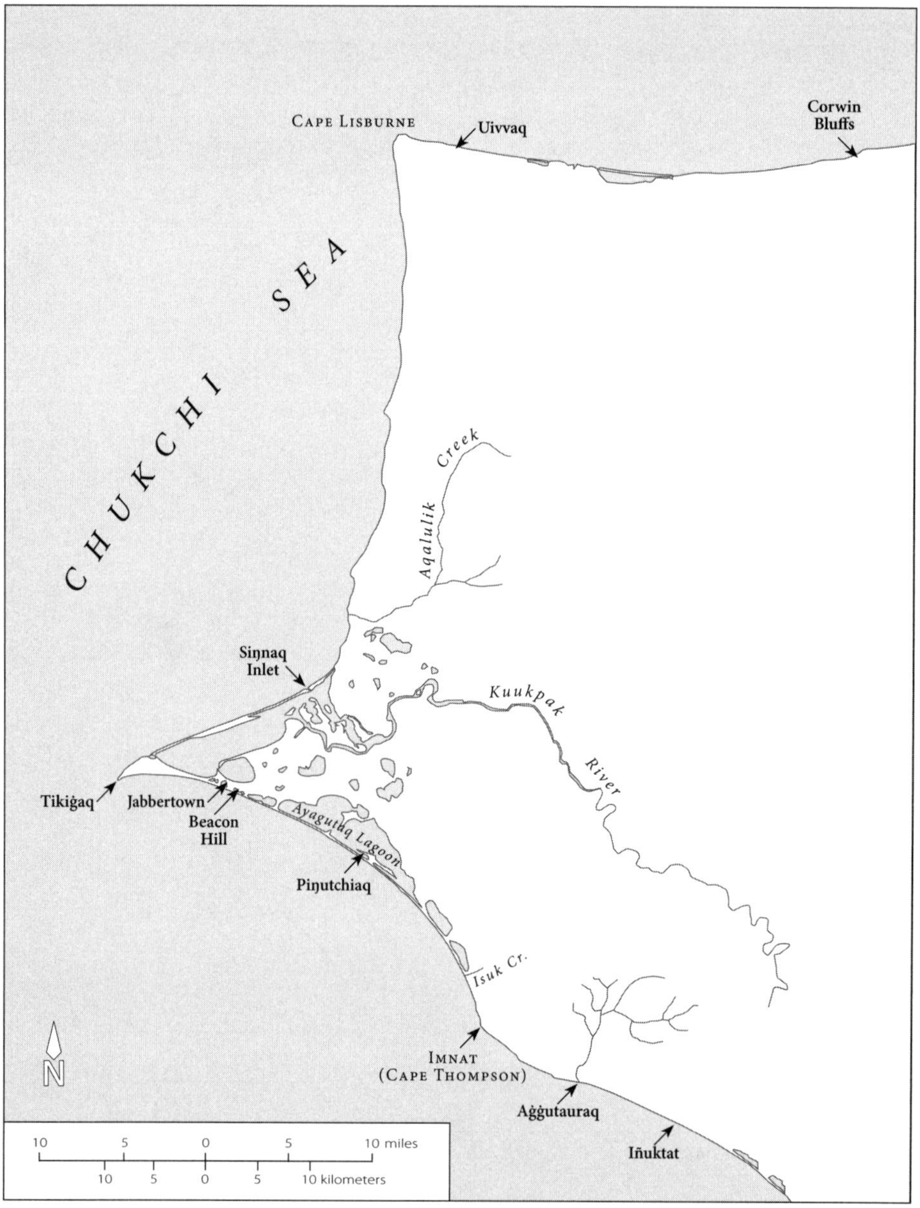

Map 2. The Point Hope Region

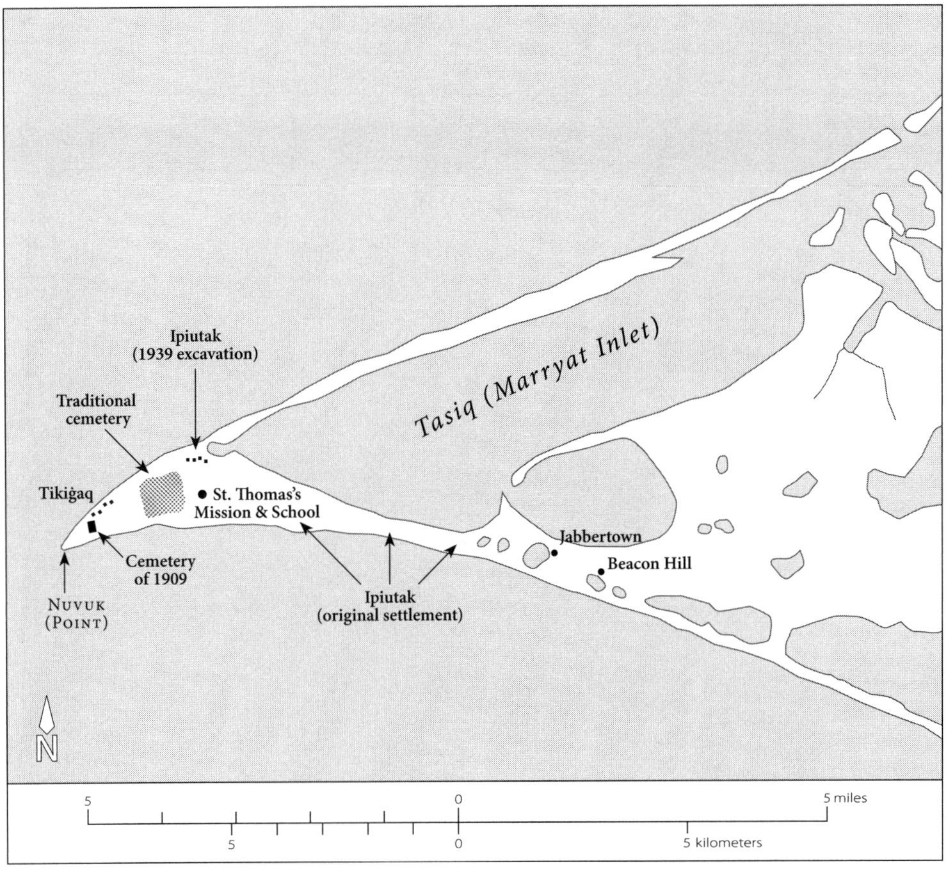

Map 3. Point Hope Peninsula

CHAPTER 1

Introduction: First Encounters

By the Cliffs Before Contact

August 5, 1826, was a mild day and some fifteen Iñupiaq people were camping on the beach about twenty-five miles from their main village in one of the most pleasant spots in their vast northwest arctic territory. Two or three iglus that had been excavated into a grassy bluff were screened from the beach by a cluster of caribou-skin tents that stood leaning away from the south wind. Nearby, a stream called Isuk rushed through a bank of last year's impacted snow and ran across the beach into the sea. Most of the families camping here would return to the village in the autumn to await the sea ice and the winter seal hunt, but a few older people would remain at Isuk, living quietly on the meat, fish, and berries they had put away that summer in storage pits sunk into the bluffs.

Surrounding the camp lay a mass of equipment: skin boats, kayaks, harpoons, fishnets, coils of sealskin rope, and a clutter of little tools for the making and mending of yet more equipment. Awaiting transformation into boots and parkas, bedding, and boat skins lay seal, caribou, and walrus skins. Sea mammal carcases—fresh, half butchered, some half putrid—lay among skeletal remains from past seasons' hunts. Higher on the bluff rose driftwood racks where the women had draped sliced meat to dry in the wind that had scarcely stopped blowing since the last Ice Age. Hunting, cutting, drying, and preserving were a daily labor. Bags made from hollowed-out seal stood by the tents filled with dried meat, eggs, birds steeped in seal oil or blueberries and meat chips preserved in caribou back fat. Tethered at a distance from this larder were the dogs that had dragged the skin boats along the inshore water from the village. And all this work and clutter was framed by heaps of gray, weathered tree trunks: spruce, birch, and cottonwood from the southern rivers which had drifted up coast and which the sea had been disgorging here for innumerable centuries.

On an August day like this, the work of hunting and accumulation was mixed with hours of enjoyment. The temperature hovered between forty and fifty degrees Fahrenheit. For Tikiġaq people, who experienced winter temperatures of minus thirty (and far lower given windchill), this was warm. The children ran barefoot, the little ones naked. Adults shed one layer of their double caribou-skin parkas and trousers. "*Siḷagiiksuq*!" ("beautiful weather"), people would murmur. Such days of sunshine and a light south wind would soon be replaced by violent equinoctial storms and then freeze-up in November.

The peace at Isuk was by no means silent. At this spot by the stream, the beach was cut off by a long stretch of cliffs—the northwestern end of the Rocky Mountain cordillera—where some fifteen kinds of seabirds nested in the summer. Half a million murres (a species of guillemot) laid their blue- and black-streaked eggs on the sandstone and limestone ledges among gulls, terns, kittiwakes, and fulmars. Puffins burst in and out of holes in the softer mudstone strata, while peregrine falcons, ravens, and golden eagles nested on the higher ledges. For the campers at Isuk, these species were a source of food and clothing. While some people swarmed up the cliffs and carefully descended with eggs stuffed into their parkas, others dangled their partners from the cliff top on ropes, or lay in the grass and scooped birds in pole nets.

Iñupiaq life was often balanced between fierce activity and contemplative periods of doing little. A visitor to Isuk might sit through hours or even days when nothing seemed to happen, and when nobody did much except scrutinize the sea and comment on the wind direction, or ramble to the cliff top, gaze inland towards the Kuukpak River, or try to make out on the western horizon the point where the main village, Tikiġaq (Point Hope), lay at the tip of its peninsula. These interludes in summer daylight hours were interspersed with haphazard sleeps, making and mending, the bloody and oily business of meat preparation, and, not least, telling stories. The men had the most freedom. The women probably worked harder. They tended the dogs, sewed, butchered animals, cleaned skins, and dealt with the children. Sometimes visitors arrived from inland or by sea. Partly they came to drink from Isuk, whose water was particularly pure and delectable. After food, goods, and news had been exchanged, there were dances and games. People scrambled up the cliff for eggs or just waited in silence.

When activity came it was often sudden. The nets stretching from the beach would stand empty for days. When belugas arrived, the thick skin mesh would thrash violently with struggling white, dolphinlike whales. When the wind changed direction, the men might take off inland to hunt caribou, returning with backpacks of meat and skins, repeating their journey to fetch what they had left and to visit the traps they had left for foxes and marmots.

Things also happened for people simply to observe. Grizzlies ambled from the ridges where they'd been browsing for berries and excavating squirrel burrows to scavenge for dead seals or walrus. A gray whale, not worth chasing for its thin layer of blubber, would swim into view in the middle distance. Or an orca would swerve inshore to attack a seal. The women would laugh, throw stones, and shout, "When you have eaten, bring us a share!" What habitually came round the cliffs belonged to a world that was known and which could be interpreted. The animals were always welcome. But people from beyond the cliffs were often dangerous.

Isuk ("End") and the Nearer Rounding Place

The south beach runs uninterrupted from the Tikiġaq Point to the cliffs, and Isuk, which means "End," was one of the last in a series of small beach settlements. The cliffs represented a multitude of sites, and twenty-five place-names between Isuk and the southern edge of the headland identify creeks, rookeries, slate and clay deposits, and trails inland. The whole promontory, which juts dramatically into the sea, was also known as the "Nearer

Rounding Place." This was because Tikiġaq boat travelers could, before arriving at this point, hug the relatively safe shoreline. On reaching the cliffs they had to take on board the dogs that been hauling the boat and navigate the deeper water round the headland.

The cliffs represented other points of transition. Beyond the headland lay southern neighbors with whom Tikiġaq traded and made war. Also, more remote in imaginative geography, were the "countries" where the souls of the animals went home to be reincarnated. When the animals returned, they had to travel past the headland until they were once more in Tikiġaq territory. Many species—birds, whales, seals, beluga, walrus—migrated to the cliffs and past them each spring. Further south, the storytellers said, was a region of mystery and rebirth. Just as the sun rose in January for the first time after its disappearance in midwinter on the southern horizon and grew daily until it appeared, from the distant perspective of the village, to be leaning against the cliffs, so the major sources of life "came round" the great barrier that screened the village from those distant southern regions.

The headland beyond Isuk also carried human implications. The cliffs had seen two disastrous battles in the early nineteenth century. In the first, a war party from Cape Prince of Wales had hidden in a cave halfway along the cliffs and ambushed a boatload of Tikiġaq people, leaving no survivors. Then, also in about 1800, at the far end of the cliffs, almost an entire generation of Tikiġaq warriors had been killed by people from the Noatak River. According to John Kelly, a commercial whaler of the late 1880s:

> At that time the growing tribe of Noatoks began pressing southward and westward... One summer, about the year 1800, a great land and boat fight took place between the Tigaras and the Noatoks just below Cape Seppings, in which the Tigaras were overthrown and compelled to withdraw from all that part of the country... So badly crushed were the Tigaras that they lost half their population, which gradually led to the abandonment of all their outstanding villages, the people taking refuge at the capital place of Tigara.[1] The sudden taking off of the leading men, the great whalemen and skillful hunters, left the tribe badly demoralised and comparatively helpless. (Kelly 1890:10)

The site of the battle was subsequently named Iñuktat, "many people killed."[2] About a decade later came another misfortune, this time to the north of Tikiġaq near Uivvaq (Cape Lisburne). Here, in 1838, a party of Russians led by A. F. Kasheverov stumbled on a horrifying bone yard: "In the ravines along Cape Lisburne...we saw human bones strewn everywhere, and, besides, there were several graves in shallow holes... In other graves there lay human bodies that had not yet decomposed and beneath each was a bow and arrows" (Vanstone 1977:64). Most of those camped at Isuk in that first week of August were in health and good spirits. But they were also survivors: a resilient people who were rebuilding the population.

Contact at Cape Thompson

While Native visitors from outside Tikiġaq territory would have been instantly identifiable, the people at Isuk might have hesitated to classify HMS *Blossom* when this British navy sloop, commanded by Captain Frederick W. Beechey, stood off from the cliffs at about 11 o'clock on that August morning.

Beechey was on the final outward leg of a voyage that had taken him from the South Pacific up to the western Arctic. Like Tikiġaq people, the British were also recovering from war: the successful but exhausting Napoleonic campaigns, which, in the aftermath of the Congress of Vienna (1814–15), had left the fleet with the leisure both to complete areas of mapping left unfinished by James Cook and to support British mercantile interests against Russian competition in the American Northwest. Beechey's assignment that summer was to support the second Franklin expedition. While Beechey had been moving through Polynesia before turning north to Alaska, his old friend Captain (later, Sir John) Franklin was traveling overland from the Mackenzie River with a view to emerging in Alaska, in an effort to prove the existence of a Northwest Passage, the longed-for shipping channel that would open a commercial lane from the Atlantic to the Pacific. Beechey's orders were to put up signals to show where he had buried provisions and, if possible, take Franklin's party back to London.[3]

In harmony with the spirit of this politically almost neutral and scientifically progressive mission came instructions from the British Admiralty on conduct towards Pacific Islanders and Native Americans: "[Y]ou are to use every endeavour to preserve an amicable intercourse with the Natives, and to caution your ship's company to avoid giving offence or engaging in disputes with them; and you are to show them every act of kindness in your power" (Gough 1973:28). Trade should be conducted fairly, "taking care that when purchases are made, an officer may always be present to prevent disputes." Mindful partly of the venereal diseases that even Cook's well-disciplined men had transmitted to Hawaiians, the officers of the *Blossom* were also ordered to be "guarded in their intercourse with the females, so as to avoid exciting the jealousy of the men." There is evidence that the sexual activities of some of Beechey's party did excite jealousy, but this, as we shall see, was the least serious outcome.

Also governing Admiralty instructions lay the knowledge that peaceful landings were often dependent on local goodwill. Beechey therefore requisitioned a cargo of "utensils," or trade goods, for distribution to Natives. Some of the goods that came ashore at Isuk included items from the following cargo:

> 50 yards of blue and red broadcloth; iron in the form of hoops and bars; 500 hatchets, nails, saws; 4 cases of beads, jewellery and trinkets of different colours but mainly blue; 500 knives; 100 printed handkerchiefs; 50 kaleidoscopes; 100 bundles of needles; 40 pair of scissors; 80 looking glasses; 36 common shirts; 1,000 fish hooks; 10 bundles of vermilion. Two double-barrelled guns had already been presented to the kings of Tahiti and Sandwich islands [Hawaii]. (Gough 1973:22)

These would be among the earliest, though not the very first, manufactured goods to reach western Eskimo people.

Beechey: "Necessary Observations"

The son of a successful London portraitist, Beechey was a Napoleonic war veteran and a future admiral of the fleet. A close observer of people and places, Beechey was also

FIG. 3. *A skinboat with Iñupiat and British sailors. These are the earliest representations of Point Hope people. Top: Two men, a woman (tattooed and wearing amulets), and a child. Bottom: Trade with British sailors. One sailor appears to engage in sexual advances. Author's photograph from Beechey 1831.*

a naturalist and draftsman, and his account of the *Blossom*'s round-the-world passage provides the first published description of Tikiġaq and its people. Beechey had enjoyed the South Pacific. He had socialized with Polynesian royalty who remembered Cook, and on Pitcairn Island he had met the last of the *Bounty* mutineers and their extraordinary mixed-race children. Cook himself had sailed past Tikiġaq in 1778, but had been too far out and low in the water to observe the peninsula. In 1819, two Russian vessels had visited the cliffs where Beechey would land and had inscribed their charts with Russian place-names—but these Beechey patriotically ignored. Also in the Russian service, the Baltic German Otto von Kotzebue had spent some days about a hundred miles south of Tikiġaq in 1816. Thus, while the people at Isuk might have had some secondhand knowledge of previous visitors, this was probably their first experience of Europeans. And just as the people of Tikiġaq would for the first time see themselves in mirrors that the Englishmen distributed, Beechey's colleagues give us our first glimpse of Tikiġaq faces.

This brave new world did not, however, seem to have provoked much wonder in either party. No Tikiġaq stories of first encounter survive, but judging from Beechey's account, they seemed neither surprised nor awed by the *Blossom*'s arrival. The British of course were not there to study Eskimo lifeways: they were focused on the business of communicating with Franklin, completing Cook's maps, and not least collecting the natural history specimens and artifacts for transportation to London.

No matter that the *Blossom* was carrying Russian maps of 1819 and that Beechey knew that another Russian expedition had three years earlier dubbed as Cape Rikord the outline

FIG. 4. *Two Point Hope skinboats approaching Beechey's party. One man holds a white fox skin; three others gesture their wish to trade. The men wear labrets and one figure sits in a bear-skin parka; some paddles show ceremonial markings. Author's photograph from Beechey 1831.*

of the headland. Beechey's first act on rounding the cliffs was to name them in English. From that day to the present, the locally known Imnat ("cliffs") would become Cape Thompson:

> On the second of August 1826, being favoured with a breeze, we closed with a high cape which I named after Mr. Deas Thomson [sic], one of the commissioners of the navy... As this was a fit place to erect a signal-post for Captain Franklin, we landed and were met upon the beach by some Esquimaux, who eagerly sought an exchange of goods. Very few of their tribe understood better how to drive a bargain than these people; and it was not until they had sold almost all they could spare, that we had any peace. (Beechey 1831:I, 262–63)

This encounter, during which Beechey identified these coastal people as Eskimos (Bockstoce 1977:16), was socially successful: goods exchange compensated for the absence of a common language. As Beechey observed, the Eskimos had some metal tools, which bespoke trade with Siberian Natives who obtained their metal from the Russians. "We found them very honest," wrote Beechey, "extremely good natured, and friendly. Their features, dress and weapons were the same as before described in Kotzebue Sound.... They had more curiosity than our former visitors, and examined very minutely every part of our dress." But from "their being frightened at the discharge of a gun, and no less astonished when a bird fell close to them, we judged they had had a very limited intercourse with Europeans" (Beechey 1831:I, 262).[4]

Lieutenant George Peard, a scientist on the expedition, also provided some vivid images, one of his journal entries containing a first description of Tikiġaq faces:

> Three [skinboats] came alongside with some women tatooed in stripes down the chin, a fashion we afterwards found common amongst them, the Men only wearing the

> lip Ornaments. This day we took a great many prawns in the dredge more like crayfish, larger and with thicker skins than our English prawns, but not so firm. (Gough 1973:152)

Peard, having viewed Cape Thompson from the water, also gave a good account of its correspondingly striped limestone, sandstone, mudstone, and shale strata, where they "found fossil shells and Tribolites and Selenites." It was Beechey, however, who supplied the best verbal sketches:

> The oldest person we saw among the party was a cripple of about fifty years of age. The others were robust people above the average height of Esquimaux: the tallest man was five feet nine inches, and the tallest woman five feet four inches. All the women were tattooed upon the chin with three small lines, which is a general distinguishing mark of the fair sex along this coast; this is effected by drawing a blackened piece of thread through the skin with a needle, as with the Greenlanders. Their hair was done up in large plaits on each side of the head, as described by Captain Parry at Melville Peninsula. We noticed a practice here amongst the women...which consisted of blacking the edges of the eyelids with plumbago rubbed with a little saliva upon a piece of slate. All the men had labrets [lip ornaments], and both sexes had their teeth much worn down, probably by the constant application of them to hard substances, of which their dresses, implements, and canoes are made. (Beechey 1831:I, 263)

Local artifacts were, however, Beechey's main focus. Both Beechey and Lieutenant and Assistant Surveyor Edward Belcher made important ethnographic collections and both had the collector's acquisitive but impersonal curiosity.[5] The Natives at Cape Thompson, wrote Beechey, had "several rude [metal] knives probably obtained from the Tschutschi, some lumps of iron pyrites, and pieces of amber strung round their neck; but I could not learn where they had procured them" (Beechey 1831:I, 263).

The British, for their part, hauled their surveying equipment up the cliffside, and this glittering brassware became the focus of intense curiosity:

> As soon as we had finished the necessary observations with the artificial horizon, to the no small diversion of our inquisitive companions, we paid a visit to the next valley, where we found a small village situated close upon a fine stream of fresh water flowing from a large bed of thawing snow... The tents were constructed of skins loosely stretched over a few spars of drift-wood, and were neither wind nor water tight. (Beechey 1831:I, 263)

Once trading and surveying had been accomplished, the Eskimos wanted to cement relations with their own great socializing contribution, food:

> The Natives testified much pleasure at our visit, and placed before us several dishes, among which were two of their choicest—the entrails of a fine seal, and a bowl of coagulated blood. But, desirous as we were to oblige them, there was not one of our party that could be induced to partake of their hospitality. Seeing our reluctance, they tried us with another dish, consisting of the raw flesh of the narwhal [probably immature beluga whale], nicely cut into lumps, with an equal distribution of black

> and white fat [the skin and blubber of the bowhead whale]; but they were not more successful here than at first. (Beechey 1831:I, 263–64)

While a food offering was one major aspect of hospitality, the dances that followed the meal represented a further expression of welcome. Of this Beechey gives an accurate, if condescending, picture:

> An old man then braced a skin upon a tambourine frame, and striking it with a bone gave the signal for a dance, which was immediately performed to a chorus of Angna aya! angna aya!, the tambourine marking time by being flourished and twirled about against a short stick instead of being struck. The musician, who was also the principal dancer, jumped into the ring, and threw his body into different attitudes until quite exhausted, and then resigned his office to another, from whom it passed to a lad, who occasioned more merriment by his grimaces and ludicrous behaviour than any of his predecessors. His song was joined by the young women, who until then had been mute and almost motionless, but who now acquitted themselves with equal spirit with their leader, twisting their bodies, twirling their arms about, and violently rubbing their sides with their garments, which, from some ridiculous association no doubt, occasioned considerable merriment. (Beechey 1831:I, 264)

Two formal greetings had thus been extended to the naval party, and many aspects of the performances described by Beechey are recognizable in the dances of today's Point Hope. The "tambourine" drum, which is the one musical instrument used by Eskimos, has likewise survived into the twenty-first century and consists of a steam-bent driftwood frame to which a whale's liver membrane is attached with caribou sinew. The drum (*qiḷaun*, "device for power") figured equally in shamanistic healing rituals and ceremonial invocations; it continues to accompany the satirical, comically erotic dances that Beechey's men enjoyed watching.

Tikiġaq Is Named Point Hope

When, on a clear day, local people stood on the cliff top and looked west, they could see a blur on the horizon, which was the tip of the peninsula where Tikiġaq's two beaches come to a point. Along the northern edge of the Point lay the main village, where in the heyday of this ancient society up to six hundred people had iglus. The name Tikiġaq is based on *tikiq*, "index finger," a term suggesting the way the peninsula extended into the sea as though pointing to the animals hunters could reach from the tip of the peninsula.[6] Tikiġaq people also had two other names for their peninsula. One was *nuna*, "land," alluding to this virtually sea-level splinter of beach gravel, which in winter was coextensive with sea ice and in summer and autumn was beaten and eroded on the north side by violent storms. Despite the apparently precarious status of this land, the term *nuna* expressed the fact that the peninsula was solid—as in fact it had been since before the last Ice Age twenty thousand years ago. A third word for Tikiġaq's Point was both a nickname and a reference to the peninsula's metaphysical identity. This was *niġrun*, "the animal," and referred to the myth in which the Raven Creator harpooned a whale-like beast which then rose to become Tikiġaq. The people, whose driftwood and whale-bone iglus were sunk into the earth of the Point, thus lived within the transformed body of that mythological creature. Since Iñupiaq names were regarded as an aspect of the soul (see Chapter 13), this three-name convergence suggested a set of meanings that lent the Point both an identity and a metaphorical life that corresponded to the lives of both humans and animals.

The Cape Thompson cliffs are today just as Beechey and other, later nineteenth-century observers saw them, and so the following description of July 1881 by the naturalist John Muir, who came through on the Revenue Cutter *Corwin*, helpfully fills in some detail:

> I went ashore this forenoon and, after passing a few minutes interviewing a group of vagabond Natives from Point Hope who were camped here to gather eggs, kill murres and loaf, I pushed on up the hillside, whose sheer scarped face forms the Cape. I found it five hundred and fifty feet high, composed of calcareous slates, much bent and contorted, and a considerable portion was fossiliferous... The slopes over which I passed today are planted chiefly with sweet fern—Dryas—with its yellowish-white flowers. A purple silene is also very abundant, making beautiful bosses of color. Phlox is present in dwarfed masses, only the stems and leaves being dwarfed, not the flowers. Anemones occur in fine patches, and buttercups, and several species of daisies and lupins... I saw no tundra here, its absence, no doubt, being due to the free drainage of the surface... (Muir 1917:144)

Here, then, Beechey stood, surrounded by his officers and surveying equipment, while below stood the grandparents' generation of Muir's "vagabond Natives." In contrast to this beautiful paragraph from Muir, Beechey provides this succinct description of the moment when Tikiġaq was inscribed on his map of Alaska: "we ascended Cape Thompson, and discovered low land jutting out from the coast to the W.N.W. as far as the eye could reach. As this point had never been placed on our charts, I named it Point Hope, in compliment to Sir William Johnstone Hope" (Beechey 1831: I, 264–65). Thus, under the casually assigned name of one of the Admiralty Lords who had commissioned Beechey's expedition, the name Point Hope entered European annals.[7]

Sexual Contact: The Voyages of Cook to Beechey

The company of the *Blossom* consisted of twenty-two officers and more than one hundred seamen, but their short stay had virtually no lasting impact. Out of a population of about one thousand, only some thirty Point Hope people met the English visitors, but even this relatively small number of outsiders would have fueled local curiosity and conversation. With the autumn approaching and the community about to converge on the Point, storytelling sessions in the ceremonial houses were due to begin. All the exotic details of Beechey's ship, the barges, uniforms, firearms, food, unintelligible language, reports from people who had been elsewhere on the coast of the *Blossom*'s itinerary, and the evidence of knives, saws, axes, needles, and scissors would have filled the season's talk and perhaps entered ceremonial house storytelling. But no Native testimony has survived. The two hundred commercial whaling vessels that were plying Point Hope waters by 1865 reduced the solitary *Blossom* to a vanishing point on the horizon of local memory.

Besides the transient presence of new material goods, the major impact of Beechey's presence was biological. Evidence for this comes from Beechey's men, but also indirectly from the writings of Cook about both Alaska and Hawaii on his final voyage. Cook and his officers were explicit on the subject. Hawaii, for Cook's men, had been a nirvana of sensuality. The young women, wrote the surgeon's mate in January 1778, were "exceedingly beautiful" and "used all their arts to entice our people into their Houses" (Beaglehole 1967:1083). To counteract this process, Cook issued orders to avert the transmission of the "'Venereal Complaint," but Hawaiian men were, within a few months, visiting the ships' surgeons. "They had a Clap, their Penis was much swelled and inflamed," observed James King (Beaglehole 1967:498). And later that year there would be encounters with women in the North Pacific. Of the Aleuts at Unalaska, Cook observed in October 1778:

> They are remarkably cheerfull and friendly amongst each other and always behaved with great civility to our people. The Women grant the last favour without the last scruple; young and old, Married or Single, I have been told, never hisitate [sic] a Moment. The Russians told us they never had any connections with the Indian Women, because they were not Christians; our people were not so scrupulas [sic], and some were taken in, for the venereal distemper is not unknown to these people… (Price 1971:248)

Cook's informants were almost certainly lying. The Russians had been in the area since the mid-1740s and had already infected Aleut and Koniag people with syphilis and gonorrhea (Fortuine 1989:242). Perhaps the discipline that Cook tried to exert on his men in Hawaii had greater success in colder latitudes. There seems to be no evidence. But given the experience of 1778, the men who came later on the *Blossom* were likely to seize the same opportunities. Evidence of this is expressed in the diary of Beechey's fifteen-year-old son Richard:

> Mr Wolfe made a present to each of the females of a string of beads. This favour being excluded from the males was not at all understood. Mr Wolfe acknowledges the favours of one of the ladies charms, who was dressed in a salmons skin gown, and whose appearance had before won the heart of one of our brother officers. (Bockstoce 1977:109)

Sexual activity is further suggested in one of the illustrations in Captain Beechey's own account of his expedition. The engraving shows a skinboat containing nine Point Hope people, four of whom are women. Two women hold up white fox skins for barter, while two men offer a bow and arrows. Also in the boat are three British seamen, two of whom are bargaining, while a third apparently engages in sexual advances.

Apart from brief landings from British vessels in the search for Sir John Franklin in 1850,[8] there are no more recorded visits to Point Hope for half a century. But relations between Iñupiat and British mariners did take place again when the HMS *Plover* visited Point Barrow in 1852–54: the most concentrated single episode of Eskimo-white contact in the mid-nineteenth century. But even at this comparatively late date, pelagic whalers had made little contact with Point Hope, and Maguire reported that a Point Hope visitor to Barrow

> did not appear to have been familiar with ships, which I account for by the lowness of the land about Point Hope, deterring the Whalers from approaching it, otherwise in the Latitude during the summer, they are in great numbers—He spoke also of his people being badly supplied with tobacco, which would not be the case on a part of the coast frequented by the American Whale Ships... (Bockstoce 1988:168–69)

Touching briefly on the subject of sexual relations, the *Plover*'s physician John Simpson wrote:

> It is said by themselves that the women are very continent before marriage, as well as faithful to their husbands; and this seems to a certain extent true. In their conduct towards strangers, the elder women frequently exhibit a shameless want of modesty, and the men an equally shameless indifference, except for the reward of their partner's frailty. (Bockstoce 1988:523)

And such relations between men from the *Plover* and Barrow women in the late 1880s were accounted for by the American John Murdoch, who wrote:

> Although, according to the account of the Natives, there was considerable intercourse between the sailors and Eskimo women, there are now no people living at [Point Barrow] who we could be sure were born from such intercourse, though one woman was suspected of being half English. She was remarkable only for her large build, and was no lighter than many pure-blooded women. (Murdoch 1892:53)

Memory of these contacts with British mariners became submerged in midcentury traditional life. In comparison with the ills of the 1880s, they became all but invisible. The British were few. American whalers would come in increasing numbers, and while commercial whaling had died out by 1900, the effects of their presence long outlived them.

One possible meeting between a British mariner and a Point Hope woman at Point Barrow remains to be explored. This encounter (if it took place at all—the evidence is tenuous) may have had a relatively long-term and not entirely negative outcome. In order to focus on the following discussion, the reader might like to view Figure 2, on page xiv. In this, the Point Hope community leader and shaman Anaqulutuq, who will appear several times in this study, is depicted with his wife and children. As observed by John Bockstoce (pers. comm. 2008), the two children to the left appear to be of mixed race. And on further inspection, it can be seen that Anaqulutuq himself, in contrast to his recognizably Point Hope wife and older daughter, also presents mixed-race features. If this prominent individual who, in 1889, was appointed "headman" by Commander Charles H. Stockton, did have a non-Native father, he would probably have been the first Point Hope Iñupiaq with such ancestry.

Anaqulutuq's initial (unpublished) appearance in written Point Hope annals was given by Charles Brower in 1885 (see Chapter 11); the other American with whom he associated was the missionary John Driggs (see Chapters 16, 17, and 20). The first time Anaqulutuq's name comes into print, however, was in Commander Stockton's article that described his landfall at Point Hope in 1889—a visit that prompted the Episcopal Church to send Driggs

FIG. 5. *HMS* Blossom *in Hawaii, 1826, before its voyage to Alaska. An image of contact in which each society is represented within the boundaries of its own marine technology. National Maritime Museum, Greenwich, England.*

to be Point Hope's missionary. Stockton's article reveals some knowledge of recent Point Hope history. February 1889 had seen the assassination of Ataŋauraq, the Native trader and tyrant; following the murder, the community had suffered yet more turmoil (see Chapter 12). Stockton's contribution to this sequence of events was to "appoint" Anaqulutuq as a new headman, a position that had recently existed only in the person of Ataŋauraq (see Chapter 9). Stockton wrote:

> The natives of Point Hope, like the Eskimos generally of northwest Alaska, have no tribal or other form of government except what exists by control of the head man, oomalik, or chief, whose superiority arises from his wealth and influence. The previous chief had lived a life that made him a terror to the community. His rule was by force alone and by the influence of the rifle, which was his inseparable companion. After a career distinguished for license, murder and robbery, he had come to a timely end by being assassinated by the brother of a wife he was tormenting to death. Since his death, up to the time of our stay in September, anarchy had prevailed. On account of the very indifferent treatment received by the survivors of the wrecked whaler "Little Ohio" from the Eskimos of Point Hope the previous winter, I determined to appoint a head man or chief who would be charged with the responsibility and duty of caring for any shipwrecked persons or destitute whites [see Driggs's experience, Chapter 16]. Anokolut [Anaqulutuq], who was appointed by me and whose appointment was afterwards confirmed by the Governor of Alaska, had married the niece of the previous chief, and was the best whaleman and hunter of the district. He had been in the

> employ of the whaling station established the previous year at Point Hope, and had been satisfactory in all his dealings with the whites. His wife was a very superior woman, and their desire for civilized usages was so great that a bread-pan of tin, some granite-ware bowls, and candles, were given and eagerly accepted as contributing to make their domestic lives more comfortable and civilized. An urgent request was made for a cooking stove, which I promised to give them if I should return the following summer. (Stockton 1890:194–95)

How Anaqulutuq enacted the role of "headman" is not recorded. He spoke some English and the manner in which he befriended both Brower and Driggs attests to both his humanity and his initiative. He was clearly also a highly intelligent man who harnessed his ability to hunting and trade. A respected shaman and *umialik*, he makes his final appearance in this volume in spring 1899 (see Chapter 26), although in April 1901, he is still active, according to Driggs's journal. The storyteller Asatchaq (born 1890) remembered him as a shaman and dancer.

Returning to the subject of Anaqulutuq's parentage, we have already seen that Point Hope had no dealings with the pelagic ships of the 1850s and 1860s. The only sustained contact took place not in Point Hope but, as mentioned above, between Iñupiat and crew members of the *Plover*, the British naval ship stationed at Barrow. In this connection, two entries in Maguire's 1853 journal give an account of a visit from a high-status Point Hope man, Samaruna, who was accompanied by two wives:

> February 7 1853 The expected event of a visit from the Natives from Point Hope 'tik-a-rams', talked of for the last few days, took place today, in the visit of Chief Sam-ma-ru-ma [Samaruna in today's spelling] accompanied by two nice looking young women his wives—he was himself a pleasing, spirited, looking man, about thirty five years of age—& had travelled as he said to see the ship, having heard at Point Hope of our being here—He recognised Dr Simpson, as having seen him before, about Hotham Inlet… I invited him down to the cabin, & made him some presents… It would appear that this man was a poet in his way, as he favoured us with a long extempore song, which included the name of the Ship & also my own—After which he stroked his stomach down several times, expression of great friendship, then he got up & came to me, fixed his forehead against mine, and used it as a fulcrum, to rub his nose several times across mine, a ceremony not very agreeable, in his heated state, from singing…
>
> February 10 The Tik-Ki-ra'me' [Point Hope] Omelik [Chief] Sam-ma-ru-ma was on board with his wife, the other having gone on the ice on a sealing excursion—He described to Dr Simpson…the barter that takes places between his people and the Asiatics, who understood him to say that he gave the Martin (Sable), Fox, Wolverine, Wolf, and other Skins & sometimes whale oil and fish for Kettles, tobacco, beads, & Knives… He said he had only one Ne'-wak or bartering friend at this place, and that he brought three kettles to barter getting wolverine skins in return—He was lively and musical today, & and sung the good name he should give the ship on his way down to Point Hope, which will be of benefit to our boats in the Summer trip to Cape Lisburne—The officers speak of these Point Hope people and of the natives to the South generally, in terms of praise, in comparison with the Natives at Point Barrow. The former express gratitude for favours received, accompanied with a degree of modesty, particularly the females… (Bockstoce 1988: 168–69, 171–72)

Fig. 6. *"Eskimo village at Point Hope, 1889." This is the earliest photograph of the village and is inscribed probably by Commander Stockton of the USS* Thetis. *The wooden structure may be a dilapidated* qalgi *over which layers of sod will be laid. Courtesy of the Archives of the Episcopal Church.*

The Point Hope party continued to visit Maguire for another five days, but when it was clear that the Captain had no wish for more extensive trade, they seem to have started their return journey on about February 16. Given that Samaruna recognized Dr. Simpson from the previous summer at Hotham Inlet (the Kotzebue region), it would appear that he had kept track of the *Plover*'s movements and that his trip to Point Barrow was most likely to visit the English for the purpose of trading. The fact that he had brought "two nice looking young women," and his mention that he had only one "bartering friend" at Point Barrow, suggests that he planned to establish formal trade relations with some man with a high status that paralleled his own. Spouse exchange, as described later in this chapter, frequently accompanied such arrangements.

Whether one of Samaruna's wives gave birth nine months later to a child fathered by an officer of the *Plover* cannot of course be established. But if this were to have happened, this isolated event in Point Hope contact history had its volitional beginning and an outcome that was, given the flourishing history of the adult Anaqulutuq, benign. For this early, perhaps first, mixed-race Iñupiaq grew to enjoy a successful career, skilled in both its traditional and transitional aspects. One further circumstantial detail remains to be added. When the storyteller Asatchaq recorded one of Anaqulutuq's dance songs for me in 1977, he appended an unusual fact. "Anaqulutuq was bald," said Asatchaq, laughing. Since hair loss, which may be inherited from either parent, occurs most commonly in Caucasian men, has its lowest incidence in Native Americans and Eskimos, and is virtually unheard of in Point Hope, this furthers speculation.

Tikiġaq Life Before Contact

Most of the people who met Beechey moved back to the village on the Point in the autumn or traveled to the nearby Kuukpak River to fish for grayling, snare ptarmigan, and hunt caribou. Up and down the coast and on the Kuukpak, there were, besides the main village on the peninsula, about twenty settlements of people who followed game from place to place and who joined families and partners in the pursuit of food and skins.

But everyone in these tiny clusters of outlying iglus belonged to a society at whose center lay Tikiġaq's metropolis: a site which, from autumn through spring, could accommodate about half the total population of around thirteen hundred people (Burch 1981:14). Tikiġaq village was the capital of a huge tribal territory which in the early nineteenth century stretched about a hundred and fifty miles along the coast from Cape Beaufort in the north to a creek at the southern end of the Cape Thompson cliffs. The inland territory comprised about forty-three hundred square miles of tundra, hills, lagoons, and a number of river systems. Besides this there were the coastal waters, which Tikiġaq hunters exploited all year. In the winter they took seal, polar bear, and white fox. In the spring and early summer there were bowhead whales, beluga, bearded seal, and walrus. Between July and October, they harvested beluga and migrating trout and salmon. Hunting on the tundra and in river valleys, they brought home caribou, wildfowl, white fish, and fur-bearing animals such as marmot, red fox, wolverine, and parka squirrel.

When the *Blossom* appeared in August 1826, most Tikiġaq residents had left the main village. The very old and disabled, along with a handful of children and other helpers, were left on the Point while the able-bodied went traveling. A few might spend the summer on the Point to build new iglus. But most people packed tents, food, and equipment into skinboats and took off along the north or south beaches until they reached their first camping places.

Here, as the sea ice rotted and separated into floes, they hunted the bearded seals and walruses whose skins would cover their boat frames the following spring. By late July, the ice had usually dispersed. Some people then traveled to more distant settlements, others moved beyond Tikiġaq's borders to meet Eskimo and Chukchi people at one of the two main inter-regional trading locations at Sisualik to the south or at the mouth of the Utuqqaq River. Here they would barter sea mammal products for meat and skins from the interior.

From July to September, almost all Tikiġaq people were scattered across their territory and beyond, returning to their iglus before the sea became too rough. Most of the village lay at the northwest edge of the peninsula. Here, built into beach ridges and with their entrances facing south, there were about one hundred iglus, each housing up to twenty people, though they would not all be simultaneously inhabited. Once an iglu had been abandoned and its wood and bone salvaged for new buildings, it was used as a cold storage pit, while the oldest houses were eventually torn away by the sea which eroded the north side in the summer and autumn.

The northwest Alaskan iglu was probably a prototype of the snow-block "igloo" that Canadian Inuit constructed temporarily on the sea ice. The Alaskan iglu consisted of two main parts. The living area was a cabin, about ten feet square, built of driftwood, with a wooden floor that was cut and smoothed with stone axes and adzes to form an inner chamber. Leading to this was a south-facing entrance passage (*qanitchaq*) built with the ribs, skulls, jaws, and scapulae of bowhead whales. Entry to the iglu was down a ladder from the top of the low-ceilinged passage; at the inner end of this, a circular hole led up to the living area. Two layers of sod insulated the iglu, which, along with the entrance passage which acted as a cold trap, rendered the house weatherproof.

Fig. 7. "Point Hope Esquimo Houses." Wooden racks rise over sod-covered wooden structures with south-facing whale-bone entrance tunnels. The peninsula beach ridges are further raised by iglu architecture. Author's photograph of engraving in Report of the Revenue Marine Steamer *Corwin* in 1885.

Edward Belcher's description below may be compared with accounts written by the missionaries Driggs and Edson in the late nineteenth century, and while comfortable to local people, they were certainly not built for large-boned Euro-Americans (see Chapter 22).

While the iglu was insulated from prevailing north and south winds it was heated and lit with two or more lamps made of stone or clay that had been excavated from the Cape Thompson cliffs, and which burned day and night using seal oil and wicks made of twisted moss or tundra cotton. Added to this heat were the half-clad bodies of the people who ate, slept, and reared children within the iglu. A membrane window and a ventilator hole in the roof provided light and air. But most iglus were dark, and European visitors found the atmosphere asphyxiating. Edward Belcher, one of Beechey's officers, who revisited Tikiġaq in August 1827, gave this description:

> I visited four or five of these [houses] and was much surprised to find so much neatness, cleanliness and comfort in the great apartment, although the entrance [passage] was much the reverse. In the principle one I examined, I had to descend from the summit of the flat of the Mound through a sort of well formed of stone, about 8 feet deep, but furnished at the bottom with two large whale vertebra for the purpose of a step. The passage then runs (about 3 feet high & the same in width)... for 6 feet...at the end of which at the top was a hole the same in height, the back half being fitted as a kind of standing bed-place, the floor remarkably clean, and of smooth fir plank, the light afforded by a square hole at the apex covered with the gut similar to the skin dresses, and in one corner stood a receptacle similar to [a font] in which was seal oil with a ridge of wick alight on its outer edge. The temperature

> was very oppressive to the lungs, and when I examined [the interior] a second time through the window at the summit [the air] was insupportable. (Belcher quoted by Bockstoce 1977:124)

Family life in the iglu centered on parents, their siblings, children, and elders. Children were, however, seldom restricted to life in the parental home, and most of them also spent time with other relatives. Marriage partnerships were usually permanent, but people sometimes changed spouses or took lovers. Spouse exchange was by mutual agreement, both within the home community and with couples in neighboring societies. These exchanges were generally formal partnerships that were often arranged to consolidate trade relations. Children born to exchange partners treated each other as kin, thus binding half siblings to intra-village and inter-regional alliances that offered protection in potentially hostile situations and at other times of need.

In addition to the iglus, there were six or seven larger houses in Tikiġaq. Belcher was shown round one of these ceremonial houses (*qalgi*) before it was opened for use in the autumn. "At length I came," wrote Belcher, "to [a house] of large dimensions, half filled with sledges, which appeared to be their dancing room, being about 30 feet square, entrance very large, and unfurnished beyond the flooring… Several frame of tambourins were hung up within" (Bockstoce 1977:124). There would also have been masks, human and animal figurines, and elaborate mobile sculptures under the floorboards, and while Belcher acquired a collection of ivory and bone carvings, he was not shown ceremonial objects that would come into play in the autumn's ritual games and dances.

The *qalgi* rituals took place before the sea froze in November and helped to reunite a majority of the population, which had been scattered during summer. But the main purpose of the ceremonies was to propitiate animals that had already been taken and to communicate with the animals people would be hunting once the sea ice froze. The rituals followed a tightly organized pattern, and while Tikiġaq had no chiefs or councils, each *qalgi* was dominated by its oldest men, skinboat owners known as *umialit* (singular *umialik*) and shamans. Women, particularly *umialik* wives, could participate in some of the ceremonials; otherwise they spent the autumn looking after children, sewing clothes and boots, and preparing food in the iglu.

The ceremonies opened with a long recitation of stories. There were the myths and legends (*unipkaaq*), which consisted of origin stories, supernatural tales, and narratives of shamans from the distant past. Then there were the ancestor histories (*uqaluktuaq*), which were stories of Tikiġaq's forebears that went back four or five generations. These might also have some supernatural content, and the older histories merged with *unipkaaq*. Alongside these narratives there were also tales known as "warning" stories, which outlined how to act in difficult ice and snow conditions and how to behave to vulnerable or threatening people such as children, old people, and shamans. Wherever people traveled, they also told stories connected to the places they passed, and thus reinforced a communal grasp of historical and mythological geography.

Viewed from the coast in summer, Tikiġaq village rose in two distinct planes, which Beechey described as he approached the peninsula tip before a strong current made it impossible to land: "As we neared [the village] we perceived a forest of stakes driven into the ground for the purpose of drying skins, and placing sledges and other articles upon; and beneath them several round hillocks, which we afterwards found to be the Esquimaux yourts, or underground winter habitations" (Beechey 1831:I, 265).

The contours of these racks and iglu mounds elevated the village above the low expanse of the Point. Village topography was otherwise dominated by the clusters of iglus belonging to extended families who lived in houses (some of them interconnected) and belonged to the same *qalgi*. Their dogs were tethered a short distance away, and these were somewhat sparingly fed seal, fish, and other meat that people had no use for. The dogs of the traditional period were larger than today's husky and were used to haul sleds on land and skinboats from the beach through inshore water. Dogs were on the whole too valuable to be risked on the constantly shifting sea ice. But dogs were never pets, and partly because they were sometimes reduced to eating feces, they were regarded as low creatures.

Tikiġaq social life was centered on the family and ceremonial house, and although children were often exposed to harsh weather, they were brought up indulgently and seldom punished. That said, Tikiġaq people did practice (usually female) infanticide during subsistence crises. The newly born, as we shall see later, had to endure a long period of perinatal confinement, and the survivors of this would thus be already tough. To encourage both dexterity and stamina, Tikiġaq also deployed a vast repertoire of games for both children and adults, while much adult time went into constructing a wonderful variety of ingenious toys. Children also practiced manual dexterity and spatial awareness in string figure games, while their intelligence was honed in the memorization and improvization of songs and stories. Physical strength and agility were also of the greatest importance to people who frequently had to run and jump for their lives on the sea ice, and long-distance racing games inculcated competitiveness (sometimes violent) as well as speed and staying power. This vigorous outdoor life was, for girls, accompanied by long application to skin sewing and animal butchering, which they learned early from female relatives. When their husbands and brothers killed seals and polar bears, it was the women who hauled them in, cut up the meat, and dressed the skins. Women were also sometimes responsible for dog handling, and often ran with the dogs as the men steered their boats through inshore water.

Boys' education was directed to the development of hunting skills, and toys often consisted of miniature weapons. As soon as he could hold a little bow or spear, a boy was off on the Point chasing squirrels, ptarmigan, and snipe, and each first capture was ceremonially marked by a distribution of gifts. Much of this early experience took place in boys' outdoor hurly-burly. But boys and girls also went out quietly to the beach in the autumn and collected tomcod that the waves had cast ashore; and later in the season they accompanied their elders who jigged for the same fish through holes in the new inshore ice.

Children were forbidden go further than this shore-fast ice. Shifting winds and currents made the offshore ice a treacherous environment. Gaps in the ice of course made it possible for men to hunt in the open water. But it took years of apprenticeship before a man had learned to negotiate wind, current, and ice conditions and was ready to hunt alone. And experience was no guarantee against being marooned on a drifting floe. Most men hunted at least within earshot of a partner. But drowning accidents and death by exposure were not infrequent.

Hunting, trading, and sexual partnerships between non-kin in the village who could share their labor and resources were of special importance in a society that lived constantly on the edge of subsistence. The land and sea were abundant, but the hunt could also fail. A family might lose its men in a boating accident or a skirmish with a neighboring society. The ice might stay locked solid for part of the winter, rendering seals inaccessible. The caribou herd could change its migration route, as probably happened in the late nineteenth century. Most crucially for Tikiġaq, a south wind might prevail during the crucial weeks of the spring whale hunt. This would force the ice pack onto the shore-fast ice and thus close the inshore open water, forcing whales to swim through channels that were inaccessible to boat crews. While the tiny hunting bands of people who lived on the Kivalina, Noatak, and Kobuk Rivers were sustained by fish and caribou, the much larger Tikiġaq population was dependent on the tons of blubber, meat, and whale skin that a single bowhead provided. As we shall see, all these societies were seriously threatened by the subsistence crises that followed contact with commercial hunters. None of them was, however, immune from starvation, even in the relatively more abundant traditional period.

Echoing the stresses of the ice and weather, against which almost every daily action represented a maneuver, were social pressures. There was competition between boat crews, *qalgi* groups, and rival families, and intra-village loyalties could lead to feuds that sometimes continued for generations. Partner relations across kin lines helped to maintain stability, as did the structure of *qalgi* organization and the shared stresses of survival during a war. War, as we have seen, decimated Tikiġaq's population in the early nineteenth century, and relations between Tikiġaq and her neighbors always consisted of an uneasy balance between hostility and the need for inter-regional trade. At the best of times, this equilibrium of necessities was profitable for all parties. People from coastal villages such as Tikiġaq were interested in meats, fish, and skins from the interior, while inland people looked forward to the sea mammal oils, dried seal meat, whale skin, and walrus ivory that the sea ice hunter provided. The reception of strangers in any Iñupiaq society thus varied according to context and season. There were relatively safe periods during trade fairs and regional feasts; in nonceremonial time, the threat from an alien society was often greater.

The life of the individual was thus richly enmeshed within extensive social patterns. Nor was the Tikiġaq community, despite its dramatic geographical isolation, alone in the larger environment. Like all Eskimo people, Tikiġaq lived within a universe of spiritual forces whose support could not be predicted. Unlike the Inuit of Canada and Greenland, the Alaskan Iñupiat did not (at least within ethnographic memory) recognize an all-powerful sea goddess. As will be outlined below, Tikiġaq's main deity was a capricious and often punishing moon spirit (*tatqim iñua)* who was responsible for providing game animals but who sometimes withheld them when he was angry. In addition to the lunar deity and the occasional influence of his sister in the sun, there were countless other spirits whose assistance could be applied for, or whose control must be sought, avoided, or neutralized according to circumstance. Animal spirits and the souls of the ancestors were similarly ambiguous. Provided that people propitiated the animals and maintained the taboos that attended almost all subsistence business, these spirits might be helpful. But relations with every supernatural power were generally unstable. The same spirit could, unpredictably, be protective or vengeful. Most accidents and illneses were attributed to paranormal influence.

At the center of these spiritual relations were the shamans (singular *aŋatkuq*), whose role it was to communicate with variously helpful or dangerous supernatural forces. The shamans' task was also to purge their clients of invasive spirits that had made them ill and to defend them against other, ostensibly hostile, practitioners. Within Tikiġaq's animistic world, the shaman was an indispensable arbitrator who identified the invisible powers that were at work at any one time, prescribed the taboos that kept people healthy and safe, and diagnosed taboo infractions which they said caused things to go wrong. It was also shamans who performed many of the rituals in Tikiġaq's ceremonial houses. Along with other practitioners—diviners, healers, and clairvoyants who had not been fully initiated into the spirit world—the shamans maintained an order within the village that was mostly respected, often feared, and sometimes resented. Two other sources of authority existed beyond the otherwise pervasive shamanistic influence. First were the skinboat owners (*umialit*) who dominated the all-important whale hunt each spring. It was their intelligence and hunting skill that led to their economic success, and this in turn lent them the status and charisma to attract men from their *qalgit* for their boat crews. The *umialit* were also central officiants in the autumn ceremonials, and this, in the absence of chiefs or councilors, contributed to a certain de facto authority in the village. An *umialik*'s influence lasted, however, only as long as he was a successful hunter. As we shall see in the history of Tikiġaq's trading chief, Ataŋauraq, the status of shaman and *umialik* could, in the context of new contact opportunities, village depopulation, and modern Americanizing expectations, be reformulated and abused, with disastrous results for both the "big man" and the community.

A last center of power and authority lay in the old people. These were the sages and long-term survivors, and it was said that *all* old people were shamans of a kind, whether or not they had undergone initiation. There was ambiguity in the status of the elders in that during times of hardship they might, as nonproducing consumers, elect to commit suicide or be killed. Testimony to this practice, which reached into the nineteenth century, was provided by Herbert Aldrich, who recorded that in the severe winter of 1885–86, "a dreadful famine ensued [in Point Hope], and resulted in not only aged persons being cast out, but also many women…there has since been a scarcity of women there" (Aldrich 1889:170). Otherwise, the old, on account of their experience and their closeness to ancestors and to the spirit world, had a near-sacred position, and they represented authoritative sources of information on both practical and ceremonial issues.

Many of these aspects of pre-contact life will be touched on in the chapters that follow. Longer and fuller ethnographies of traditional life can be found in works cited in the bibliography. For the time being, this brief introduction to life in old Tikiġaq should serve as a basis against which to compare the cultural and historical changes that this book now attempts to describe.

Notes

1 Kelly was probably the first to create this near-accurate version of the name Tikiġaq, and "Tigara" became a standard semi-Anglicized term for the village. Before Iñupiaq became standardized, this spelling was also used by local people. A photograph of a Point Hope country-and-western group from the mid-1950s shows a placard reading "Tigara Play-Boys" (Vanstone 1962: plate V-B).

2 Three local stories recall those wars. In one, some inland Eskimos ambush Tikiġaq hunters at Iñuktat. In two other stories warriors from Cape Prince of Wales or Kivalina hide in a cave at Imnat and ambush Tikiġaq boat travelers (Lowenstein 1973–89).

3 Franklin reached his furthest point west on Alaska's north coast on August 18 before turning back. Five days later, Thomas Elson, second surveyor to the *Blossom*, and traveling in the *Blossom*'s barge, reached Point Barrow, just 146 miles west of Franklin's point of return. There was, despite this narrow miss, no way Franklin could have reached the *Blossom,* which was now hundreds of miles to the southwest.

4 In 1816 Otto von Kotzebue used a musket to kill a snipe at the Shismaref Inlet, remarking that Natives there were "wholly unacquainted" with firearms "and they could not get over their astonishment" (Kotzebue 1821:I, 229).

5 Beechey and Belcher materials remain in the British Museum's Department of Africa, Oceania, and America and the Pitt Rivers Museum, Oxford. See Bockstoce (1977).

6 Martin Frobisher recorded this word on Baffin Island in 1577, making it one of the first in the Eskimo language to be transcribed. Beechey also made a "collection of words," mostly from the Kotzebue Sound, including *teg-heya* for "first finger" (Beechey 1831:II, 622).

7 In 1821 the Russians Vasiliev and G. S. Shishmarev had bestowed the name Golovnin on Tikiġaq, also naming the "Cape Thompson" cliffs Cape Rikord (Bockstoce 1977:17). The men who planted messages for Franklin also called the cliffs "Bottle Cape," but this did not stick. Beechey's party also came close to naming the Tikiġaq peninsula "Dungeness," since it projected "into the Sea like Dungeness [point]" on the southeast coast of Kent (Gough 1973:152).

8 These included Richard Collinson's HMS *Enterprise,* Captain Henry Kellet's *Herald,* and T.E.L. Moore's *Plover.*

CHAPTER 2

Contact with Chukchi and Europeans up to 1854

Traditional Contact with Siberian Chukchi

Regular contact between Native Alaskans and Euro-Americans started when commercial whalers found their way through the Bering Sea in the 1850s. But north Alaskan Eskimos had seldom been wholly cut off from the Eurasian continent and they had traded with Siberian Chukchi, a mere fifty miles across the sea, at least since the early eighteenth century. It was from the Chukchi that Alaskans had obtained the metal goods that Beechey observed in August 1826. Trade did not, however, imply friendship between Eskimos and Chukchi, and there were occasional hostilities. Individuals who lost their way and landed on the wrong continent might be killed or enslaved, and people on both sides told fantastic stories about each other. According to the Chukchi, the "tooth people" (Alaskan Eskimo men with lip ornaments) walked barefoot and wore tails (the latter probably referred to animal and bird amulets sewn on their parka hems). In the mid-eighteenth century, Russian Chukchi Daurkin quoted a Siberian story about "Tikegen," "which in strong weather is moved several versts out of place, but when there is no such weather it stands in the usual place" (Daurkin 1765, quoted by Krauss 2005:173).

A story from the Alaskan side describes how a hunter from Point Hope, Miġalliq, was captured by the Chukchi and survived to return home and pull a remarkable deceit on his former hosts. Miġalliq, according to this story, had discussed with his captors the virtues of Siberian copper against Point Hope's slate harpoon points. Before Miġalliq started to walk home over the ice, some Chukchi women tattooed his back with a map of the Siberian coast as evidence that he had been to the other side.[1] Before leaving, the Eskimo invited his hosts to visit Cape Thompson cliffs, where they could gather slate. The next summer at the cliffs, Miġalliq received the Chukchis and exchanged slate for copper goods. Historically true or not, the story eloquently expresses how trade goods moved between the continents despite the insecurities that attended such transactions.

Another good indication of intercontinental hostility was reported in 1816 by Otto von Kotzebue. When Kotzebue showed some drawings of Alaskan Eskimos to a Chukchi man, the Chukchi exclaimed, "If I meet such a fellow with two bones [lip ornaments] I shall pierce him through" (Kotzebue 1821:I, 262). Similarly, an Alaskan story tells of a local hunter who spontaneously killed a Siberian who had drifted to Cape Dyer about thirty miles from Point Hope (Lowenstein 1973–89). In the late nineteenth century, the Point Hope shaman Asatchaq conducted a supernatural duel with a Siberian rival, eventually killing him with amulet magic (Asatchaq and Lowenstein 1992:147). But as Burch has pointed out, it was relative numbers and degrees of strength that often determined whether Native groups would engage in hostilities (Burch 2005:55). Much the same consideration often applied to relations with white visitors, though as we shall see, individuals such as Charles Brower and the missionaries Driggs and Edson were made welcome in Point Hope.

The Tobacco Trade

It was tobacco as much as metal that dominated the Chukchi-Alaska trade. The Chukchi had acquired a taste for Chinese leaf from Russian traders, and by about 1750 they were exporting it to Alaska. This endpoint to a round-the-world traffic had started in the sixteenth century on the East Coast of North America, and it was a lucrative business. Alaskan Eskimos paid, in skins, enormously inflated prices for tobacco, and the Chukchi redoubled their profits when they sold Alaskan furs to the Russians. Smoking, intensified by deep inhalations from tiny pipe bowls until the smoker passed out, was both exhilarating and addictive. The Alaskans probably had no previous experience of intoxicants. Alcohol was unknown. And while some Siberian reindeer herders consumed psychotropic mushrooms, Alaskan Eskimos avoided them (the name of *Amanita muscaria* fungus, which grows in the birch woods of the subarctic, means "makes your hands drop off," a reference to the tingling capillary shrinkage that follows *Amanita* poisoning). But by the late eighteenth century tobacco had caught on in coastal Alaska. To give one example: in the absence of Euro-American whalers one winter, a Point Hope man made a dangerous journey across the ice to visit Siberian Chukchi simply to ensure he would have plenty to smoke that season (Asatchaq and Lowenstein 1992:139ff).

The Chukchi were regular summer visitors, but they spent as little time as possible in Alaska, returning home as soon as they could. The most important market was at Sisualik, a sand spit a hundred and twenty miles south of Point Hope, near modern Kotzebue, where Iñupiat would meet annually in mid- to late August. The trade fair was an important event. Tents, kayaks, and skinboats lined the shore in territorial groupings. And in addition to inter-regional trade, communal beluga hunts, fishing, and berry-picking, there were competitive games, shamanistic seances, storytellings, spouse exchange, and a certain amount of inter-regional conflict.

While the Alaskans traded sea mammal products and the rich furs of the subarctic river valleys, the Chukchi brought tobacco, metal, glass beads, and the reindeer skins that became increasingly important to Alaskan Eskimos when, in the late nineteenth century, the caribou herds had diminished. This intercontinental trade was tense and impersonal:

> The Chukchi first comes, and lays some goods on the shore, and then retires; the American comes, looks at the things, puts as many skins near them as he thinks proper to give, and then also goes away. Upon this the stranger approaches, and examines what is offered him; if he is satisfied with it, he takes the skins, and leaves his goods instead; but if not, then he lets all the things lie, retires a second time, and expects an addition from the buyer. In this manner the dealing seems to me to continue without

> speaking, and there is no doubt but the Chukchi obtain here the skins for the Russian trade. (Kotzebue 1821:I, 228)

And as hinted at elsewhere by Kotzebue, the furs that the Chukchi sold to the Russians included wolverine, muskrat, marmot, black bear, foxes, and lynx from riverine Eskimos, while the coastal hunters would offer sea mammal products.

The First Guns and Liquor

It was guns and liquor that most drastically changed Alaskan society in the later nineteenth century, and while the Russians defensively kept firearms to themselves, small American traders had by 1815 brought small quantities of firearms to the Bering Strait. The Pacific Fur Company from Fort Astoria (in today's Oregon) traded guns and ammunition with Iñupiat as early as the 1820s (Bockstoce 1988:2). As mentioned, the Iñupiat who watched Kotzebue discharging a musket at Shismaref Inlet could not get over their astonishment, while possibly the first recorded Native gunfire in northwest Alaska, directed at a Russian ship in the Hotham Inlet, erupted in the summer of 1820. But while muskets with ammunition were, in very small numbers, known in this area, the Point Hope Eskimos who met Beechey in 1826 appeared not to have encountered firearms. A tripartite agreement between the British, Americans, and Russians in 1824–25 that outlawed firearms or alcohol trading with Alaskan Natives was impossible to enforce in this vast area, and in complicated inter-regional and inter-ethnic trade maneuvers of the 1850s, a limited number of guns were coming into the hands of Point Barrow Natives from Vunta Kutchin Indians, who had acquired them from Hudson's Bay Company traders at Fort Yukon (Bockstoce 1988:82).

Captain Rochfort Maguire, who spent 1852–54 off Point Barrow on HMS *Plover*, identified one musket owner in that village (Bockstoce 1988:129), while three years earlier, in August 1849, one of the *Plover*'s boats on an exploratory trip west encountered a group of some eighty Iñupiat among whom there was a man, probably the same individual described by Maguire, who demanded gunpowder for his solitary musket (Bockstoce 1988:29).

This was one in a mere handful of guns in Native possession at the time and according to Maguire, only a few people were "acquainted with destructive nature of fire arms and," having witnessed the ship's gun, "were consequently more afraid of those weapons which they knew we possessed in abundance" (Bockstoce 1988:253). Muskets were also coming from the Hudson's Bay Company post at Fort MacPherson on the Peel River (a branch of the Mackenzie), and again Maguire describes an individual who had procured "a gun at Barter [Island] which has the date of 1850 on the lock, two other men have also got guns so that there are now at Noovook [Barrow] four guns all H.Bay's—at present there is a pressing demand for gun powder" (Bockstoce 1988:275). Later, however, Maguire humorously describes how the musket owner was unfamiliar with its mechanism and had not yet learned to operate it (Bockstoce 1988:287).

Maguire's residence at Point Barrow coincided with the arrival of American commercial whalers in 1854, and this unleashed a regular flow of firearms and liquor all the way up the coast from the Bering Strait. By the 1880s, ownership of new and sophisticated guns had become commonplace, an account of whose features was outlined by John Murdoch: "The chief change resulting from their intercourse with the whites has been the introduction of firearms. Nearly all the Natives are now provided with guns, some of them of the best modern patterns of breech-loaders, and they usually succeed in procuring a supply of ammunition" (Murdoch 1892:53)

And possession of firearms, suggested Murdoch, had two major consequences:

> This is in some respects a disadvantage, as the reindeer [caribou] have become so wild that the Natives would no longer be able to procure a sufficient number of them for food and clothing with their former appliances, and they are rendered dependent on the ships. On the other hand, with a plentiful supply of ammunition it is easier for them to procure abundance of food, both deer and seals, and they are less liable to famine than in former times. (Murdoch 1892:53)

Whether the scarcity of caribou in the later nineteenth century followed overhunting with firearms or that the herds were reduced or diverted from their earlier migration routes, or both, remains uncertain. The impact of firearms on seal hunting had clearer results. In the winter, seals could be shot in small areas of open water, but then must be retrieved in newly invented one-man umiat and/or with floating grapples. In the spring, basking seals and walrus could be taken in larger numbers with fewer problems of retrieval (see Chapter 20).

Liquor also arrived in the early 1850s: first to Bering Straits people, next to the Arctic. Captain T.E.L. Moore exaggeratedly described Natives as "constantly drunk," though in 1852 he wrote more sympathetically: "the whalers have [done] the Natives a vast deal of injury, by the introduction of…spirits" (Collinson 1889). The drinking habit was similarly observed in 1854 by Richard Collinson, who observed how Kotzebue Sound Eskimos requested rum as a principle exchange item.

From these isolated beginnings in the 1850s, a current of liquor grew which swept through the north Alaskan communities, leaving no village untouched. By 1880, the alcohol problem had been compounded by epidemics brought by American whalers and, at the end of the century, by miners and gold prospectors. The European bacteria that since 1492 had laid waste to millions of Native Americans from the Caribbean to California had finally arrived in the northwest Arctic.

Notes

1 Whether this map reflected the influence of contact is an open question.

CHAPTER 3
The Alaska Purchase and the Russian Period

The Alaska Purchase

The period of high contact in the late nineteenth century was so much a part of American frontier experience and so typical of Native-white intercourse that it takes effort to remember that when the missionary John Driggs arrived in Point Hope in 1890, the purchase of Alaska had occurred only twenty-three years before. By the 1880s, the controversy over the $7 million deal in 1867 had died away, but Alaska was still largely unexplored by white men, and mainland Americans remained indifferent to their far northern district.

The United States had grown prodigiously in the nineteenth century. California had, in 1850, been the last state to be added, and Alaska now completed the Pacific boundary. The value of the annexation had been strongly argued by William Seward, Secretary of State to President Lincoln. Alaskan fisheries offered a vast potential resource; even the ice, which some imagined could be transported to the American mainland, exerted a certain lure. Not least, American control of trade on the Pacific rim would exclude commercial and political competition from Russia and Great Britain. The annexation was nonetheless generally viewed as a farce, and the new territory was dubbed "Seward's folly," "Seward's ice box," and jokingly by Bret Harte, "Walrussia."

That Alaska was cold and distant posed a first alarming problem. Its sheer scale and inaccessibility compounded the challenge, and most of the great land would remain out of reach until the mid-1880s. Public indifference was matched by administrative apathy, and Alaska's political status remained vague. Neither state nor territory, the land was construed simply as a U.S. district. A handful of military outposts in southeast Alaska were established in the late 1860s, but these soon dwindled to just two, and the army was withdrawn in 1877.

Since the military, and from 1877, the U.S. Revenue Marine, were the only representatives of U.S. power in Alaska, there was no civil or judicial authority until 1884, when the first governor was appointed. Few civilians wanted to travel beyond the reach of American courts even in southeast Alaska, and there was plenty of scope for enterprise on the mainland, where the economy was booming, without having to broach the uncertainties of Walrussia. A few settlers did venture north to the old Russian capital, Sitka. But for almost two decades after 1867, Alaska's Euro-American population fell below that of the Russians of the previous eighty years.

The Russian Period

The Russians had first arrived in Alaska with the voyage of Vitus Bering in 1741, and by 1743, hunters were moving from Siberia to the Aleutian Islands in pursuit of the sea otter, whose fur brought fabulous prices in China. This early episode of European enterprise represented the most oppressive chapter of Native-European contact. The early Russian hunters (*promyshlenniki*) did not come to settle: they arrived simply for the hunt, and in pursuit of this they massacred and enslaved large numbers of people along the northern Aleutian chain.

It took almost a generation of the *promyshlenniki*'s crimes before Russian government interest in Alaska was aroused. In the early 1780s, partly to appease Catherine II's moral disquiet, a semi-regulated company was formed to pursue the fur trade in "Aliaska."

In 1784 Grigor Shelikhov and his wife, Natalia, with a force of some two hundred men established the first Russian colony on Kodiak Island. The invaders met fierce resistance from the Koniag Eskimos, but this was brutally put down and the Shelikhovs established both a trading settlement and a school, which in 1794 was taken over by Russian priests—a pattern of educational management that the American churches would follow in the nineteenth century. Management of the fur trade was by then in the high-handed control of Aleksandr Baranov, whose excesses were, to some extent, restrained by the Orthodox clergy (Fitzhugh and Crowell 1988:75).

Russian atrocities in the Aleutians were most famously documented by Joseph Billings, an Englishman who had served with Captain Cook in the 1770s, and in 1789 was engaged by Catherine II to explore western Alaska. Following an expedition in 1790 Billings described the methods of the *promyshlenniki* in horrifying close-up:

> The hunters were accustomed to act as follows: Upon arrival...at an inhabited island the (chief of hunters) sent an armed boat...[to] take from the Natives all furs and valuable articles they possessed and if the least opposition was made they were silenced by the muskets of the hunters. Wives were taken from their husbands and daughters from their mothers... They used, not infrequently, to place men close together and try through how many the ball of their rifle-barrelled musket would pass. (Sauer 1802:56)

One positive result of reports like this was the creation, in 1799, of the Russian-American Company, to which Emperor Paul I granted a monopoly over the Alaskan fur trade for the next twenty years. With the partial absorption of the *promyshlenniki* into Company employment and the relatively humane educational work of missionaries, relations between Russians and Native Alaskans were moderately improved. By 1867, there were forty-six Orthodox churches, some with lay Native clergy, and many Russian missions provided at least a modicum of defense against Baranov's agents. Here was the original pattern of church intervention which would be repeated by the Americans, whose missionaries, in theory, would offer Natives some protection against the abuses of commercial whalers.

Russian relations with the Tlingits in southeast Alaska were predictably terrible, though an uneasy symbiosis of Tlingit and European interests evolved. Trade was transacted in a guarded courtyard of New Archangel (modern Sitka); the Russians taught the Tlingit to garden; and in return for vegetables and fish, the Tlingit received metalware (but not guns), dry goods, and blankets. To the fury of the Russians, British and American freebooters in small, fast boats sold the Tlingits firearms and ammunition in return for sea otter pelts, and this complicated local tensions. Here, we should note a foreshadowing of the illicit open-sea trade in firearms and liquor that would take place between Iñupiat and American whalers in the nineteenth century, which the Revenue Marine would be powerless to regulate.

With the growth of British and French interest in the Pacific after the Napoleonic wars, conditions improved slightly. Competition between the British and Russians was particularly intense, and this brought in the Russian navy, with the result that the Russian-American Company, monitored now by a government agency at first hand, passed to the military. Baranov was cashiered. There followed a period of minor reforms. A limited medical service was established, and from 1818, smallpox vaccination attempted—though this did nothing to check a ferocious epidemic in 1836. There were also measures to regulate fur prices and Native wages; and in 1828, the otter and seal populations were to become partially protected by hunting quotas.

But for all the social and environmental depredations of their eighty-year presence, the Russians occupied only a fraction of Alaska. Traders came and went, and beyond the Orthodox priesthood and a population of new Creole or mixed-race Alaskans, particularly in the Sitka region, there were no Russian settlers. The shore-based whaler-traders who settled temporarily at Point Barrow and near Point Hope in the late nineteenth century followed a similar pattern, and their impact on Native life was, as we shall see, comparable.

Russian exploration of the territory was also limited. The state had, after all, not laid claim to Alaska. The Russians "took" Alaska in the spirit of enterprise to exploit the fur harvest, and Company traders seldom left their compounds. There was no love lost between Company regulars, Native hunters, and middlemen. The Company negotiated no treaties with the Natives whose land they occupied, and they achieved their limited aims by sheer armed superiority.

By the mid-nineteenth century, the Russians were ready to pull out. Like their counterparts on the beaver creeks and the buffalo plains of mainland America, they had virtually exterminated an entire species—the sea otter. Worse, they had enslaved whole Eskimo and Aleut populations. And both the Aleuts and Koniag Eskimos had been decimated by contact diseases.

Early American Occupation of Alaska

Alaska's white population numbered about six hundred when the Company withdrew, and the few Russians who stayed were priests. Back in 1766, Catherine II had conferred Russian citizenship on the Aleuts and the Alaskan peninsula people, but this amounted to little more than entailment to Company demands. With the American purchase, the "uncivilized tribes," as they were described in the treaty of annexation, were thus literally "bought in" by the Americans. Native Alaskans were offered no rights in exchange for their inclusion within the American ownership of their territories.

In practice, American ownership meant virtually nothing. Six army posts were established in southeast Alaska under the theoretical supervision of Major General Jefferson

C. Davis, a protégé of General Sherman, presently engaged in wars against western Indian tribes. Five years previously, Davis had murdered a U.S. general during a hotel fracas, and in Alaska, he was responsible for a number of crazy actions, including the Kake War of 1869, in which he destroyed an entire Indian village (Hinckley 1972:101). Early the next year at Fort Wrangell there was a brawl between some hard-drinking soldiers and their Indian hosts. Howitzer shells were fired into the township, but fortunately, there were no casualties.

With virtually no transport at their disposal, the U.S. Army neither occupied nor administered the territory. In Washington, a weary desire for disengagement and contempt for the army runs through the Congressional record. By 1877, the military had been reduced to three companies. Ten years had passed since the purchase of Alaska, and these six hundred thousand square miles of American property were being run by 108 soldiers. Even the monthly steamer that hauled supplies from Oregon was hardly worth the expense, and where the question of the liquor trade was concerned, the War Department conceded that policing this could well be transferred to the U.S. Revenue Cutter Service, an arm of the Treasury Department that was later enfolded into the U.S. Coast Guard. And thus it was that troops were withdrawn and the Revenue Cutter Service given responsibility for law and order (Mitchell 1997:53).

The Alcohol Trade

With respect to the trade in alcohol, the Intercourse Act of 1843 was the single American law of any relevance to Native welfare. This law, "to regulate trade and intercourse with Indian tribes," was extended to Alaska in 1873 and prohibited selling liquor or breach-loading, repeating firearms to Natives. According to Major James B. Campbell, "the strictest provision of the Indian Intercourse law will hereafter be rigidly enforced" (Hinckley 1972:98). But this was a vain boast, and in the decade following Purchase, the anti-liquor law became impossible to enforce in Alaska. The whites were mostly small-time fur hunters following trade routes set up by the Russians. Traveling in small craft from Portland, San Francisco, and Seattle, these operators made liquor a major currency of exchange, wreaking havoc with Native communities as far as St. Michael.

In April 1877, William G. Morris, a special agent to the Treasury Department, issued a graphic, although possibly exaggerated, warning:

> During the summer the Alaskan coast swarms with small vessels and canoes navigated by desperate and lawless men, bent on smuggling, and trading in ardent spirits. A [customs] collector is stationed at Wrangell, but he will be utterly powerless to enforce the revenue laws and stop the smuggling from British Columbia. (Morris 1879:28)

Morris also contested a previous Smithsonian report that claimed "perfect peace" between Natives and whites. It was the miners and traders en route to gold diggings upriver from Wrangell who represented the danger. If this transient population came to overwinter in Wrangell, they would surely bring liquor with them. Permanent white residents, by contrast, posed far less danger—as was most often the case in and around Point Hope later in the century.

At the heart of the danger lay an issue that had haunted white society in more than three hundred years of contact with American Natives. Not only were the tiny U.S. garrisons powerless to control commerce in liquor, but the army itself was implicated in the trade. "From the first sergeant to the drummer-boy, soldiers taught the secret of distillery to the Natives" (Morris 1879:23). The technology was simple. To distill from a fermenting mash of molasses, flour, dried fruit, and other cheap dry goods, all that was needed was fire, a tin teapot, and a condensation pipe—in southeast Alaska this latter was supplied by the fleshy tubes of kelp that swarmed in the waters of the inside passage. It was from the name of the village of Kootznahoo, near Sitka, for the venomous rotgut, dubbed *kootzenoo,* then *hochenoo,* that the term "hooch" entered the American language. And fundamentally the same method of distillation was introduced in Point Hope in 1887. In the absence of kelp, rifle barrels were utilized, with deadly effect, as condensation tubes.

Both expense and a constitutional dislike for military rule within U.S. territory led to the withdrawal of the army. A third factor was continued indifference to possession of Alaska and its "uncivilized tribes," who were vaguely identified as "wards" of the government. There was, in addition, virtually no northward emigration from the American mainland where postwar reconstruction was taking place along the railroads and in cities. Early hopes of Alaska's commercial possibilities were thus shelved. Refrigeration technology, for example, made the export of ice superfluous, and Alaska's fish canneries were developed only after 1900. The rule of the army thus came to an ignominious end, leaving law enforcement to the U.S. Revenue Cutter Service for the time being.

CHAPTER 4

The U.S. Navy, the Alaska Commercial Company, and the U.S. Revenue Cutter Service

The officers of the U.S. Revenue Cutter Service (USRC) were no strangers to southeast Alaska. As early as 1867, the Revenue Cutter *Lincoln* began a series of exploratory trips, its mission being to enforce U.S. law along the coast, suppress smuggling, and begin a systematic reconnaissance of Alaska's seaboard. The Cutter's job was also to protect and regulate commerce inside the three miles of Alaska's territorial waters—most importantly, the fur seal hunt on the Pribilof Islands, which had long been plundered by the *promyshlenniki*. But now the Russians had gone, and the field was open to American and British entrepreneurs.

Both the Aleut people, lured by the promise of liquor, and a diminishing seal population were at risk. But the territory was so vast and Revenue Cutter Service personnel so limited, that the hunt was impossible to regulate, and by the end of the century, the fur seal approached extinction.

Another big player came to harvest the fur seal, though in a more regulated manner. This was the Alaska Commercial Company (ACC). Formed in 1868, the ACC obtained a monopoly on the Pribilof Islands fishery in 1870. The slaughter of seals on the Pribilofs had, by this time, become national news. In 1868, a quarter of a million skins had been taken on one island alone, and the push towards an ACC monopoly with its annual catch quota of one hundred thousand also derived from an effort to conserve the seal population.

Given that the San Francisco–based ACC would soon deploy big, fast steamers and receive a renewable lease from Washington, DC, the company soon squeezed out most of its competitors. Unlike its smaller rivals, the ACC also established a permanent station on the Pribilofs. Here the company provided housing, school, and medical care to its Aleut employees and tried to operate a ban on the liquor trade.

As the USRC had done intermittently since 1867, the company also collected "Native curios," many of which it displayed in its own museum in San Francisco (these are now housed in the Hearst Museum in Berkeley, California). But the ACC contract was renewed only once. The company's enormous profits became a target of criticism and envy, and in 1890 its charter was withdrawn and the ACC went into a decline.

In its high period between 1870 and 1890, the ACC represented the major arm of U.S. commerce, law, and culture in Alaska. One sympathetic observer was Captain Charles H. Stockton, who was cruising north on the U.S. naval ship *Thetis* in summer 1889. Stockton, who would also make a significant visit to Point Hope that summer, disembarked at the

Fig. 8. *The Revenue Cutter* Bear, *built in Dundee, 1874, as a sealer. Fitted with a steam engine, its oak ribs were reinforced with steel plates to withstand sea ice. While the Revenue Cutter* Corwin *patroled the western Arctic between 1880 and 1885, the* Bear *undertook annual arctic cruises from 1886 to 1925. U.S. Coast Guard image.*

ACC station at Unalaska and wrote an enthusiastic article about it for *National Geographic*. Touring the ACC storehouses on Unalaska, Stockton was impressed by a luxurious sea otter fur, which, he mused, might fetch £170 at "the great fur market in London" (Stockton 1890:174). The company storehouse was, wrote Stockton, like Macys of New York. The paradoxical relations between "outposts of progress"—where Native peoples were employed to harvest what had previously been their own—and the commercial center that they served is neatly expressed here. Stockton's orders lay primarily in "looking out for the commercial and whaling interests of the United States in the Bering Sea and the Arctic Ocean" (Stockton 1890:172)—a task that would soon fall entirely to the Revenue Cutter Service, which had long policed only southeast Alaska and the Bering Sea. In the 1880s, the service started to increase its range, and dating from 1880, when the *Thomas A. Corwin* made its first far-northern cruise, the arctic trip became an annual fixture. Stockton's single journey to the Arctic, as we shall see, had particular impact on the future of Point Hope.

Two great Revenue Cutters, the *Corwin* and the *Bear*, played a vital and often dramatic part in the history of the next two decades, and they had two distinct, if opposing, constituencies. First were the commercial hunters, who had been whaling and trading along the Alaskan and Siberian coasts since 1850. Second were the Natives, who were nominally under the protection of the American government and whose well-being was being compromised by whalers who shared Revenue Service protection with them.

The commercial hunters consisted almost entirely of whale hunters: John Bockstoce has identified twenty-seven hundred annual whaling cruises to northwest Alaska between 1848

FIG. 9. *Michael A. Healy (1839–1904) with his parrot. The most celebrated captain of the Revenue Cutter Service, he commanded the* Thomas Corwin *and then the* Bear *from 1886 to 1895. Following an 1896 court-martial, Healy was reinstated to various other ships, 1900–1903. U.S. Coast Guard image.*

and 1914, and to the U.S. Treasury under which the Revenue Service operated, their activities were of high economic interest (Bockstoce 1986:15). The skippers and crews of the Yankee fleet consisted largely of Euro-Americans, Cape Verdeans, and African Americans, engaged, in the Euro-American view, in economically beneficial, if hazardous, work.[1] In the service of this enterprise, the *Corwin* and *Bear* did much to protect property and life, while also helping to promote early scientific work in the Arctic. As we shall see, both John Muir and the ethnologist Edward W. Nelson traveled on the *Corwin* in 1881. And the following years saw two significant coast guard expeditions up the Kobuk and Noatak Rivers.

The brilliant, controversial "Hell Roaring" Mike Healy, who took over the Scottish-built ex-sealing barkentine *Bear,* saw his mission on an even more heroic scale. "The *Bear* is more than just a famous ship," he wrote, "she is a symbol for all the service represents—for steadfastness, for courage, and for constant readiness to help men and vessels in distress." Another way of putting it was that "Bear ship," as some coastal Eskimos called her, was capacious enough to carry back south "the motley and numerous gang of castaways, federal agents, destitutes, missionaries, lunatics, scientists and prisoners" who wanted, or had, to get out. Two such Americans from the Point Hope shore-based whaling station were taken onto the *Bear* in July 1891, and as described by Alaska's missionary organizer Sheldon Jackson, they were archetypal cases of Healy's "motley":

> A deserter by the name of Wm Brown was received on board for medical treatment. Last winter through exposure he froze his feet and hands necessitating the amputation of his left hand. He was covered with vermin and greatly reduced in strength by starvation and neglect. Stephen Cushin, a miner, crippled with rheumatism was also received on board for treatment and taking out of the country. (Jackson 1890–1900: Entry for July 31, 1891)

FIG. 10. *Iñupiaq boys, possibly from Point Hope, on board a Revenue Cutter ship. The arm gestures of the two boys (right) and foot position of the boy at center show they are dancing. The officer (left) is learning the steps. A rare moment of culture contact which involves no goods exchange. Alaska Historical Library.*

The other major constituency were the coastal Eskimo who were being irretrievably changed by the presence of the white men the coast guard came to protect. Irving C. Rosse, the ship's surgeon who traveled with Captain Calvin Leighton Hooper on the first arctic cruise of the *Corwin*, wrote that he and Hooper were both filled with "enthusiasm for humanity," and described how he "examined and treated suffering Eskimos by the hundreds" (Rosse 1883:16–43). Dr. Samuel Call, medical officer aboard the *Bear*, was similarly active among Natives, and it was partly his popularity as a physician that led Healy to claim that shamanism was in decline.

The Alaskan missionary organizer and Commissioner of Education Sheldon Jackson, whose work is described in Chapter 14, and who traveled regularly on the *Bear*, responded to coastal Native distress with this dramatic assessment:

> ...hundreds of human beings starve and die—their fate unheeded by the world outside. To the starving Native of Siberia there is always the possibility of [Native reindeer herders] coming to their relief. But on the Alaskan side, where are no herds of domestic reindeer...there is nothing left the people but to starve and die. (Jackson 1890–1900: Entry for July 1891)

Assumptions of American cultural superiority sometimes nonetheless accompanied these expressions of concern. On the same journey, Jackson, Healy, and Dr. Call fell into conversation with some Eskimo shamans, and persuaded them to provide a demonstration:

> Captain Healy bantered one of the shamans who was visiting the ship to show his powers. He replied he could do nothing on shipboard. So we proposed to accompany him on shore, the captain making him a present of some powder, lead and tobacco. With the rattling of a Native drum, and monotonous Hi-yah-Hah chorus of women he pretended to suck from the flesh of Dr. Call, ship's surgeon, a piece of sponge, and a second time a piece of walrus hide. He then attempted to have two men strangle him with a rope, but could not make it work. A blind shaman then took up the performance and held an inch board 18 inches by 3 feet in size to his mouth by suction. He also allowed his hands to be manacled behind his back and then worked his hands through or in some way got the manacles off. It was a rather lame affair. We returned to the ship about 11 o'clock P.M. to find that Mrs Healy had spread a pleasant lunch for us. (Jackson 1890–1900: Entry for July 1891)

It was all too easy to ridicule "fakiristic" aspects of the tradition by encouraging displays to be performed out of context. But it was the mission of Jackson and his collaborators to introduce a "rational" Christianity, supported by scientific medicine, American law, and English-language education; it thus came naturally to discredit the indigenous systems it was the American project to replace.

But the Revenue Cutter's presence would always be intermittent and its influence a hit-and-miss business. Extended along an immeasurable coastline in heavy seas between early July and mid-September, the Revenue Service could do little more than occasionally touch base among Native communities—many of whose inhabitants would be scattered on hunting and trading journeys. And while the Revenue officers closed almost daily with whaling vessels, most of these could easily slip out of range if they wanted to avoid being searched for liquor. While Healy claimed that his *oomiak-puck-pechuk-tonika* (the big ship that was against "tonic") had almost entirely suppressed traffic in "Pain-killer," "Jamaica ginger," "Bay rum," and "Florida water" (Rosse 1883:20), the liquor trade went on more or less unchecked. One profound, if historically minor, irony lay in the fact that Healy—celebrated not without reason in 1894 by the *New York Sun* as the Greatest Man in America—was himself a victim of alcoholism; less to his credit, he indulged in at least one drinking binge with Point Hope's murderous "trading chief," Ataŋauraq (see Chapter 10).

Notes

1 For a list of other nations represented, see Aldrich (1889:17) as quoted in Chapter 6 below.

CHAPTER 5

The Commercial Whale Hunt—1

The disasters that overwhelmed nineteenth-century Iñupiat were not the result, as on mainland America, of white settlement or land appropriation. The Iñupiaq calamity was entirely on account of the whale men who arrived midcentury, infected the Natives with diseases, introduced them to alcohol, and plundered the sea until the whale and walrus populations were all but exhausted. As with the destruction of the buffalo, which happened on the mainland during the same period, Natives and white men were in pursuit of identical game, though for quite different purposes. While Native Alaskans needed whale and Indian hunters required buffalo both for subsistence and in order to sustain their cultural integrity, the white man's hunt was for short-term gain. In that the most useful manufacture to which baleen was put was the construction of umbrella frames, the whale hunt was also a culturally frivolous endeavor.

Numerically, the whale men who penetrated the Chukchi and Beaufort Seas of Alaska were few and their ships were scattered over thousands of miles of coast. The white man also came and went seasonally, seldom spending more than two or three years on a single voyage. Nonetheless, it took a remarkably small number of Euro-Americans with sophisticated harpoon technology to slaughter almost two entire populations of sea mammals and, during brief periods of contact with Natives, transmit infections which would—particularly in the summer months between the 1880s and 1905 when most contact happened—spread through whole villages.

The western arctic whale hunt started in July 1848, when Thomas Roys, in his bark *Superior*, entered the Bering Strait sea ice and found his ship surrounded by innumerable oil- and baleen-bearing bowhead whales. These were larger, fatter, more plentiful, and more docile than the Greenland right whale, which they closely resembled, and within just four weeks, Roys took eleven animals. Laden with huge quantities of oil and baleen, Roys headed south to Honolulu, where he published an account of his voyage in a church journal that circulated in the Pacific whaling community. Faced with a decline in both sperm and right whale populations from the North Atlantic to the South Pacific, the American industry was electrified by Roys's news. (Roys himself did not return. Not unlike Melville's Captain Ahab, who had lost a leg to Moby-Dick, Roys blew off his hand with an exploding rocket harpoon.) Within two seasons of 1848, some two hundred whalers had followed Roys's path into north Alaska, and the chase would continue until the bottom fell out of the baleen market in the early twentieth century.

The pursuit of the Alaskan bowhead corresponded in time and in impact with the massacre of the Plains buffalo. And while tortured Ahab was nothing like the bluff Roys, Melville was writing his novel precisely at the time of Roys's push into the Arctic. This historical coincidence chimes interestingly with a passage in *Moby-Dick* where Melville laments the destruction of both whale and buffalo, but then proceeds to a prematurely cheerful prophecy that "whalebone" whales would survive predation:

> concerning these last mentioned [baleen] Leviathans, they have two firm fortresses, which in all human probability, will for ever remain impregnable...the whale-bone whales can at last resort to the Polar citadels, and diving under the ultimate glassy barriers and walls there...and in a charmed circle of everlasting December, bid defiance to all pursuit from man. (Melville 1851:Ch. 105)

That the whale and the buffalo should have been hunted to near extinction during the same decades was no coincidence to Melville. Just as the Civil War (which postponed the buffalo kill for five years) represented the first fully mechanized modern war, so the pursuit of buffalo and bowhead whale exemplified the industrialization of hunting. Before turning to the whale hunt, it is therefore worth looking briefly at what happened to the buffalo.

Melville was on the East Coast when, in 1851, he published *Moby-Dick*, and the buffalo roamed the territory nine hundred miles to the northwest. The northern herd had been under pressure from white hunters since the 1830s, but in midcentury Plains Indians were still taking them freely. By 1840, however, Indian hunters were responsible for only part of the national kill: buffalo-skin robes and pemmican for shipment to Rocky Mountain trappers accounted for about one quarter of the northern Plains hunt (R. White 1994:246). The whale populations, on the other hand, were already in crisis, and Melville's conjecture that this species would retreat to the "ultimate glassy barriers" was accurate only in the short term, because later in the century steam-powered ships would be pursuing them into the high Arctic.

If the buffalo had been pursued simply for robes, the herds would not have been pushed to extinction so fast. At this stage, even the commercial hunt remained seasonal. Hides with short, new hair were worth taking only in the autumn and winter, and then they had to be cleaned and tanned by Indian women—which further slowed export from the Plains and meant that the process remained partially in Native hands.

It was not, however, the demand for robes and pemmican, but the demand for industrial components that doomed the bison. Untanned skins, easily transportable from new western railheads, now had a special utility as machine belts, and in the pursuit of this, four million animals in the southern herd would be slaughtered by 1874.

The hunt then turned north, and by 1883, once a railroad had been driven into the Dakotas, the northern herd was entirely destroyed. "The slaughter was so thorough and so quick," wrote historian Richard White, "that not even the [white] hunters could believe what they had done. In the fall of 1883, many outfitted themselves as usual. But there was nothing to hunt except piles of bones bleaching in the sun and wind" (R. White 1994:248). These statistics postdate Melville's prophecy, and while the "speedy extinction" that he predicted came slower than he had anticipated, when the moment arrived it also dealt a coup de grâce to the Plains Indians.

The fate of the Alaskan Iñupiat was similar to that of the Plains hunters in that their principal game animal was plundered and they were infected by diseases and incapacitated by the alcohol and tobacco that the hunters brought with them. But in significant ways, the Iñupiat were more fortunate than mainland peoples. By 1915, commercial hunters had left and the Iñupiat retained, as they do today, an almost unhindered freedom to hunt whales, albeit since 1978 on a quota. Equally significant, the white man of the nineteenth century had no interest in the Arctic beyond the goods he could export. Northwest Alaska therefore remained within Native control, and the handful of adventurers who chose to settle there made no appreciable impact on Native lands.

The Bowhead Whale

The herds that Roys met north of the Bering Strait were bowhead whale (*Balaena mysticetus*), though the taxonomy remained uncertain for over a century. The bowhead is a toothless species, and like the right whale (*Balaena glacialis*), which it resembles, it feeds through the black feathery plates of baleen (or "whalebone") that hang from each side of the upper jaw for trapping plankton, which its tongue then sweeps into the throat. The right whale was well known to Americans. It had been hunted very hard in Greenland and the eastern Arctic, and by 1850, yet more had been killed around Kamchatka and the Gulf of Alaska (the "Kodiak ground"). When Roys traveled beyond these waters in 1848, the plentiful species he located, which was fatter and more docile even than the right whale, at once rekindled the whale-hunting enterprise.

A few bowhead would have been present in southwestern Alaska, but its subtle differences from the right whale had gone largely unnoticed. The American whalers were of course interested in the largest whales they could capture: at first both for their oil and baleen, but soon just for the latter. A large adult might carry up to six hundred baleen plates weighing thirty-two hundred pounds. At the end of the commercial hunt, this harvest fetched $5 a pound, making the baleen from a single whale worth up to $16,000. When Roys sailed back through the Bering Strait after a month of hunting, he had captured eleven bowhead whales, filled his hold with sixteen thousand barrels of oil, and carried a stupendous quantity (unrecorded) of baleen. In the seventy years that followed this exploit, twenty-seven thousand separate whaling expeditions penetrated the same waters, 150 ships had been lost, and almost twenty thousand whales from a population of thirty thousand had been slaughtered.

The ice-loving bowhead whale of the high northern latitudes belongs to the Mysticete (moustached) group of baleen whales, and is the only species to inhabit the northern circumpolar seas. Between thirty and sixty feet long in adulthood, and weighing about one ton per foot, the bowhead gets its name from the contour of its head, which is molded by the arch of its upper mandible. The lower jaw, with its white chin markings (which help distinguish it from the right whale) and scattered white bristles, is supported by two vast, curved jawbones. Barnacles and lice that mark other baleen whales are absent from the thick, smooth, gently furrowed, blue-black skin of adult bowheads.

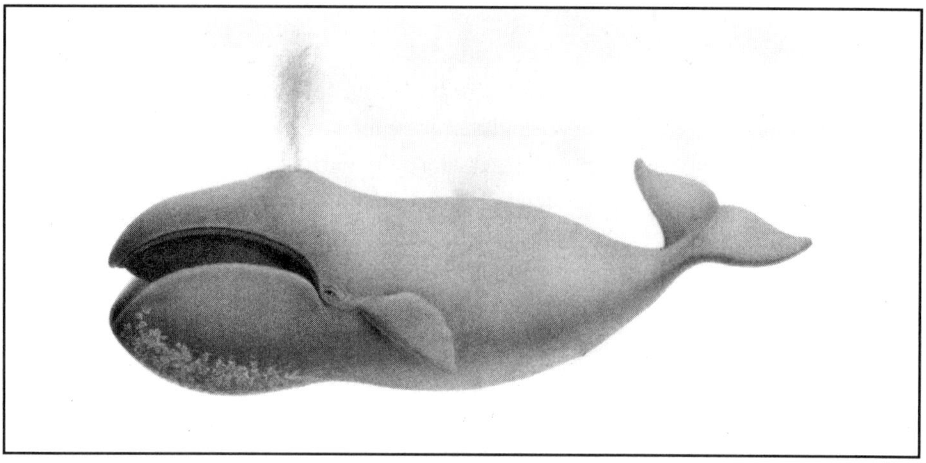

FIG. 11. *The bowhead whale. Engraving published during the period of the Alaskan commercial hunt. Author's photograph from C. M. Scammon,* Marine Mammals of the Northwestern Coast of North America *(San Francisco, 1874).*

Surrounding the body between the inch-thick skin and the inner tissue lies the whale's insulating fat. Ivory white or pinky yellow, depending on the whale's age, this fibrous blubber covers the whole body in a layer of varied thickness. The lips are virtually all blubber; the lateral flippers and tail flukes have a thinner covering. Elsewhere the blubber may be ten or twenty inches deep. The head of the bowhead is reinforced with flesh, fat, and cartilage for breaking young ice in frozen waters, and added to the head's weight are more than three hundred baleen feeding plates that hang from the two sides of the upper jaw.

In average years, the bowhead spend autumn and winter in the central and western Bering Sea. Native people say that when the bowhead are ready to migrate north, groups of age- and sex-related whales gather and move north together. Scientific observation suggests that the spring migration consists of three main "pulses" of whales of increasing size and age (Sheldon and Rugh 1995). As the ice shifts, closes, and opens again, a new group forms and waits for its moment.

Ice breakup in the Bering Sea starts in March and early April when temperatures rise and the wind shifts from northeast to south or southwest. Broken and haphazard lanes of water open through the Bering Strait and run north into the Chukchi Sea. These channels trend directly past Point Hope and up to the north Alaskan coast. But this shifting maze of lakes and alleys also includes a single, well-defined near-shore lane of water that lies between the land-fast ice and the older, thicker, and unstable ice pack. It is through this channel that the whales, with their preference for coastal waters, migrate. Females with a yearling and a guardian adult frequently migrate together. Smaller groups like these are followed every few miles by individuals within the same run, and such groups and individuals often surface in the same locations. Iñupiat report that whales leave patches of bubbles where they have risen, thus marking a trail for the herd to follow and ensuring that bubble patches are repeated.

Not much is known about how the bowhead subsists in the winter. There is little to eat on the northward migration, and half of the bellies of whales taken in the spring and examined by scientists in 1986–87 and 1993 were empty (Sheldon and Rugh 1995). A Point Hope man found a single gastropod in one; a few tiny fish and mollusks have been found in others. Bowhead migrate for one overwhelming purpose: to reach their summer feeding ground in the Beaufort Sea of northwest Alaska and east toward Amundsen Island at the boundary with Canada. There for four months they continuously forage, returning south along the coast of Siberia in the early autumn.

The main prey of the bowhead is the current-borne zooplankton: tiny, shrimplike crustaceans with the high fat content upon which whale blubber depends. Like other baleen whales, the bowhead feeds by swimming through plankton and gulping drafts of food and water. As the whale's mouth extends, the baleen packed on either side spring forward. Each baleen plate is edged with a fringe in which the plankton gets tangled. When the whale's mouth closes, the water is forced out and the tongue mops the plankton out of the fringes. This skimming and scooping action accounts for most of the process. In heavy banks of plankton, groups of whales sometimes feed together, often swimming counterclockwise a few body lengths apart. But mud streaming from the mouths of breaching bowhead, and pebbles and mollusks found in their stomachs, suggest that whales are occasional bottom-feeders.

The bowhead swim with long, deep dives interspersed with short periods at the surface. When they come up for air they exhale, inhale, and then dive shallowly several times in succession. As the whale rises, the head with its prominent blowhole appears first, followed by a V-shaped exhalation. This cloud of condensation is mixed with water from the grooves of the blowholes; the nostrils clap shut audibly when the blowhole submerges. The whale's roll continues as the rounded back emerges, and once the back has come up, the head again submerges. The back goes down, and a few seconds later the head reappears to begin the next roll, again blowing and inhaling. Each sequence involves some fifteen ventilations. At the last breath in a series, whales arch their backs, dive steeply, and swim deep underwater for up to twenty minutes before rising again for breath.

The channels that make migration possible are not always open. On any one voyage the whales inevitably reach places where the pack ice and shore-fast ice have come together, or where water has been newly frozen. But the bowhead has evolved to live in the sea ice, and its prominent nostrils allow it to breathe through cracks and tiny airholes. Its streamlined back gives it mobility beneath the ice, and whales sometimes find pockets of air between the ice and water.

In the absence of airholes, the bowhead can smash vertically through at least one foot of sea ice. Rising slowly and exerting pressure with the top of its head, the whale lifts the ice to make holes within ice hummocks for later whales to use and keep open. If the ice is too thick or completely impassable, whales backtrack and circle in ponds and small channels to stop the water from freezing. They can, in extreme situations, swim an hour or more beneath the sea ice. But when they rise from long periods under the water they are slow and exhausted.

For most of the spring the whales simply travel. The small, plump variety of bowhead known in Point Hope as *iŋutuq* enjoy loafing in calm bays. But while whales appear seldom, if ever, to sleep, they do play, mate, and calve during migration. For most of the year the adult whale is solitary, and migration provides pairing opportunities. Around cooperative females, males are competitive and jostle one another. They occasionally lash out with their tail flukes or get slapped; beyond this level of sexual rivalry they are a nonaggressive species.

Calving takes place between late winter and early summer; most calves are probably born around April. Born large, they suck from two nipples on the mother's belly. Calves are still with their mothers in April, May, and June, suggesting that they were born on the previous migration. Neonates swim close to their mother, often grasping the mother's flippers with their own; or they travel in a pool in the dip of her back, half swimming, half riding. As they grow bigger they play around and circle, swim back and forth, practice breaching, meander off, and then give chase to their mothers. In six months they are weaned. Besides the occasional killer whale, a rare visitor to the high Arctic, their only predators are humans.

Baleen and Its Uses

When the first ships entered the Bering Sea, they were in pursuit of both whale oil and baleen. By 1880, however, while the price of baleen had reached four dollars per pound, oil was fetching a mere sixty-five cents a gallon. In the light of this discrepancy in value and the long, messy labor of rendering oil onboard ship, the ship owners abandoned their trade in oil to focus almost entirely on the baleen harvest (Bockstoce 1986:260). So for the next three decades, until the hunt petered out at the turn of the century, only a tiny proportion of the bowhead's bulk was utilized. Once the head had been stripped of "bone," the whale's body was abandoned. And given that this hunt took place mostly on the open sea, relatively few carcasses lay in reach of Native boats. Like the elephant hunt in Leopold II's Congo of the 1890s, which was pursued for the European luxury furniture market, so baleen (and, to a smaller extent, walrus ivory) also fed into a relatively short-lived commerce. Iñupiaq people also had their uses for baleen. Sliced into strips, it is phenomenally strong, and this made it useful as lashing material and for fishing line and nets. It also made good sled runners and, once softened in hot water or urine, was used by Natives to make spoons, buckets, shovels, and mask pendants. But baleen was never central to Native technology. Sealskin rope and sinew made bindings that were easier to handle; ivory and bone made serviceable runners. Compared to stone, wood, skins, and sinew, baleen had only marginal utility. Thus, while commercial whalers deprived Natives of subsistence from the whale, the loss of the baleen itself involved no deprivation.

Interest in baleen in nineteenth-century Europe and America was short-lived. Demand arose in about 1850, prices climbed through to the 1890s, and the market collapsed around 1900, when the whales had all but gone and whalebone was anyway being replaced by spring steel. And just as baleen had been an Iñupiaq side item, so this product was a somewhat trivial accessory to Euro-Americans. Unlike oil, which had a civic utility for lighting and as a lubricant, the whale hunt was dedicated mainly to the acquisition of corset stiffeners, dress stays, and a small number of other minor products. In the first decades of the nineteenth century, the North Pacific whale hunt had been dedicated to the pursuit of bone for hooped skirt cages. Then from the mid-1870s until about 1900, the fashion for wasp waists arrived, and this called for corsets that were larger and more constricting. It was primarily this demand that took the whale hunt into the higher Arctic, and which drove the price of baleen to its dizziest levels. There were other uses for baleen. The thick end of the "bone" could be twisted into canes. Umbrella and parasol ribs, whips, children's hoops, billiard-table cushions, fishing rods, penholders, paper knives, shoehorns, and even the hilt inlays for ornamental Japanese swords were made from baleen, while brush manufacturers mixed baleen shavings with pig bristle and upholsterers stuffed mattresses with the peelings. These were the end products of an industry that, over thousands of dangerous miles, sent thousands of men in pursuit of animals, twenty thousand of which they uncomprehendingly destroyed and whose stocks could only, with luck, be naturally replenished. Few constituencies of producers and consumers—the roughnecks who pursued the whales and the mixed urban folk of the Euro-American fashion market—could have lain further apart. And concealed within this separation was a Native American population that was dying simply on account of being caught in their own territory.

The Native Hunt

Shifting sea ice surrounds the Point Hope peninsula all winter. When a channel in the sea ice opened in the spring and the bowhead migrated past the village, Point Hope hunters launched their skinboats. The beginning of the hunt was usually signaled in mid-April by duck, goose, and murre migrations, while the snow buntings, which came to nest in hummocks and iglu ruins, arrived as harbinger of the whale migration. Point Hope's *umialit* were poised for this moment. First to come rolling through the open water were herds of white belugas. The bowhead migration followed.

Whaling was done from skinboats, which held crews of up to eight men. The boat consisted of a sturdy wooden frame latticed with steam-bent wooden struts that were covered each year by seven or eight new skins from the bearded seal. A skinboat was a Point Hope family's most valuable possession. Each boat was managed by a married *umialik* couple; together these were the skinboat owners. The male *umialit* were the leading hunters in the village, and they used the meat and skins they had accumulated since the previous summer to pay fellow members of their ceremonial houses to work in their boats. The *umialik* woman's part was separate but equal. While her husband hunted on the ice, she, for the most part, stayed in the village. Her ritual passivity was vital to her crew's achievement, and some informants say that the woman was the real hunt leader.

In contrast to the partly hidden role of women, men's work was visible and strenuous. Their first labor was to cut a path for the skinboats from the south shore to the open water. Once a trail was ready, the boat sleds were man-hauled to the ice edge. At this stage each crew's main challenge was to maneuver the boat into a good position. Competition for propitious places in the bays and headlands of the ice lasted all season. As the crews spread the length of the shore to the cliffs at Cape Thompson, *umialit* looked for quiet water where the whales might loaf. If they could harpoon a whale as it rose to breathe, the crew would be saved an exhausting chase.

The moment the crew arrived at open water, a long vigil began. Each man was dressed in new clothes: two-layer caribou skin parka and trousers, lined boots, mittens, and

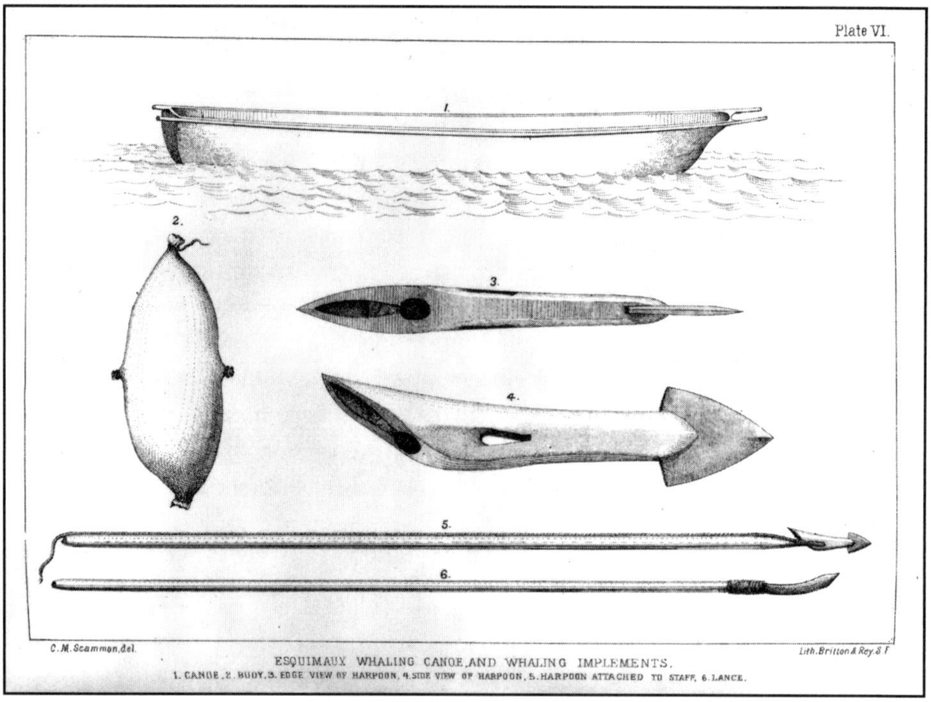

FIG. 12. *Iñupiaq skinboat with whaling equipment. The sealskin functioned to buoy a whale following a strike. Harpoon heads are fitted with blades carved from Cape Thompson slate. Author's photograph from C. M. Scammon,* Marine Mammals of the Northwestern Coast of North America *(San Francisco, 1874).*

eyeshades. In addition to the *umialik*'s share of harpoons and lances, the men brought their own equipment: knives, bows, bird spears, slings, and seal-hunting tackle. For about six weeks, and so long as a north wind kept a water channel open, they would stand on the ice, intermittently hunt the smaller species, and watch for bowhead. When the wind was from the south, the pack ice came in and the hunt was suspended.

Camping places were seldom secure for long. The same winds and currents that made the hunt possible could always shift the ice and force crews back to land-fast ice or to travel the coast until they found fresh harbor.

The hunters ate little. Each *umialik* provided a large wooden bowl of frozen meat, replenished daily by a boy—nicknamed "little muskrat"—who ran between the camp and the *umialik*'s iglu. When the whales were not running, the men hunted seal, harpooned belugas, and threw slings into the flocks of eider and murrres that swept overhead to their breeding places. Cooking on the sea ice was forbidden, so all meat taken on the hunt was pre-cooked in the village or eaten raw. The men drank melted snow from seal-flipper skins, which they kept under their parkas.

Bouts of hunting were scattered between long periods of inaction during which crew waited silently or listened to stories. The older men and shamans told the primordial myth that described how the Raven Man harpooned the sea monster whose body had become the

Point Hope peninsula. This story was talismanically associated to the hunt and its recitation was propitious. The owners of magical songs would hum and sing to bring whales to them. If the crews made other noises or fouled the ice—waste was dumped well north of the whales' path—the whales would be offended and go elsewhere. The cold, monotony, and hunger were exhausting. Some men took quick naps on the boat sled while the rest of the crew kept watch. If a whale rose by the skinboat, the harpooner's reaction had to be instantaneous.

Most whales came less easily. When lookouts saw a whale's back or its forked plume of steam, the skinboat was launched silently. The *umialik* climbed in last and pushed off with his steering paddle while the crew strained the boat through choppy water, sometimes closing in, then losing track, always watching as the whale dove under ice floes, often to disappear completely. When the skinboat closed on a whale, the outcome lay with *umialik* and harpooner. As the harpooner stood up in the bow, the *umialik* kept the boat's position steady from the stern. Everything depended on timing and coordination. The harpoon pole was heavy and the whale enormous. The harpooner aimed at a "death place," at the heart or between cervical vertebrae. Following his strike, the blade on the foreshaft disengaged. The harpooner laid his shaft in the bow. The sealskin line paid out across the gunwale; if it tangled or snagged, the boat would capsize and the whale might drag it under.

The harpoon was central to all sea mammal hunting. The five main parts were a wooden hand shaft, fore shaft, slate blade, sealskin line, and three drag floats made of whole inflated sealskins. The largest of these was fitted with a mask called *Iñuġluk* ("strange person"). Following his strike, the hunter twisted and detached the hand shaft; as the harpoon blade cut down and aslant into the animal's body, the fore shaft uncoupled. Detached from its shaft the harpoon line ran free, dragging the floats across the water; these slowed the whale's escape, helped drag it to the surface, marked its position, and buoyed it up until hunters could reach it.

Ideally a kill would be quiet and simple. But chasing a harpooned whale might take hours or days, and sometimes took the boat miles out to sea. All this time, *umialik* and harpooner sang. They sang songs to lure the whale, and songs to bring the whale floats back to them. The songs were also a signal to rival *umialit* that the first boat needed help. As supporting crews arrived, the men planted spears and harpoons in the animal. The boats finally lined up and towed the whale to the nearest firm ice. Here they tethered its body and the successful *umialik* sent a runner to the village to call his wife.

At this point the woman *umialik* entered a new ritual phase. Her whale had been taken. Dressed in new clothes, she led her crewmen's wives to the sea ice, where she offered the dead whale a drink from a ritual pot, using water from ice taken from a pond near the village. Once she had slaked the whale spirit's thirst—all sea mammals were believed to be thirsty for freshwater—she thanked the whale for coming, and welcomed it to her family.

The senior *umialit* now supervised the butchering. This had to be done before the ice shifted, otherwise the whale might be lost. The first *umialik* to fix a harpoon took the major share. In their order of arrival, all the boat crew got portions. The whole whale was dismembered and hauled back to the village on sleds, while the head was tipped back into the sea to ensure that the soul would be reincarnated.

After a whale had been butchered, the men spread out again and continued to hunt for as long as the ice allowed. Around the end of May, the ice became too rotten to support the boat crews, and most of the late migrants were too heavy and fast to make hunting possible.

Back in the village, the women started to prepare for a three-day festival to celebrate the *umialit* who had taken whales and to thank and propitiate the souls of whales that had been killed. After the skinboats had been drawn up on the beaches and the women had prepared food, the whole village circulated through the grounds of each ceremonial house. Here everyone was feasted with soured raw meat, boiled skin and blubber, tongue, heart, liver, and intestines. Finally, the flukes were sliced up and distributed.

On the last day of the feast, the dancing and singing came to a climax in a skin toss game (*nalukataq*). The young men of each ceremonial house cut handgrips round the edges of a walrus or sealskin blanket that had been removed from one of the skinboats. As the old men beat their drums, protected from the wind by upturned skinboats, *umialik* men and women leapt in turns onto the blanket and were tossed above the heads of the singers. As they leapt the women threw gifts of food and skins into the crowd, and the younger *umialit* competed in aerial acrobatics.

By now, the spring had brought new game into Point Hope waters, and as soon as the whale feast was over the community hurried to hunt seal and walrus from the south shore. Surplus whale meat was stored underground and left to freeze for use in autumn and winter.

Commercial Whalers: San Francisco to the Arctic

Driven by a need for oil after the Revolutionary War and the War of 1812, New England whalers launched explorations in the early nineteenth century that took their ships to the furthest southern waters and eventually back across their own continent to what, in the 1850s, was still Russian Alaska.

The Nantucket whaling industry had depleted America's own waters by 1820, and the most lucrative grounds until midcentury would be in the North and South Atlantic and South Pacific. These voyages, which lasted up to two and a half years in pursuit of sperm whale, were principally for its oil yield, but also for what Melville, in *Moby-Dick*, described as the "absolutely pure, limpid, and odoriferous" spermaceti, which had become a major resource in the candle-making trade. Like Melville's *Pequod*, whaling ships traveled down the coast of Chile, around the Galapagos Islands, to Hawaii, and finally east to the "Japan grounds." Honolulu, from the 1820s until about 1860, became the principal harbor in which the Americans refitted and took on fresh supplies—and it was from here in 1848 that Thomas Roys announced his breakthrough into the Bering Strait. Hawaii remained the Americans' refitting center until the sperm-whale population crashed in the 1840s; by the 1860s, once sugarcane became Hawaii's main industry, scarcity of labor and increased harbor costs drove the whalers to a more viable home base. Thus, from the late 1860s and 1870s, San Francisco slowly evolved into the business and supply center for the Alaskan enterprise.

Until that point, however, the journey from ports on the American East Coast even to southwest Alaska remained complicated and arduous. One problem was that by the 1850s, New England crews had become hard to recruit, and the ships had to stop in the Azores and Cape Verde islands to take on hands: these were mixed-race Portuguese (a number of whom would settle at shore-based stations near Point Hope and at Barrow). The next major stop after rounding Cape Horn during the southern summer was Hawaii ,where, like the sperm-whale hunters, Alaska-bound ships were refitted, and where they took on men and supplies. Like Cape Verdeans, the Hawaiians (known as Kanakas) and other South Pacific crewmen also became regulars in the western Arctic over the next fifty years.

On the basis of Roys's news, one fifth of the entire world's whaling fleet converged on the coast of Alaska in 1849, and then continued another thousand miles into the Arctic. Whaling logs for 1849–50 record "whales aplenty," and the ships reaching Hawaii that autumn averaged well over a thousand barrels of oil per vessel. But after three fat years, the rest of the decade was commercially dismal: despite some catches north of Point Hope, the word went round that the whales in the arctic ground had been exhausted. Why whalers initially struck rich and then suddenly failed is hard to explain. One twentieth-century theory suggests that the first rush of hunters may have wiped out a whole southern Chukchi Sea bowhead subpopulation. A nineteenth-century view corresponded to Melville's: the whale, observed one captain, was no longer the slow and sluggish beast that they originally found it. The bowhead had sensed danger and had grown very shy. Whatever the truth of the matter, the whalers had, by 1856, killed seven thousand animals out of a population of twenty or thirty thousand. And by 1914 they would bring the total slaughter to twenty thousand (Bockstoce 1986:15).

The 1860s and 1870s were marked by a series of further disasters. First, in the mid-1860s, whaling was interrupted by an astonishing military adventure, when the Confederate steamer *Shenandoah*, commanded by a grimly fanatical James Waddell, was dispatched in the summer of 1865 to the North Pacific with orders to "greatly damage and disperse the Great American whaling fleet" (Bockstoce 1986:106). Waddell burned, sank, or ransomed some thirty-eight ships, took over a thousand prisoners, and in the process dented the Union economy to the tune of $1.5 million. As news of the offensive spread, sixty whalers evaded the Confederates by sailing north to Cape Lisburne. This bizarrely efficient but anachronistic Civil War maneuver made whaling impossible. But by the time the *Shenandoah* was steaming south through the Bering Strait, Waddell learned that General Lee had surrendered. Only the party of Natives who extracted some harpoon irons from burned-out whalers reaped any advantage.

Limited commercial success returned in the late 1860s. But a combination of a depressed U.S. economy, falling oil prices, and terrible sea conditions in the 1870s almost completely put an end to the industry. The years 1871 and 1876, when fifty-seven vessels went down (with heavy loss of life in the latter season) were the most catastrophic. In 1876 there had been only twenty whaling ships in arctic waters, and these now concentrated on hunting walrus, whose population they severely reduced. A flood of new industrially manufactured substitutes for whale oil—particularly kerosene, which had gone into widespread use as early as 1856—made the pursuit of whale oil an insupportable expense. This diminution of the commercial hunt took pressure off Native subsistence, while a decrease in contact also had direct health benefits for coastal Alaskans (Larson, personal communication 2006).

But two new factors sustained the commercial whalers for another three decades. One was the sudden new demand for corset bone. Second, steam-powered ships could go at nine knots (in contrast to six under sail), and could thus maneuver faster and more safely in sea ice. Steam was expensive in both men (requiring trained engineers) and materials. But this was offset by baleen profits in the 1880s and 1890s, and these in turn made longer tours possible. Increased power enabled whalers to reach the high north Alaskan coast by August, hunt in the Beaufort Sea, follow whales along the Siberian coast in the autumn, or steam east to Herschel Island, where they could spend the winter. Between 1880 and 1914, these became settled patterns, and it was during this period that longer and more destructive contact with Alaskan Natives occurred.

The early 1880s marked the beginning of San Francisco's domination of the industry. Coordinated by the dynamic and far-sighted ex–merchant skipper Josiah N. Knowles, the Pacific Steam Whaling Company became the business hub of an increasingly industrialized commerce. In 1869, San Francisco had finally been connected to the East Coast by the railroad. Now, to equip and provision the almost 350 voyages that started and finished in San Francisco in the 1880s, the wharfs became a busy outfitting and warehouse center, where baleen and oil were also processed and then packed by rail to the manufacturing heartlands. But steam, which tripled the cost of a voyage, by no means drove out sailing schooners. More than half the vessels that left San Francisco for their eight-month tour south in November and December and then to the Arctic continued to travel with the wind (Bockstoce, personal communication 2005).

The Final Years

The last phase of the hunt was set in motion by two men with Point Hope connections. Charles Brower had spent some months of 1885 whaling with Native people in Point Hope (see Chapter 11). In 1888, when Brower had set up his own whaling outfit in Barrow, he commissioned the Welshman Joe Tuckfield (who was later to live in Point Hope) to explore the Mackenzie River delta, where there had been Native reports of whales. Tuckfield returned to Barrow to announce a whale population that was "thick as bees," and that Herschel Island just west of the delta offered that rare thing in the western Arctic, a good harbor. This news reanimated the industry, and on Josiah Knowles's initiative, ships, warehouses, and supplies were dispatched to the tiny island from which the Pacific Steam Whaling Company would launch its last—and briefly most profitable—push.

The two seasons of 1892 and 1893 were spectacular. In 1892, the *Mary D. Hume* steamed into San Francisco after twenty-nine months at sea carrying baleen from thirty-seven whales (Bockstoce 1986:266). The following year saw a gigantic slaughter of 286 whales. But as in the 1850s and 1860s, these successes were followed by a run of poor seasons.

Fifteen ships nonetheless overwintered at Herschel in 1894, and fifteen hundred people were resident on the tiny island. The American whale men had worked out how to bank their ships up in the ice and snow, but they could not hunt seal or caribou or sew the skins they needed for the winter. Even the community of local Inuit and Gwichin Indians who converged on Herschel were insufficient to service their needs. The whalers therefore began taking onboard Natives from the northwest Alaskan villages they passed en route to Herschel Island. During 1894 and 1895, virtually the entire population from Barrow and almost a hundred Point Hopers traveled to Herschel. In the desperate search for subsistence, Point Hope people were also moving north to work at the shore-based whaling stations run by John Kelly and Charles Brower at Point Belcher and Point Barrow. Brower visited Kelly's station in December 1891 and described his house as "filled with Point Hope people" (Brower nd:417–18), though it is possible that some of these were ex-Jabbertown employees from the Noatak River (Cassell 2000:279; Larson, personal communication). By 1893, Kelly was again whaling in Barrow with a probable mix of Point Hope and Noatak crew members (Bockstoce 1986:238). The result of these northerly migrations was to make Point Hope's remaining population more and more dependent on the local whaler-trader society at Jabbertown, a community that, as we shall see, resembled the larger Herschel Island enterprise.[1] These events are described in more detail in later chapters.

As the price of baleen sank to all-time lows, these last twenty years of commercial whaling paradoxically represented the highest points of culture contact at Point Hope. The whales had largely gone—fifteen hundred whales, or 80 percent of the total kill, had been taken as early as 1883—but the white man, in the hope of a sudden upturn in luck, and resorting increasingly to the fur trade, stayed on in the Arctic through to the turn of the twentieth century. At Point Hope, these were the men of the Jabbertown whaler-trader station, a number of whom, like Joe Tuckfield, married into the community and also provided both employment and subsistence opportunities. Reduced by hunger and disease, a large proportion of the coastal Inuit, Point Hope people included, were thus driven into the service of these Americans.

Notes

1 A further rationale for the Point Hope migrations may have been that people under taboo following death in the family could still participate in the hunt for *commercial* ventures without directly compromising their relatives' hunting at home in Point Hope (Mary Larson, personal communication; Larson 2004:79).

CHAPTER 6
The Commercial Whale Hunt—2

> I can not speak too highly of the skill, seamanship, courage, and endurance of the whaling masters. They are a fine body of American seamen.
>
> —Commander Charles H. Stockton

> Consider that, with the majority of them, the very fact of their being sailors, argues a certain recklessness and sensualism of character, ignorance, and depravity; consider that they are generally friendless and alone in the world; or if they have friends and relatives, they are constantly beyond the reach of their good influences; consider that after the rigorous discipline, hardships, dangers, and privations of a voyage, they are set adrift in a foreign port...consider that by their very vocation they are shunned by the better class of people, and cut off from all access to respectable and improving society; consider all this, and the reflecting mind must very soon perceive that the case of sailors, as a class, is not a very promising one.
>
> —Herman Melville, *Redburn*

The Men of the Whalers

With the notable exception of this accolade from Commander Stockton, a chorus of denunciation of the commercial whalers arose from the late-nineteenth-century churchmen who deployed missionaries to counter the physical and moral depredations that the whalers inflicted on Alaska Natives. They were, according Sheldon Jackson, of "an inferior grade...emphatically hard cases" (Jackson 1890a). The whalers, he continued, were "the vilest men," pursuing a destructive profession that wreaked heedless injury on the people of the northwest Arctic—never mind that crews themselves endured exceptionally foul and dangerous conditions.

The rapacity of the commercial hunt was partly driven by the system of "lays" or shares in net profits—a system that coincidentally paralleled the Native distribution of shares in an animal kill (Aldrich 1889:107). In lieu of wages, an entire whaling crew, from captain to cabin boy, was paid in accordance with the success of a voyage: one third of profits going to the ship owners, one third to the ship's maintenance, and one third divided among the men. According to rank, the officers received shares of between 1/8th and 1/100th, while ordinary seamen made do with "long lays," which generally amounted to 1/160th or 1/250th. Given that many in the crew would start a voyage in debt for their clothes and equipment, and that once under way they could only buy from the ship's slop chest on credit, many would be "paid off" on return to San Francisco, their debt written off, with a single "iron dollar." The whale and walrus hunt was thus charged with ferocious determination for the maximum kill. And since every whaling voyage was also a trading journey, it is scarcely surprising that many officers left port with trade goods, which, besides legitimate cargos of dry foods, bolts of cotton cloth, tobacco, and metalware, included illegal but lucrative liquor and firearms. An ordinary seaman's journey would most likely be futile unless he could also do similar private trading. The very economic structure of the enterprise therefore contained, in the shape of the liquor trade, a built-in destructive potential even before it reached the Arctic. Given that the men would also spread diseases and abuse Native women, the human cost of trade relations was incalculable.

These inauspicious circumstances for both the commercial crews and the Natives were compounded by terrible shipboard conditions. Diet was a regime of hardtack, potatoes, salt meat, molasses, and occasional servings of duff (a boiled pudding of flour and molasses), with the more wholesome and anti-scorbutic addition of fresh meat when available. Life belowdecks was atrociously crowded, dirty, and violent, with discipline correspondingly brutal. Whaling vessels, wrote the American naval historian Samuel Eliot Morison, "brought out the worst traits of human nature [and were] more efficient schools of vice than reformatories… [S]kippers, who on shore passed as pious church members, were cold blooded, heartless fiends on the quarterdeck…blackguards who degraded the flag and the name of America" (Morison 1921:324).

The competence of the crews, many of whom had been entrapped in San Francisco by company agent "land sharks," was very poor and deteriorated in the later nineteenth century (Bockstoce 1986:326). And the fleet Stockton vaunted was in fact only "American" in that it was run by Yankees, served an American urban economy, and drew largely on white men for its senior officers. Subordinates and ordinary seamen were otherwise an assortment of multinational casual workers. When the journalist Herbert Aldrich left San Francisco on board the *Young Phoenix* in March 1887, the thirty-four crewmen consisted of Portuguese, Scandinavians, Germans, Spaniards, Englishmen, Irishmen, and a handful of American ranch hands (Aldrich 1889:17). The most skilled of these, as we have seen, were Hawaiians and mixed-race African-Portuguese Cape Verde islanders. But with the completion of the transcontinental railroad and post–Civil War opportunities opening in the far west, there were relatively few American-born volunteers who would be interested in a dead-end whaling lay (Hunt 1975:127). The result, according to one writer, was "a set of miserable scamps who sold the vilest whiskey to the Natives and robbed them of their hard-earned furs, whalebone and other trade" (Stuck 1920a:177). In January 1887, a *San Francisco Chronicle* journalist expressed the same opinion in greater detail:

> Crews are of a most heterogeneous description; men who have never seen the sea… some of the greatest drunkards…jail-birds and thieves. The majority who can pull a boat are Kanakas…the lower grade officers are nearly all colored men…from the Cape

> Verde islands... The mates and masters are, with few exceptions, Americans, hailing from New Bedford or other Eastern whaling ports... When it is remembered that the crews of whaling ships are composed largely of a useless set of men and partly of a lot of vagabonds who speedily demoralise the others, it must be conceded that a strong hand is required to keep order and preserve discipline... (Bockstoce 1986:226)

It was this culture of "scamps" and "beachcombers" (men who had jumped ship) that Sheldon Jackson and his colleagues set out to challenge in the 1890s in their push to evangelize some of the coastal villages. Dr. Driggs at Point Hope was notable among missionary voices for not waxing morally hysterical about the whale mens' failings; accounts of both Point Hope's more decently behaved whaler-traders and their "roustabout" colleagues, along with their relationship with the missionaries John Driggs and Elijah Edson, who both befriended and defied them, are given in Chapters 16, 18, 20, 21, and 26.

Contact with Coastal Iñupiat

The Native and commercial hunts were geographically separate from each other and were usually conducted at different times of the year. The Natives pursued the whale from the sea ice in April and May while the white man had to wait until July, when he could maneuver in open water. Until shore-based stations at Barrow and Point Hope were established in the late 1880s, there was thus little contact between coastal Iñupiat and white men until the summer, when Native boats approached the ships on the open sea or the Americans came ashore for fuel, food and water, and to recruit Eskimo labor. Since almost all whalers were also traders, July, August, and September were when goods were exchanged, liquor entered the villages, and the seamen transmitted diseases to Native populations. Earlier we saw how confidently both Northwest Coast Indians and Eskimos approached the ships belonging to Cook, Kotzebue, and Beechey, rightly assuming that the Europeans would, like themselves, be traders and that their goods would be useful. Despite new and ongoing abuses by Euro-Americans in the late nineteenth century, the Iñupiat kept coming, for the white man's imported goods became increasingly valued as replacements for the game resources the commercial hunters had depleted. Flour, biscuits, rice, and other dry goods entered the Iñupiaq diet to supplement or substitute for the missing whale and caribou. A decline in the caribou population in the 1880s, possibly through cyclic change, but perhaps on account of overkill with repeating rifles, left coastal Natives also with insufficient skins for clothing, and the bolts of calico that traders brought with them could never be an adequate substitute. Game shortages similarly rendered the harpoons, arrows, lances, and other traditional equipment inefficient. The men needed rifles, ammunition, steel knives, and harpoon irons. Women required metalware to cook non-Native food such as flour soup and bannock, and scissors and steel needles to work manufactured cloth for both their families and visiting whalers.

As we saw in Beechey's trading inventory, the white man had also long ago created a market in accessories such as mirrors, glass beads, buttons, colored yarn, and combs. The repertoire in new trade goods was increased in the later nineteenth century to meet modified hunting needs and increased dependence on cooked food and a liking for hot drinks. In addition to nonrepeating gun and harpooning equipment, the traders brought steeltraps, axes, saws, hammers, nails, and other hardware. The new staple foods included tea, coffee, sugar, beans, macaroni, canned meat and milk, and dried fruit. All these goods—in addition to alcohol and tobacco—were welcomed by the coastal populations. In consequence, many Natives on the coast north of the Seward Peninsula partially adapted their subsistence activities so as to coincide with the arrival of ships both on the open sea and at regular places of landfall. Henry Woolfe, who traded near Point Hope in the mid-1880s, described this scene on the Seward Peninsula coastline:

> The coast Natives abandon the winter houses as soon as the first sign of thaw is visible, erecting tents at convenient spots along the coast where seals can be shot amid the shore and ground ice, and there await the arrival of vessels whose masters are known to be inveterate traders... Everything in [Native] possession, from large bundles of whalebone to the smallest ivory or bone carving, is brought for trade by these people. (Woolfe 1893:14)

Like coastal Alaskans, the Siberian Chukchi at Plover Bay prepared for the arrival of whaling vessels. One American captain wrote:

> [We were met by] a dozen skin canoes with eighteen to twenty-five people in each. In fur bags were boots, mittens, coats, caps, and pants, all made during the winter out of seal skin to be traded with the whalemen... [In return] some would want tobacco; others asked for cartridges. Again some wanted tea, while many would demand powder, lead and primers. If there many [Native] articles grouped together to warrant paying the price, then a trade would be made by giving a rifle with some cartridges. Not unfrequently they catch a whale, the bone of which is divided among those who take part in the capture, to be traded along with the ivory that they save from the walrus for the useful articles to be found aboard ship. (Cook 1926:145)

This observation is one of many commenting on the ways Native people adapted to the presence of American commerce. The Iñupiat themselves had always been traders, but as we have seen, trade between Natives often alternated with warfare, and this same tension found expression in relations between Natives and white men. Regional trading was formal and had, as I have noted, the aim of achieving a balance of exchange between contrasting products. In addition to fulfilling subsistence needs, this barter had a socially useful aspect. If the homeostasis achieved through barter was temporary, formal partnerships between individuals in different societies, often handed down to the children of the participants, provided relatively safe avenues of communication and reduced the frequency of trade dissolving into conflict. Each Native generation also grew up with personal and inherited memories that gave trade and warfare a historical context. The individual acted both within collective expectations of home and away Native groups, but also as a participant in a continuum that had the justification of ancestral practice. In these ways, people pursued both trade and war in a state of mutual, if uneasy, comprehension.

Relations between Natives and white men were filled with the same uncertainties, but with less mutual understanding. Captain Rochfort Maguire, whose HMS *Plover* wintered off Barrow from 1852 to 1854, and who himself took an unusual interest in Native life, gives a good overview of an indifference that was shared by the American whalers:

> [F]inding no prospect of any understanding being come to with the Esquimaux and our Men, the latter in fact seeming to have no wish or inclination to renew their visit to the Village—I found if we pursued the system of holding back, a valuable opportunity would be lost of acquiring a knowledge of their habits and customs which it would be interesting to know. (Bockstoce 1998:145)

The interests of Natives and Euro-American whalers nonetheless coincided on two things: trade and sexuality. And if Native people saw sexual relations as a means of getting something that required relatively little effort, in trade both parties were intent on getting as much as they could for minimum output. Native traders were often on their guard against parting with their best objects, and knowing the premium on baleen and ivory, they could often hold out for satisfactory bargains. Sometimes they would tease the white man. "The Eskimo," wrote one observer, "never show all the trade they have, but carry a little here and a little there, perhaps producing something new every day, but always keeping something behind." But then the same Native might grow weary of this game. "An article," wrote the same observer, "will be taken to one ship and Native will refuse to part with it. He will then approach another vessel and gladly exchange for a fraction of what has just been offered for it. A [metal] louse-comb…will frequently purchase what a knife would not have bought an hour before" (Wrangell 1980). This behavior, which the wife of one whaling captain described as "sly" and "cheating," speaks of canny self-protection in transactions between vigorous protagonists, in which, nonetheless, the white man would ultimately win. In contrast to the well-comprehended arena of inter-regional Native trade, the tendency often arose in the end to throw things away and grab the white man's offer. But some Native traders also established good partner relations with white men, as did Samaruna at Point Hope in 1888 (see Chapter 9): each side trustingly fulfilling orders made a whole year ahead. The trade in alcohol, however, on which many Euro-American traders depended in order to secure advantageous deals, represented the most destructive element in all these transactions; and it was largely this which so demoralized the societies that fell in with it.

The Liquor Trade

> Illicit traders, taking advantage of this northern craving for drink, have of late years been in the habit supplying the most villainous compounds, in exchange for small quantities of which the improvident Eskimo gives his choicest furs. Some captured specimens of these prohibited articles, bearing the respective labels of *Bay Rum*, *Jamaica Ginger*, *Pain Killer*, and *Florida Water*, with a view to defrauding the revenue, proved on examination to be nothing by cheap alcohol of a highly inflammable nature to which a little colouring matter had been added. (Rosse 1883:20)

Native life over many centuries had achieved continuity on account of strong social codes and the discipline of people's responses to environmental demands. Every capable member of a community played a part in the work of subsistence, and every village economy was shored up and extended by an inter-regional trade that could usually make up deficiencies in the local system. Nothing was ever perfect, and there were famines and Native wars. But nineteenth-century Alaskan Iñupiat continued to live in equilibrium with their environment and in relatively stable relationship with their neighbors. Early European visitors such as Cook, Beechey, Kotzebue, and Maguire did comparatively little to destabilize Iñupiaq life, although even Cook's officers, as we have seen, took advantage of Eskimo women (Fortuine 1989:266). But apart from perhaps introducing venereal disease and reinforcing a Native taste for tobacco, they left coastal societies intact. The impact of the commercial whalers was of a different order: there were more of them; they were less disciplined, stayed longer, destroyed essential game, and spread disease; and most of them traded in alcohol.

Not everyone was a liquor trader, and the probity of certain individuals stands out, in particular those like Captains John A. Cook and W.A.P. "Little" Porter, whose wives overwintered with them onboard ship. Also there were characters such as Captains Barney Cogan and Ebenezer Nye, who expressed insight into the impact of the commercial hunt. "Captain Cogan has taken very few walrus," wrote his colleague Nye in 1879; "he says that for every one hundred walrus taken a family is starved, and I concur in his opinion. I don't want or need money bad enough to go for walrus" (Cook 1926:74). Nor did the big operators, such as the Pacific Steam Whaling Company and the earlier Alaska Commercial Company, engage in alcohol trading. But as E. W. Nelson pointed out, many Native people expected liquor in exchange for their goods: "At many places from Point Hope to Point Barrow we were offered whalebone, ivory [and skins]. Whiskey and cartridges seemed to be about the only articles desired by these people in exchange" (Nelson 1899:231). This was a position that some shore-based traders, such as Heinrich Koenig (an ex–Salvation Army man) at Point Hope, who refrained from the liquor trade, stood out against. At the other end of the spectrum there were criminally unscrupulous operators such as James McKenna and Edwin Newth, the two most notorious liquor merchants of the 1890s. Of them the Jabbertown trader Jim Allen wrote:

> I have heard it said that Captain McKenna would have the bottom of the hold of the *Fearless* lined with barrels of liquor to trade with Natives along both the west shore and the east shore. Captain Newth was another who did that sort of thing, but not on so large a scale... In 1896 at Herschel Island, [McKenna's] ship was called the hock shop. (Allen 1978:140–41)

And among the shore-based villains was Peter Bayne, who both traded in liquor and taught Point Hope people how to distill their own spirits.

The impact of alcohol on arctic peoples was immediate and dramatic. Possibly on account of a genetic intolerance shared by many Native Americans and Asians, the effect on Iñupiat was like that of a swift-working poison. Anti-social behavior, murder, suicide, and abuse of women and children were the overt results of drinking. Less visible were fetal alcohol syndrome, cirrhosis, and depression. And in varying degrees, this entirely new and dangerous set of behaviors weakened resistance to disease, disrupted family and community life, and often rendered men incapable of hunting.

While Euro-Americans could not understand how they were spreading epidemic diseases, clearly they did comprehend the evil associated with alcohol. One Iñupiaq loan word for liquor was *taaŋŋaq*, from the English "tonic"—as in "Here's my tonic!"—though the term probably had its origin through contact with the Chukchi when Captain T.E.L. Moore wintered HMS *Plover* just south of the Bering Strait in 1848 (Bockstoce 1988:23). But even the morally neutral rum in the captain's locker could represent a poison when given to people with no tolerance for it. The issue was scarcely new. Like the diseases to which Native people had no resistance, liquor had been ravaging Native populations across the continent since the early sixteenth century.

Observers who witnessed the effects of the liquor trade were unanimous in their condemnation. But despite the Congressional Act of 1873 prohibiting a liquor trade with Natives, whisky continued to pour into the Arctic. Partly in an effort to protect the coastal populations, at least during the summer months, the U.S. Marine Revenue Cutter Service began widening its tour of duty. The mandate of the Service was also, somewhat paradoxically, to protect legitimate American commerce, which included the whale hunt. But as Captains C. L. Hooper and Michael Healy made clear in their reports through the 1880s and 1890s, at least there was an intention to protect Native populations by attempting to enforce the federal law.

Hooper's initial comments emerge from his first far-northern tour on the U.S. Revenue Cutter *Corwin* and paint a dismal, though exaggerated, picture of life on St. Lawrence Island:

> [T]hey make houses, boats, clothing etc., of the skins of walrus and seals, and sell the bone and ivory to traders for rum and breach-loading arms. So long as the rum lasts they do nothing but drink and fight. They had a few furs, some of which we tried to buy to make Arctic clothing, but, notwithstanding their terrible experience in the past, they refused to sell for anything but whiskey, breach-loading rifles, or cartridges. (Hooper 1881:10)

The naturalist John Muir, who was traveling on the *Corwin* on the same visit, described a crowd

> of "jolly, well-fed people" who came alongside the steamer, offering ivory, furs, sealskin boats etc., for tobacco and ammunition. There was much inquiry for beads, molasses, and most of all for rum and rifles, though they willingly parted with anything they had for tobacco and calico. After they had procured a certain quantity of these articles, however, nothing but rifles, cartridges, and rum would induce them to trade. But according to American law, these are not permitted to be sold. (Muir 1881:24–25)

Later in the decade, the young journalist Herbert Aldrich gave this far more alarming account of Point Barrow in 1887:

> I went ashore and while there, saw the sad effects of the pernicious habit of treating, to which Americans are so wedded. Over half the adult Natives, regardless of sex,

> were intoxicated. Some of the men were paralysed with liquor, and one old fellow was frothing at the mouth in his crazy delerium [sic].
>
> A little tonga—liquor—would have purchased the whole settlement and everything in it... [L]iquor is not directly traded, but now and then a captain...lubricates a bargain for whalebone with a few drinks of whiskey, and perhaps makes a makes a present of a bottle or so to the unfortunate possessor of the bone. Such cases are few but they could be fewer... (Aldrich 1889:114)

At Point Franklin, Aldrich found two whole settlements of fifteen iglus each deserted:

> These habitations are all tumbled in. The Natives do not know how long the settlements have been deserted... Probably the decay of these settlements dates from the great wreck season [1871]. Not only were many Natives poisoned by drinking the contents of the medicine chests at the time, but many more moved away out of superstitious fear. The population here has been reduced at least one half in twenty years. (Aldrich 1889:148)

Seventeen years later, Aldrich noted how drunkenness led men to miss the hunting season:

> [S]tarvation was thus brought on. But other agencies have also aided in this decimation. The introduction of firearms has driven most of the [caribou] far inland, many Natives following. White men have...driven the walruses from the shore, and greatly reduced the numbers of hair seals and whales. All these causes, together with the great mortality among children, much of it natural, but in some instances, intentional, have caused several years of famine and death. (Aldrich 1889:141)

The Revenue Cutter's summer patrol provided at least some incentive for the traders to practice subterfuge, and by the mid-1880s they had complicated the traffic by offloading their rum on the Russian side, thus providing Siberian Natives with extra supplies, some of which they would occasionally bring to Alaskan Iñupiat with value added. Healy apparently believed he was making progress in eliminating the trade, claiming that:

> whiskey traffic in northern Alaska has almost ceased. The beneficial effects of our annual cruises are apparent in the changed condition of the Eskimos. Sickness has decreased; the people are better clothed; more attention is paid to their boats; food is plentiful; fur, bone, and ivory for trade are abundant, and the large number of healthy young children in every village dissipates former fears that the race might become extinct. (Healy 1889:17)

Sheldon Jackson was overconfident in the Revenue Cutter's effectiveness.

> Healy's record record is...brilliant in protecting half-civilized Natives from...the demoralization that comes from the white man's rum... The temptations were very great when a bottle of whiskey would purchase $500 worth of furs, and the profits were a thousand fold. Through the vigilance, firmness and tact of Captain Healy, this trade has been almost broken up on the American side of the coast: so that no part of the United States is today so free from intemperance as Arctic Alaska. (Jackson 1891)

Jackson was of course deluding himself that the liquor trade had been "almost broken up," and his optimism was at odds with Jim Allen's analysis of Captain Healy's behavior, discussed earlier.

Confirming Allen's account, Herschel Island's visiting missionary Isaac Stringer wrote in February 1895:

> There are fifteen ships wintering at Herschel Island and some of the captains appear to be more unscrupulous in the matter of giving whiskey [than previous ones]. It is my opinion that the Revenue Cutter 'the Bear' is a sham and a fraud. The Commander—Captain Healy is supposed to inspect all the whaling ships and allow only one case of rifles and a limited quantity of liquor on each ship for the men's use. And yet it seems that an unlimited supply of both rifles and liquor are brought to Herschel Island. If this goes on for a few years the Eskimos and Indians will be completely demoralised. (Stone 1981:110)

On a smaller scale, there were incidents of gross deceit which a few drinks rendered invisible to the Native trader. One such transaction on the Siberian side was described in some detail by Captain William McIntosh of the whaler *Charles Hanson*:

> When his bone and skins had been brought on board the Native was invited into the cabin where a bottle and glass were set before him, and in a short time his two hundred pounds of bone was bartered for twenty gallons of alcohol, say $10, a box of hard biscuits, $1, and a sack of low grade flour at fifty cents, in all $11.50, while the two hundred pounds of bone will realize in the eastern market $6 per pound, or in other words the captain got $1,200 worth for $11.50. (*Nome Nugget*, July 5, 1905)

Edward J. Knapp, who worked in Point Hope in 1905, similarly pinpointed how easily Natives could be enticed:

> Knowing full well the appetite which the Natives have for intoxicants and the value of the pelts and whalebone which they accumulate...the whalers come supplied with cheap whiskey, which they peddle out in exchange for the Native commodities... The whaling fleet always arrives in the Arctic before the Revenue Cutter does, and in cruising northward succeeds in keeping the cutter outdistanced. The whaling fleet usually arrives off Point Hope...two to three weeks before the cutter appears there. (*Seattle Post-Intelligencer*, September 24, 1905)

And when Natives were invited aboard for legitimate trade, alcohol could lead to the abuse of both women and the children they had brought with them:

> [T]he ship was more or less crowded with Natives night and day...and a large proportion of them were often girls, ranging from 12 to 20 years of age. These young girls were enticed into the officers' quarters and into the quarters of the seamen and there debauched, the unfortunate creatures receiving for their favors a handful of hardtack or an old shirt... [T]he same conditions exist on the American side from the Cape Prince of Wales up to Herschel Island. (*Nome Nugget*, July 1905)

Alcohol took deeper root within the social system when Natives learned to distill their own spirits. As with Indians of Russian America and at Point Hope in 1889, people all along the coast took enthusiastically to constructing their own stills, which, given a small quantity of dry goods, yielded plentiful supplies of lethally intoxicating liquor. Sheldon Jackson identified one of the difficulties within this situation. To eradicate home distilling, it would be necessary to withhold useful and legitimate trade items such as flour, sugar, molasses, and dried fruit—and this was not going to happen:

> The white men furnish the intoxicating liquors or materials from which they can distil liquor, and then when the white man's life is endangered by drunken Natives, they appeal to the Government for protection from their own crimes. Indeed at every [reindeer herding] station visited, the white men would speak of the smuggling of liquors with the attending drunkenness and demoralisation that was going on at the adjoining trading post. (*Nome Nugget*, July 1905)

Knowledge of distilling technology was also spread by Native people themselves. The missionary teacher Vene Gambell on St. Lawrence Island despondently described how a Point Hope woman came ashore from a whaling ship in 1887 and within hours had shown villagers how to ferment a mash of flour, molasses, and dried fruits. The distillery itself was an assemblage of contact goods and animal parts:

> Suspended over the lamp was a large, sheet-iron can, the contents of which were boiling and simmering with a singing noise. To the nose of the can was affixed a rude kind of gooseneck, contrived from a large, hollow bone; and from this the barrel of a gun projected to one side, passing through a kind of pan which Hoonakia was heaping up with bits of ice. At the far end of the gun barrel, on the other side of the pan of ice, was a little copper kettle, into which dripped a tiny stream of liquor.... During the next three weeks the village was in an uproar night and day.... Many of the men had altogether ceased to hunt, and even exchanged their walrus lances and guns for liquor. (Mitchell 1997:137)

But liquor was only one of the poisons that the white man carried in. Unwittingly or not, he also brought diseases. The course of these epidemics will be described below in Chapters 13 and 22.

CHAPTER 7
Jabbertown: Point Hope's Shore-Based Whaling Station

> In 1897–98 there were thirteen different whaling stations between Point Hope and Cape Seppings, owned and run by white men, and connected with these outfits were forty white men—that is, there were forty men who were not Eskimos, for in that country every man who is not a Native is called a "white man", whether he is an American, Japanese, Portuguese, or negro, and this whaling colony on Point Hope included all these nationalities and many others.
>
> —Second Lieutenant E. P. Bertholf

Background to the Trading Stations

At the heart of Point Hope's contact history were the commercial whaling stations that came and went on the south shore between 1887 and about 1910. During that period, between twelve and fifteen outfits established themselves, some of them briefly; others, such as the one that Heinrich Koenig operated, ran for nearly two decades. The existence of these stations at this crucial period of history intensified contact between Point Hope people and a wide variety of Euro-Americans and rendered the presence of non-Iñupiat on home territory familiar. As we shall see, several of the Jabbertown men settled in the area for many years and in the process at least enabled local people to focus on these strangers as individuals. Nearest to Point Hope, four and five miles from the village, stood the Jabbertown and Beacon Hill stations, while the most distant was at Cape Thompson, which probably lasted for only a season or two. The significance of the name Jabbertown is discussed later; Beacon Hill was named for the high ground on which the station was built, and where in the 1890s the missionary Driggs made a wood pile to serve as a beacon for summer ships.

In comparison to the big San Francisco companies, these shore-based businesses were minor affairs. Equipped with the latest harpoon technology, as were many of the local Eskimos, the most successful outfits could launch up to five small boats; others made do with a single dory and a combination of modified Native harpoons and lances. But since everything depended on a six-week season, whaling was a precarious business for everyone in the Point Hope region. With luck and skill, these weeks would yield enough baleen to

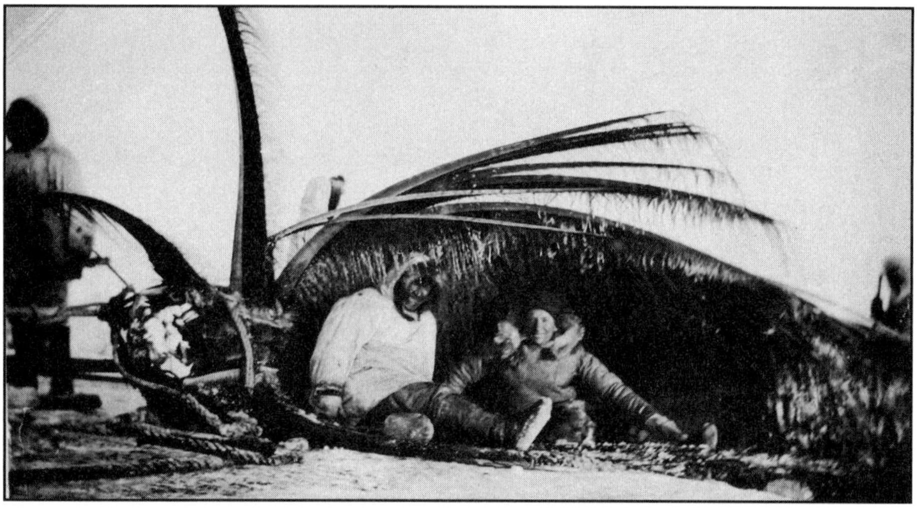

Fig. 13. *Two men, Iñupiaq and Euro-American, celebrating the capture of a large whale. The whale has been butchered on the sea ice and the upper mandible cut away to free the baleen. They pose from within the "bone" in a manner that suggests an upturned skinboat being used as a windbreak. Alaska State Library, Historical Collection.*

keep the station men going with credit until the following season, while for the rest of the year they hunted, fished, mined coal at Cape Lisburne, and did some laborious but futile gold prospecting. Since the main center of activity was at Jabbertown and nearby Beacon Hill, this whole shore-based enterprise will, for convenience, be described as the "Jabbertown culture." Most important, however, as we shall see from Heinrich Koenig's journals, Jabbertown also became a trading post where Point Hope people exchanged baleen, skins, and ivory for food, tobacco, metalware, cloth, and other goods, and where traveling prospectors could acquire provisions. Thus, during the two decades of Jabbertown's existence, Point Hope hunters increasingly became trappers and participated in a process in which local goods acquired an increasingly Euro-American economic identity.

The trading post had long been an institution on the American frontier. Traders, trappers, soldiers, and missionaries had, since the seventeenth century, created communities in the wilderness territories, and the next centuries saw them moving west. But whereas trade on the American mainland was almost invariably accompanied by land appropriations, the Jabbertown men had no territorial ambitions. This factor may not, of course, have been apparent to the Point Hope people as they watched the new community grow. But economically powerful as they were as local traders, the Jabbertown operators were more like the solitary mountain trappers of Oregon and the Rockies than the agency and fort traders of the Great Plains and along settlers' trails. While Jabbertown was new to Point Hope, it also represented the northwestern endpoint of a continuum.

The effect of contact with the big ships has been outlined already. But at Jabbertown the white man, hitherto a transient presence, came ashore. The growth of the Point Hope stations intensified many destructive contact processes, but also neutralized some of them. On the negative side, the new residents on the peninsula were in competition with

FIG. 14. *Jabbertown in late spring or summer. Two Euro-Americans and one Iñupiaq hold up baleen from a large bowhead. Right: Iñupiaq men sit on upturned sleds along whose sides they have laid baleen for cleaning. In the background are sod-insulated living quarters and a flagpole. California Academy of Sciences.*

Point Hope hunters for whale, walrus, and caribou. There was also the ruinous influence of Jabbertown's worst characters. The core of the Jabbertown population was, however, unusually well behaved. Several men, such as Koenig and Hachmann (both from Germany [Westphalia]) and Joe Tuckfield and Jim O'Hare (from Wales and Ireland), married Point Hope women, assimilated into Native society, and moderated the influence of roustabouts and scoundrels.

Shore-based commercial whaling was not a new idea. The English, Dutch, and Basques had run outposts at Spitzbergen in the early seventeenth century. Closer to home, there were whaling communities in eighteenth-century New England, and in the 1850s a dozen or so operations on the beaches of Northern California. The Pacific Steam Whaling Company (PSWC) set up a station at Point Barrow in 1884 to boost its deep-sea effort. But the inspiration for this Barrow enterprise came partly from Native whalers, who carried their boats over the ice each spring to hunt in open water. Commercial hunters based near Native villages realized, in the 1880s, that they could launch a similar effort, and in this way lengthen the whaling season by up to two months. Instead of waiting until July, when the ships gained the open sea, small crews of whalers in lightweight whaleboats and *umiat* could operate like local hunters from April to June, and also employ Natives to boost their numbers.

This first commercial endeavor in Barrow failed. In 1885 the Natives caught—by an inexplicable fluke—an astonishing twenty-eight whales while the white men took none, causing the PSWC to abandon its experiment, leaving their most enterprising employees, some of whom had married Iñupiaq women, to set up on their own. One major troublemaker in the party had been Peter Bayne, a quarrelsome individual from New Brunswick, who in 1885 retreated to San Francisco to drum up funds for his own outfit.

Surprisingly, Bayne won the support of the S.H. Frank Company, a minor player in the Alaskan business, and in 1887 arrived at Point Hope with a team of seven white men and an African-American man.

Peter Bayne and His Legacy

Bayne at first tried to establish a base in Point Hope village, but he was quickly seen off by Point Hope's self-proclaimed "chief," Ataŋauraq, so he settled on the south shore about five miles from the village. In the light of a physical clash between Bayne and Ataŋauraq (described below), it is worth noting that Ataŋauraq's hostility towards Bayne and this first commercial station was initially supported by other Point Hope men, a fact that raises an interesting conundrum. On the one hand, there was straightforward antagonism towards a white competitor. According to John Kelly, who spoke Iñupiaq and once ran an operation at Jabbertown, "The Natives under the leadership of Ataŋauraq kept the station in a constant state of alarm by menaces and threats for more than a year" (Kelly 1890:10). On the other hand, it was possible that Native support for the hated Ataŋauraq in his quarrel with Bayne might provoke the latter into killing the chief for them. As we shall see, Bayne tried to suborn Kelly and Charles Brower to kill Ataŋauraq, and rumors of this may have reached the Native village. As Kelly concluded (he was a reliable informant even though his pamphlet was edited by Sheldon Jackson), "In June 1889, the Natives made friendly overtures of peace"—which supports the possibility that Ataŋauraq's death in February that year removed the *casus belli* (Kelly 1890:10).

Bayne settled on a stretch of bluffs among some recently abandoned iglus and erected the first frame houses on the Point Hope peninsula. There would be Euro-American cabins and storage houses on the south shore until about 1910, when the white men abandoned them, and many of these would be purchased by Natives and moved to Point Hope, where they rose among the iglus. These were the forerunners of later twentieth-century architecture in the village, and about six of them were still inhabited until the late 1970s. As we shall see, Bayne also built himself a house in Point Hope, and his Danish employee Christian "Charlie" Klengenberg describes this and other Jabbertown architecture after his own arrival in 1892:

> [The trading] posts were built of thin planks of dressed lumber from San Francisco. The planks made the inside lining for the building, and against that there was banked a thickness of about eight feet of the mossy sod, with a thicker rampart of mixed tundra and sand sloping outward so as to bear the pressure from above, and resist any assault from the sea. The post I was in was about thirty feet square, and about eight feet to the ceiling.... [The] roof also was built of thin planks, over which several plies of tarred paper were laid, with tundra over that, and all seeming tight and compact.
>
> The post inside was like a cave, stored with various supplies. There was but one door for entry, and there was a skylight in the centre of the roof. We had no windows. There were six cubicles for sleeping quarters, with wooden partitions reaching within a foot of the ceiling. We depended upon kerosene lanterns for light, and we had three iron stoves such as the Russians use, in which the driftwood which we gathered would burn all away to a white ash. Apart from the door, the accommoda-

> tion and lay-out of the post was like an ordinary Eskimo house, or igloo.... But into an igloo one enters by a tunnel, which makes for more warmth and thicker air.... (MacInnes 1932:57)

Bayne's enterprise encroached on two important Native activities. First, when he took his boats out in the spring, Bayne would compete with Point Hope whalers. Then in June, when Point Hope families set up seal hunting camps on the south beach, Bayne and his men would be on site, building and repairing boats and cabins and competing for seal and walrus. Bayne's conflict with Ataŋauraq stemmed partly from these intrusions. And after his brawl with Ataŋauraq would come Bayne's most injurious legacy to the village: his instructions on how to build distilling equipment.

Bayne was a recognizable type of frontier tyro who moved restlessly between Barrow, Point Hope, and Nome until his death from a prematurely exploding whale bomb in 1900. His irascible temperament is expressed vividly in a note to "Cooper" Koenig on the subject of a dog:

> Coopr [sic]: I send you your Dog Chain, But I did not shoot your Dog, or take the chain of him. I told Washuk I thought it was your chain but Washuk said it belonged to one of Mr Smiths Natives the one that was lost on the ice... Chekourouk or Washuk Shot the Dog and took of the chain P. Bayne. (Koenig Collection nd)

This delightfully short-tempered growl—Bayne on best behavior here—well expresses the man's confrontational, material preoccupations. The recipient, Koenig, had in fact been one of Bayne's original crew of 1887, but Koenig left him the next year and traveled to San Francisco to buy equipment for his own whaling business.

But Bayne was not such a fool as to antagonize everyone. He needed Point Hope men in his whaling crew and, anticipating the needs of his first season, he ingratiated himself with some conspicuous hospitality. The eighteen-year-old Point Hope man Kiḷigvak, for example, used to walk down the south beach and visit Jabbertown where Bayne entertained him and other potential crewmen with rice and molasses (Lowenstein 1973–89).

Bayne did succeed in employing a few Point Hopers, and to Ataŋauraq's fury, somehow managed to catch a whale off the north shore. In spring 1889, after the death of Ataŋauraq, Bayne brazenly recruited some men to build him a house in Point Hope. As though aspiring to Ataŋauraq's "chiefdom," Bayne wanted to be on hand to cash in on the baleen from the Native catch, and in a curiously patronizing gesture, offered a prize to the first *umialik* who caught a whale. But Bayne's fundamental misreading of local etiquette lay in his miserliness. Instead of supplying his crews with meat on their sea ice vigil, Bayne offered each of them just two pilot crackers and a cup of tea or coffee for each meal. Ataŋauraq, as *umialik,* had himself only narrowly survived violating the taboo against cooking on the sea ice in 1885 (see Chapter 11). There was no way that Bayne could successfully flout this other, most serious obligation of generosity, and eventually he saved his skin by providing his crew with meat. While his original base remained at Jabbertown, Bayne—surprisingly—stayed in the village and again launched his boats from there in 1890. It was here that he established a relationship with a Point Hope woman, but as a note from the missionary Edson in 1895 mentioning "Captain Bayne's former squaw" suggests, this was a temporary arrangement.

Following a fatality from a whale bomb in 1890 for which he was held responsible (there is no further information), Bayne fled to Barrow in 1891, where he entered into a conspiracy to defraud the owner of a schooner that Bayne had intentionally ran aground. In a complex series of maneuvers, Bayne and an accomplice attempted to slander John Driggs, whom they accused of buying molasses for distilling (Bockstoce 1986:311). Bayne was arrested but escaped conviction and returned to Point Hope.

Whatever the implications of the bomb explosion, Bayne had already shamelessly got away with having corrupted the Point Hope social system with the liquor recipe he had given Ataŋauraq. The technology was easy to reproduce, and by the early 1890s only three Point Hope households were without their own distillery. Perhaps in the end Bayne felt some small remorse, or else sought credit with the Revenue Marine. Almost ten years later, he busied himself in helping Second Lieutenant Bertholf of the *Bear* to root out the stills whose poisonous gut rot continued to undermine Point Hope well-being. "With regard to the illicit distilling of spirituous liquor by the Natives," wrote Bertolf in July 1898, "I have visited and searched all the Native houses from Cape Thompson to Point Hope, discovering and destroying six stills and about ten gallons of mash nearly fermented. On two of these trips I have been assisted by Capt. Peter Bayne and Mr Henry Koenig who have whaling stations on this point" (Bertolf 1899:21).

Bertholf, who apparently knew little of Bayne's prior involvement, continued:

> After telling them about the bad effects of whisky, I recited instances where whole villages had been depopulated by indulgence in the liquor. The Natives appeared very much impressed, and they showed us many places where the stills had been hidden, buried in the snow, and in a short time we had destroyed nine stills, which had been concealed in all sorts of snow banks. The following day a Native woman came all the way down here from Point Hope, a distance of 15 miles, and gave me five still pipes, the tubs of which she said she had broken up. (Bertolf 1899:21)

Bertholf was doing what ship-bound officers on the *Bear* could never achieve. And while many people resented his interference—an Iñupiaq term for the coast guard was *aŋuyaqti* ("harmers")—many people, in particular the elders, were glad of the intervention. Bertholf well understood, as did many Eskimos, the impact of drink on hunting, but as Ataŋauraq's history will show, he was mistaken in his analysis of its social effects: "The Natives being as a rule good natured, the liquor generally has the effect of making them hilarious and voluble, but they are not usually vicious when intoxicated. Of course when a bad-tempered Native gets drunk he become very bad, but this is not often the case" (Bertholf 1899:30).

Altogether, Bertholf destroyed twenty stills that winter. But even Koenig, a Salvation Army man and an abstainer, continued to trade in sugar and molasses. In response, Bertholf made a mash containing molasses to which he experimentally added seal oil to prevent fermentation. The liquor troubles of the 1880s and 1890s seemed to die down in the new century, but there is no record that any of the traders adopted Bertholf's recipe. Bayne, having quarreled

FIG. 15. *Jabbertown in late spring. Koenig and Hachmann sit on a boardwalk; bundles of baleen, awaiting transport to San Francisco, tower above them. Behind, a Native woman walks with a child on her back; another is at a doorway. Tom George smokes a pipe. The boat frame and drying racks with seal skins evoke both commerce and subsistence. Koenig Collection, Archives, University of Alaska Fairbanks.*

with Koenig (who for some reason filed a complaint about him in San Francisco), spent the rest of his career buccaneering between Point Hope, Nome, and Barrow.

The next main players in Jabbertown were Koenig, Hachmann, and the American George Marlin, who in May 1889, with two men in San Francisco, signed an agreement "to establish a station and give their entire time to whaling and trading with the Natives" (Koenig Collection). Each partner furnished $500 or "what money they can raise," and then Koenig, Hachmann, and Marlin traveled back to Point Hope to build on what Bayne had somewhat shakily established. The contract for their company, Geo. Marlin & Co., bound them to their common enterprise until October 1892. After Marlin sold out in 1894, Koenig maintained the enterprise under the name of Beacon Hill Whaling Station (Larson 2004:57).

Heinrich "Cooper" Koenig

Koenig's career raises an unanswerable question: Did this mild-mannered but persistent German damage Point Hope's culture and environment to a lesser degree than the venomous but less effective Peter Bayne in the slaughter of the animals the Iñupiat needed? On the one hand, Koenig dominated the Jabbertown whale hunt for twenty years. On the other, his dealings with Iñupiat were honorable and fair-minded, he married into Point Hope society, and his journals reveal him as a sensitive character.

Born in Westphalia in 1851, Koenig served in the Franco-Prussian war of 1870 before he immigrated to the American East Coast, where he worked as a cooper, rope maker, and smith. In 1875, as Henry F. King, he joined the Pennsylvania National Guard, but left the following year for whaling voyages in the South Atlantic, followed, in the mid-1880s, by journeys into Alaska. A diary written in the Atlantic provides a flavor of Koenig's rather gentle nature: "6 January 1884 Nice wetter. Little cold. Seen nothing. It's my job feeding the Geese wish gives my lots of plesure watching them." The games—innocent of Coleridge—reflect a boyish Ancient Mariner: "11 December Good wether. Light wind. Cought 3 Albatros. The bigest is 10 feet from tip to tip and 3 feet hight. We had lots of fun with this birds. Fasten a blue Ribbon around his Neck and sett them free again."

When Koenig reappeared in the northern hemisphere, he was barrel making in a lime kiln in California, when he ran into his cousin Johannes Hachmann, and in 1887 the two men shipped from Oakland to the Arctic. For the next two decades they would be business partners: first in Jabbertown, later as farmers near Seattle.

Koenig was thirty-six when he came to Alaska and his diary still buzzes with a young man's enthusiasm. "I wrote Eskimo language" reads one ambitious entry, while another describes walruses seen, the skins acquired, the bustle of boats in midnight sunshine, whales rising through ice, and Natives aboard ship who had come to do business. In August 1887 Koenig signed an agreement to stay, and his last diary entries, edited here from twelve different places, contain valuable accounts of days at Bayne's station:

> Employed stowing store room… Get up the coal from the beach... Shot three ducks. Employed carpentering, blacksmithing in house, storing coal... Building shed over coal, meat and Bread out of the old foresail... Making false keel for boat and getting drift wood for fuel… Clear pleasant weather… Keeling boat, coal tarring roof of house, getting driftwood, making sled, boat cover, geese, partridges, five ducks. Bought thirty-six lbs deer [caribou] meat… Various jobs about house. Finish boat cover... Lashing sled... Set lantern... Putting gravel in front of the house…George Marlin deer hunting...

Carpentry, metalwork, weatherproofing, and hunting were routines that would continue. Daily effort continuously went into combating the weather.

Heinrich Koenig & Co.

Koenig's first years were not successful. He was a sociable man but the station was isolated and insecure. His annual supplies and mail failed to arrive three times in the early 1890s, and in 1893 his San Francisco partners dropped out of sight. Still, Koenig wrote optimistically: "You will remember that I said that we perhaps would not make anything in 3 years but would do well if we stuck to it… I am going to stick to our agreement." But his partners had retreated and Koenig was now on his own.

Situated as they were, the Jabbertown communities could not, however, be self-sufficient, and one essential resource was the labor of Iñupiaq women. In most practical matters, the Euro-Americans could get by. But without warm and adaptable skin clothes, they would be in severe difficulty. In contrast to the men on visiting ships, the Jabbertown men needed boots, parkas, skin socks, and mittens all year round. And given that these new whaling stations were also trading posts, there was a flow of raw materials that could be turned to their own advantage.

As described already, many of the seamstresses employed on the ships were also taken in as sexual partners. Sexuality had often been a feature of inter-Iñupiaq trading relations.

FIG. 16. *Heinrich Koenig and his wife Pausana outside their trading station at Beacon Hill, 1906. Pausana wears a reindeer skin parka with white fox trimmings. Koenig's buttoned sealskin jacket, trousers, and skin boots recall Ataŋauraq's culturally mixed attire. Koenig Collection, Archives, University of Alaska Fairbanks.*

Men and women, even from mutually hostile groups, often established trading partnerships and swapped spouses, and the "shared" children could perpetuate their parents' friendships. These quasi-sibling partnerships meant that people could travel beyond the home territory and find support among relative strangers.

Men on the whaling ships could not of course fulfill what Native exchange partnerships offered. The outcome, instead, was the exploitation of women—some of whom willingly entered into relationships in the expectation of a quasi-traditional outcome. Sexual relationships at Jabbertown, with notable exceptions, as described in the next paragraph, represented a number of different aspects of these subordinations. Economically powerful white men represented a lure. On the assumption that women would send food and goods to their families, Native families sometimes encouraged these liaisons. And since Bayne's arrival in 1887 predated Ataŋauraq's murder, there was an added reason for women to escape the village, since the "chief" was himself a dangerous sexual predator. But whereas women on the ships would often be separated from their families, relationships with Jabbertown men at least had a local context. Sexual and economic power remained with the white man. But women could, if they wanted, simply walk home.

Mutual compatibility also inspired a number of Jabbertown marriages, and some Jabbertown men—among them Koenig, Hachmann, Tuckfield, Jim O'Hare, Jim Allen, Charlie Klengenberg, Marx Lieb, Joe Ferreira, and Tom George—married or lived with Iñupiaq women in long-term partnerships. Pausana ("Daisy"), who became Koenig's companion in 1889, was an intelligent and enterprising woman, and was probably the first Point Hope Iñupiaq to manage a Euro-American house. Over twenty years of residence, Koenig's property at Beacon Hill grew into an elaborate cluster of cabins, storage sheds, and the two-story building that became home to Pausana, Koenig, and their five children. Adapting to a context that was both newfangled and local, it was Pausana and her married neighbors at Jabbertown who laid the foundations of women's domestic life in the houses that began to replace Point Hope iglus in the early twentieth century.[1]

The year 1900 was a watershed for the Koenigs. On New Year's Day, Pausana gave birth to John, who would eventually become a teacher at Lincoln High School in Seattle. Later that year, the Koenigs sent their oldest boy, Frederick (Fritz), to live with the Hachmann family on their Washington State farm. By 1909, the whole household had followed, and the Koenigs scaled down their lives from the whale hunt to rabbit breeding. A letter of 1907 from Koenig (still in Point Hope) to Frederick expresses the affectionate and moral sentiments that seem to have informed the Koenig family life:

> My dear Son Fritz: Tonight I feel a longing to see you. And in the mind, I do get your dear face before me. I hope and trust that you are well and happy. I, your mother and brothers are all well. We are longing for the time when we all shall be together again, if God wills it…
>
> Be a good boy, attend church and learn all about God and his love for you. Be always kind to smaller children than you and polite to them what is older than you. Look after [Hachmann's children] Diedrich and Bertha and live in peace. Billie is learning to read and write. We have school every day [at Jabbertown]. Qukuq, Nuvuk and some more attend it. We have get a little snow and some ice but only by Joe [Tuckfield's] house at the Inlet are there seal. Bill set one trap and caught one owl…
>
> You have seen many thing since you left here that was strange to you. Keep out of all the mischief and live so that the people amongst whom you are living will like you, and make the most of your time learn all you can of the good things and leave all evil alone. Do not smoke or drink so we can be proud of you for a son and brother when we come down. God bless you in the wishes of your father and mother and brother. Cooper, Puiny [Pausana], Billie, John, Josie.

Frederick and his brothers John and William followed their father's advice, and they all pursued professional careers in Seattle. A letter from little Fritz to his mother, who was still in Point Hope with her brother Nuvuk in 1905, reads as follows: "Dear Uncle Noovook and Mamma: Whan we went over to the ranch the last time we was plowing and planting the potatoes. I saw Dr. Driggs over there. He asked me if I wanted to go up to Alaska."[2]

Pausana eventually joined Koenig. The boys never again saw Point Hope and Nuvuk Koenig (possibly the first Point Hope man to adopt a surname) remained as his brother-in-law's agent. The unanswered question remains. Koenig and Hachmann were clearly better liked and trusted than operators such as Bayne. But what difference did Koenig's honorable character make to the impact of his business? Certainly, as Elijah Edson describes in Chapter 20, Koenig and his close colleagues restrained some of the scamps and roustabouts. And in the process of bringing together stable Iñupiaq–Euro-American families and "domesticating" the contact experience through consistently decent behavior towards

Native people, Koenig and his partners quietly brought home what was anyway coming and perhaps mitigated the crude impact of visiting whalers. As we shall see later, the missionary societies were alarmed at the fact that Native Americans were encountering only the worst representatives of the so-called civilized world. Here was an illustration of the way that a wilderness trader could also do more than simply profess Christianity. This may perhaps be balanced against twenty years of relentless commercial hunting.

Jabbertown Trade: Koenig's Journals

Koenig's trade journal extends from 1889 until 1895, and while it itemizes many hundreds of transactions, the goods, both Native and imported, were relatively limited in range. Two categories of goods left Point Hope for Jabbertown. First there were skins, clothing, and meat for the station men's personal use. Second were the baleen, ivory, and saleable skins that could be exported, and it was the commercial value of these which supported Jabbertown's economic motive. But since Jabbertown had neither the time nor manpower to procure the meat and skins on which their survival depended, this trade would have been impossible without support from Natives. In contrast to the shore-based California stations, an outpost in the Arctic without access to local goods through Native labor would have been unworkable.

Koenig's journal details his first six years of trading, but there were many unrecorded transactions, particularly informal barter between the traders themselves. Just as there were limits to the white man's goods, so local materials that flowed into Jabbertown were relatively few in kind, the principal ones being, according to Koenig:

> *Raw Materials*: Whalebone, ivory, sealskin, seal and caribou meat, beluga skin, caribou skins, caribou sinew, white fox skins, squirrel and dog skins.
>
> *Processed Materials*: Ready-made parkas, trousers and boots, mittens, underclothing, seal line, canoe frames.

These regularly traded goods were supplemented by seasonal or occasional materials:

> Skins of red fox, polar bear, brown bear, wolverine [which they called badger], lynx, marmot, mink, fishnets, grindstones (local stone).

Going the other way, the main goods leaving Jabbertown for Point Hope were:

> *Foods*: flour, sugar, hard tack (biscuit), molasses, bacon, rice, tobacco, tea.
>
> *Manufactured goods*: cotton print (calico) for women's parka covers, cotton drill (for men's parkas), harpoon irons, rifles, gunpowder, cartridge cases, primers, lead shot, knives, saws, metal cups, thimbles, matches.

Koenig as Employer

As Jabbertown grew, it became a center for Native employment. Work in return for trade goods was not a novelty to Point Hope people, some of whom had experience of work on whaling ships. Employment possibilities broadened once the Jabbertown stations arrived. First there were new housekeeping roles for women who went, either as wives or housekeeper companions, to live with Jabbertown men, while men found occasional employment year-round. Community relations were sometimes uneasy and the existence of Jabbertown would always be dependent on Native cooperation if not always goodwill. In the spring this meant crewing for the white man, although this work was refused by the majority. And as Point Hopers migrated north in the 1890s, those who remained preferred to launch their own boats so that the commercial whalers were forced to recruit southern Iñupiat crews. Over the next decade, Jabbertown would attract a varied community of both single men and families from Shishmaref, Cape Prince of Wales, Kivalina, the Kobuk and Noatak Rivers, Kotzebue, and Little Diomede Island. Collectively this changing population came to be known as Yappataunmiut, or "Jabbertown people," though the term also incorporated the non-Iñupiat.

Other work was occasional: house building, transport of goods by boat and dogsled, ice-cutting from nearby ponds and sea ice, longshoring from supply ships, and subsistence hunting. The character of this latter work was ambiguous. White men viewed Native hunters as their employees. On the other hand, the nexus represented a trading relationship in which the Native hunter saw himself as a partner among equals who had the power to give, withhold, and demand a fair exchange. Jabbertown traders presumably understood this, and those who had married Point Hope women lived comfortably within such reciprocal relationships. Given the extreme weather and the precarious nature of Jabbertown's lines of supply to San Francisco, this was *force majeur.*

Towards the end of the century, when baleen and caribou skins were scarce, the Jabbertown enterprise focused increasingly on furs, so that Native men were engaged more and more in trapping. That white men were interested in furs came as no surprise. Iñupiaq people themselves valued wolverine, lynx, and muskrat, and there were traditional stories about competitive trading for the better items. Wolverine, marmot, muskrat (often received from riverine Iñupiat), white and red fox, and squirrel could all be taken in Native snares and later in steel traps that must in themselves be acquired in trade. However successful any one hunter or trapper, economic power nonetheless resided with the whites, and the stores of goods from which they could make payments gave Jabbertown the advantage.

The main items of exchange leaving Jabbertown for Point Hope were hunting equipment, cloth, flour, and tobacco. Koenig's 1891 journal gives the following examples of payment to each of nine men for some work (unspecified by Koenig): "1 rifle, a box of 1.44 cartridges, 4 lbs tobacco, 30 lbs flour and 5 yards of cotton print." By 1893, Koenig was employing at least thirteen Natives, presumably not all of them from Point Hope, for "working on flaw" (hunting from flaws in the sea ice). A representative four of these received severally:

> 283 lb flour, 28 lb tobacco, 28 lb lead, 7 lb gun powder, 3 boxes of primers, 2 knives, cotton drill, cotton print, .50, .45, .150 and .44 cartidges, 1 trap, 1 axe.
>
> 60 lb flour, 5 lb tobacco, 6 lb lead, 1.5 lb powder, 1 box primers, drill, print, cartridges.

98 lb flour, 9 lb tobacco, 13 lb lead, 1.5 lb powder, 1 box and 190 primers, cartridges, 1 saw, 1 gallon molasses.

180 lb flour, 11 lb tobacco, cartridges, 3 lb powder, 350 primers, 3 gallon molasses, 1 thimble, drill, print, 11 lb lead.

It is worth noting that white cotton drill and patterned cotton print (for men and women, respectively), which were made into "snow shirts" to cover skin parkas, were semi-decorative articles that added little to their insulating properties but which perhaps represented a token of assimilation into Euro-American habit. Tobacco and molasses were also luxuries for which Iñupiat would pay inflated prices. Trading on the often-destructive tastes that the Euro-Americans had themselves created would continue to be profitable.

Exchange Values: Koenig's Journal

Koenig's character as both patron and participant comes through in his journals. While in 1889 Koenig is trading with anonymous "Natives," by 1890 he is transcribing Iñupiaq names. Koenig notes some eight hundred separate deals, and their number and continuity suggests that he was a fair enough trader. Given their increasing dependency, the Iñupiat had little alternative. But they had experience of almost three decades of trade with white men, and participated in setting exchange values. Prices for the major items were often established at Port Clarence each spring and broadcast from there—though rates of course changed, and the white men often complained that Natives reserved their best goods until the end of the season. Given year-round trading opportunities at Jabbertown and the white men's knowledge of local economic and cultural values, maneuvers such as these came to be replaced by more straightforward, though often tense, processes of barter.

While Natives well understood the commercial worth of ivory and animal skins, they also knew that it was weight and length that mattered in the baleen trade. The bigger the whale, the heavier and more valuable the "bone." The "dock price" that whalers received in California fluctuated dramatically over the second half of the nineteenth century. When the arctic trade got under way in the 1850s, values rose sharply and continued to grow through the decade as whales grew scarce. In 1863 the price shot from 82 cents a pound to $1.53, with an average value in the early 1870s of 70 cents a pound, with another jump by 1877 to $3.50. As the whale population dropped, the price of baleen rose further, reaching an average of $5.38 a pound, and in 1891 a high point of $6.70. Values thereafter fell, the average from 1891 to 1900 being $2.50 to $4 (Bockstoce 1986:205–8). None of these fluctuations were reflected in what Alaskan Natives received. As noted in the journal extracts below, the white man paid so little for his trading outfit that his profits—whatever he received at dock—would be disproportionate to his most generous exchanges. But an additional factor makes Native receipts from the baleen trade difficult to calculate: this lay in the fact that by 1890, baleen had become only one of many exchange items. Traders had always been interested in furs. But just as silver and other minerals rose in value in the wake of declining Californian gold finds in the 1850s, so furs took their place as the whales declined. Thus the Point Hope hunter who, as the the Euro-American market demanded, had turned semi-commercial whaler, would by the early twentieth century be further transformed into a trapper.

But baleen did remain an item of Point Hope–Jabbertown commerce until at least 1910 (and in a very small way until the 1930s), and continued to represent a somewhat mystifying focus of interest. Given that "whalebone" was marginal to local economic life, it was strange to Iñupiat that it was this and not meat that the white man wanted. The paradox lay in the fact that the traders gave food and other goods in exchange for an almost worthless commodity, but that in the process they depleted a crucial supply of food.

But the white men's baleen hunger, which they transmitted to Native people, continued undiminished, and the first entry in Koenig's journal reveals why. Twenty-five pounds of whalebone received from a Point Hope Native was exchanged in July 1889 for 425 pounds of flour. This modest whalebone bundle would have yielded Koenig roughly $125. The flour with which he paid cost him perhaps five cents per pound—a total of about $1.50. So long as whales survived to sustain it, this represented a stupendous profit.

The following table, from the first page of Koenig's July 1889 journal, shows some representative deals in both raw goods such as baleen (whalebone) and sealskin and manufactured Native goods like boots, canoe frames, seal line, and fishnets. In the context of the subsistence crisis, it is also worth noting how small were the quantities of goods that Point Hope people let go. But as we have seen, white men had long noticed that Iñupiaq traders portioned out their goods with the greatest foresight, and this trickle of often smallish items may suggest long-term strategies with visits from better provisioned pelagic whalers in view.

September–October 1889

Received from natives	Given by Koenig
Whalebone 25 lb	425 lb flour
Dogskin boots	1 yard drill
1 large, 1 small sealskin	250 primers
1 coil seal line	5 yard drill
6 deer skins	50 lb flour
whalebone 16 lb, 5 foot long	75 lb flour, 4 (?) drill (?)
seal line	1 package matches
whalebone 18 lb, 5 foot long	25 lb flour, 100 .44 cartridges
1 seal skin	1 small tin cup
frame for small canoe	100 lb flour, 5 yards drill
large sealskin	25 lb flour
1 larger seal skin	25 lb flour, half lb tobacco
1 white fish skin [beluga]	25 lb flour
2 small seal skin	150 primers
1 spotted seal skin tanned	3 lb flour
1 seal poke [bag]	1 lb flour
9 lb small whalebone 5 foot long	3 lb powder, 250 primers, bar lead

3 lb whalebone ditto	200 primers, 2 bars lead
1 pair deerskin boots	half lb tobacco
1 deer skin	5 bars lead
underclothing	3 bars lead
6 lb whalebone [skullbone & jaw]	20 lb flour, 11 b (?)
1 white fox skin, 1 mink	5 yards print
110 white weasel skins	6 yd print, 5 bars lead, 1 tin cup
1 dog	25 lb flour, 1 qt molasses
2 deer skin artiga [parka] & 2 pair pants	75 lb flour + (?)
1 pair deer skin boots	half lb tobacco
(?) seal line	4 lb tobacco
2 fishnets	25 lb flour 1lb tobacco
1 red fox and 1 white	2 knifes [sic]
1 white fox	half lb tobacco
1 squirrel skin shirt	(?)
3 white fox skins	half lb tobacco
1 coil seal skin line	3 lb lead
1 dog	some sugar and little other things [illeg.]

The journal for the next two years is relatively uniform. Then, inexplicably, in 1893, the food inventory got more varied and sophisticated, though the same trend was not consistent through the century. In addition to the staples already noted, Koenig now itemized lard, baking powder, potatoes (fresh and dried), syrup, dried tomatoes, apples and peaches, bacon, butter, oatmeal, cornmeal, ginger, and cinnamon. One drawback contained within this commerce was the increasing place of carbohydrates, low in fat and vitamins, in an otherwise nutritious Iñupiaq meat diet. A second drawback lay in the trade for tobacco, to which many Iñupiaq had become addicted, and which remained a major exchange item. A third was that throughout the 1890s, much of the disingenuously traded dried fruit, sugar, and flour would go into liquor production. As if the disease situation were not bad enough, even the most innocuous commodities were capable of being turned into poisons.

While trade goods drew Point Hope people towards mainstream American culture, the use of money, first recorded in a transaction of 1905, signaled the beginning of Point Hope's economic assimilation—although barter remained the principal means of payment up to the 1930s (Larson, personal communication 2006). By the turn of the century, some Point Hope *umialit* were operating their own boats, trading independently, and working on a mixed system of barter and wages for the white man. The Jabbertown whaler Jim Allen provides a vignette of this latter status in a letter to Koenig on behalf of one of his own Point Hope employees: "Kayuktuq has nothing to pay you with, but if he gets a whale he is willing to pay you either in bone or money."

This fragmentary note illustrates the precarious situation of even a relatively successful *umialik* who was still occasionally catching whales. Like most of his contemporaries,

Kayuktuq was trapped. Despite depopulation, the weakened condition of the home community, and competition from white men, he had assembled a crew. But whereas in the past he would have fed his men on meat, he now had little to offer them. Given Koenig's flour—and note here Bayne's previous inability to maintain a crew with hardtack—he could launch his skinboat and repay the white man with the proceeds of a baleen sale. But the odds were against a successful outcome. It was a convergence of the most undesirable uncertainties.

Notes

1 Among about ten houses transported to Point Hope was one of Koenig's cabins, bought by George Omnik, who lived there until about 1977.

2 On one of his rare trips south, the missionary Driggs visited his old Jabbertown friends. See Chapter 27.

CHAPTER 8

Jabbertown: The Transformation of Point Hope's South Shore

The South Shore's Transformation

Heinrich Koenig & Company (HKC) held out for almost twenty years, but the very continuity of Koenig's business in itself represented change. For two thousand years, the south beach had been in the possession of a people whose iglus, storage pits, meat racks, and camping places extended to Cape Thompson. By 1890 the south side had been overtaken by new populations. Traders' houses surrounded by cabins, boat sheds, warehouses, barrels, piles of lumber, and the general detritus of commercial and subsistence life now dotted the twenty-mile shore. The year 1890 had been a disastrous one for Point Hope, and a hundred fifty people had immigrated to Point Barrow to work for commercial whalers. In the absence of sufficient local manpower, as mentioned already, the whaler-traders recruited Eskimos from the south, and by 1893 there were up to a hundred of these immigrants in and around Jabbertown, many of them living in new iglus. The combined presence of Euro-American and Eskimo newcomers, who intermarried among themselves and with Point Hope returnees, had a lasting impact on the future composition of Point Hope society, and the new demography of Point Hope—which had already been changed by the wars and subsistence crises of the early 1800s—included the descendants of commercial whalers and Iñupiaq incomers.

The depopulation of the Point made it difficult for Point Hope people to protest the growth of these settlements. By the end of the 1890s, when Alaska's first legal system had been established, the dispossession was at least theoretically sanctioned by law. The same period coincided with the Seward Peninsula gold rush of 1898; the speeding up of communications between Barrow, Jabbertown, and Nome; and the growth of Nome's state courthouse. Hunting and trapping now comprised just one branch of commercial possibility in the Arctic. Several Jabbertown traders joined the stampede to the beaches at Nome, some carrying with them proxy claims for friends and leaving hastily drawn-up powers of attorney for the conduct of continuing business (Koenig 1880–1914).

Complicating the process, the exchange of news on Nome's golden beaches encouraged some disappointed prospectors to travel north to try their luck in the hills behind Point Hope, stopping off at Jabbertown for provisions and equipment. This increasingly interconnected territorywide activity reinforced a formal legalistic trend at Jabbertown, and

FIG. 17. *Sketch of Jabbertown, at Beacon Hill. Background: Koenig's residence and storehouse with canvas tents to right. Against the bluff, upturned boats and sail canvas provide Iñupiaq-style cooking and working shelter. The man on the beach is an Iñupiaq newly arrived in search of work and trade. Drawing from* San Francisco Chronicle, *April 12, 1898. Franz Heinrich Koenig Collection, Archives, University of Alaska Fairbanks.*

Koenig's papers are full of locally drafted employment contracts, affidavits, and claims. In 1898, for example, Koenig commandeered a thousand feet of land that he staked out round his station.[1] And in March 1900, the whaler John Kelly sketched out maps of coal-mine claims near Cape Lisburne and, with gold in view, at a tributary of the Pitmigeaq River. Fifteen white men, including the missionary John Driggs and two Point Hope Iñupiat, put their names to these, though the basis for their claims is unclear.[2] Just as whales and walrus were being taken as the legitimate property of hunters from outside, so land was appropriated by people who, paradoxically, had no long-term plans to stay. Joe Tuckfield and John Driggs, whose names were on both claims lists, were the sole exceptions and continued to prospect until 1914.

"Gentlemen of the Point"

With these genteel words in a letter from Missouri, John Kelly broadcast to his Jabbertown colleagues the news headlines of 1900: "Britain has annexed the Transvaal... Kruger visited Europe... Galveston, Texas was wiped out by a tidal wave... McKinley beat Bryant... The battleship *Kentucky* was sent to Manilla." Later that year Kelly would arrive at Point Hope on a gasoline schooner "about the middle of July and hope to see you all, my Native friends, the men, women, boys and girls... I see by the papers that a man who went to Point Hope says he discovered gold at Cape Dyer" (Koenig 1880–1914).

Kelly was the most mobile of the Jabbertown entrepreneurs, and his connections in the lower States also made him a sought-after guide and interpreter. Granted that the inform-

FIG. 18. *Rustan [Rustam] Nelson, one of five African-Americans who worked as Jabbertown whalers between 1887 and 1910. He managed John Kelly's whaling station at Piŋutchiaq on Point Hope's south shore, and was a close friend of Driggs at the turn of the century. Drawing from* San Francisco Chronicle, *April 12, 1898. Franz Heinrich Koenig Collection, Archives, University of Alaska Fairbanks.*

tion about gold was false, the fact that it came from a published source, that prospectors were arriving from Nome, and that Jabbertown men were so hungry for news were marks of a modern American world in which the Arctic had its expanding place.

Jabbertown's own makeup had, correspondingly, always been cosmopolitan. There were of course white Americans at Jabbertown, but they constituted a minority. Of these, the Pennsylvania Dutchman George Marlin and Tom Randall were in Koenig's immediate circle, and the names of about sixteen other white Americans are recorded.[3] Among the five African-Americans were the well-loved Tom George and Rustam Nelson, who co-managed Kelly's station. Anton Bethancourt and Joe Ferreira were two among the Portuguese and mixed-race Cape Verdeans, and the names of three Scandinavians and two Jewish immigrants are also represented. Two other long-term residents were the Irishman Jim O'Hare and the Welshman Joe Tuckfield. Among about fifty others were occasional Japanese and Hawaiian sailors. Thus, while Jabbertown brought to Point Hope a culture that served mainly white consumers in mercantile America, only about ten of the recorded thirty-two names were ethnic representatives of this constituency, while about fifteen (including Bayne) were immigrants. Point Hope people, half of whom were now frequent migrants, had a reasonable notion of these identities. They called black men *taaqsipak* ("very dark"), Portuguese and Cape Verdeans *pulgi* (the Iñupiaq rendering of "Portuguese"), and white men were *naluagmiut* ("people of the bleached sealskin"). But as

R. L. Stevenson also noted in Polynesia, even the black people of those islands who were associated with traders were characterized as white men.[4]

Shared Territory and the 1893 Storm

With a population in 1893 standing at about one hundred twenty, Point Hope, with the majority of its population in Barrow and Herschel Island, was at its lowest ebb. But there were so many new people on the peninsula that with the exception of Alaskan Indians and Aleuts, it was as though a profile of the whole territory was represented. Ataŋauraq had put up some resistance to Jabbertown's presence. Otherwise, there were no hostilities beyond routine individual disputes and arguments over ownership and employment. Besides, there was a balance of community weaknesses, with no interest group powerful enough to exercise a destabilizing threat. There was competition for resources, but there were also motives for cooperation: Jabbertown required Native labor, meat, and skins. Point Hope, in return, needed compensatory supplies. A simple pragmatism dictated a degree of collaboration between the communities. Both were oriented to survival, and both were exposed to the same weather and environment.

A dramatic example of the leveling effect of the weather is provided by a violent storm that blew across from Siberia in October 1893. Accounts of this were recorded by both the missionary Driggs and the whaler Christian Klengenberg, who described the storm's impact on both communities. Driggs's account in a letter of June 8, 1894, is relatively phlegmatic:

> On the 13th of October 1893 during a very severe blizzard, the sea came breaking on the land, driving the Natives out of the village and forcing me to desert the mission. At first I did not intend doing so, but a wave burst in the door and I thought it best to leave. After an absence of seven days I returned home and found that no damage had been done to the house outside of the bursting in of the door, a damage that was easily repaired with a few screws.
>
> Out of doors everything looked desolate. Along the ocean front the land had been cut away to a considerable extent and all the snow had been thoroughly saturated by the ocean water and spray that no suitable snow could be procured to melt for household use. On the night of my return another big storm arose and the following evening I thought it best to desert the house before I was again forced to repeat my former experience of dodging waves and wading though ice water and slush, an operation I did not care to repeat, as I had slightly frozen my feet and had been forced to thaw them out with snow.
>
> That night I slept alongside of a dog sled, with a few clumps of snow thrown up as a wind break and then continued my trip back to the mountains, where I met some belated Natives and stayed with them in a sort of brushwood shelter living on uncooked, frozen fish. By the first of November there having been an abundance of snow storms and the weather being sufficiently settled, I again returned home and opened school for the second time...A young woman who had been a pupil at the mission was overtaken by these blizzards, while on the cliffs on the mainland and is supposed to have been blown off into the ocean; she and a young man with her, have never been heard from...

The cheerful Klengenberg, who chose to sit it out at Bayne's station, seems to have almost enjoyed the event: "We laughed when the first waves came splashing over our roof…their force broken by the long outward slant of our wall towards the sea. And when the first of the bitter cold salt water came through the skylight and soused Captain Bayne…that was a good joke" (MacInnes 1932:66ff).

The sea eventually smashed through the skylight, and Bayne's men withdrew inland, returning when the storm subsided three days later. Klengenberg continues:

> The Eskimos quickly realised that it was beyond the ordinary… They all ran for high land, the men and the stronger women staying only to carry such belongings with them as they could. One group even managed to carry a whale boat far inland, and it speaks well of the health of the villagers that all of them were able to get away quickly in the night except one… Sick man, wife carries him to iglu on higher ground…
>
> Driggs packed some supplies in a large sheet, making a bag of it on his shoulder, when he set off through the dark storm, but he was blown down so often on his way to Jabbertown that he lost half of what he had in the sheet, part of which was afterwards found and restored by the Eskimos. He had his feet frozen. (MacInnes 1932:66ff)

Even by local standards, this was a freak storm and it cut several meters of land from the north side, but while Driggs's house and all the Point Hope iglus were flooded, no lasting damage was recorded beyond the two deaths noted by Driggs. It is possible that one or two of the ceremonial houses (*qalgi*) may have been badly hit. This was a ritual season and the storm may have contributed to the erosion of the cultural life centered in the *qalgit*.[5]

Intercultural Assimilation: Joe Tuckfield, Nuvuk Koenig, and the Jabbertown School

The hunting, socializing, fuel gathering, prospecting, and goods exchange in which Iñupiaq and white men were often equally involved are recorded in accounts of deals, routines, and mishaps. Some of this interaction can be appreciated most vividly against the backdrop of new settlement topography. In addition to the south shore stations, there were white men on the north side, extending from the Mission House, Marryat Inlet, and the coal-bearing cliffs beyond Cape Lisburne. But all this was fluid. Kelly and Nelson, who ran a station at a huge bluff called Piŋutchiaq, were both itinerants. And paradoxically, while Kelly employed southern Iñupiat at Piŋutchiaq, he attracted Point Hope people to his Barrow operation. Joe Tuckfield, on the other hand, had started his early career at Barrow and then moved to Jabbertown. From there he decamped for part of each year to a boat at Marryat Inlet.

Tuckfield was undoubtedly the man who most successfully assimilated into Native society. While remaining close to Koenig, he married a shaman and also entered local custom by adopting a young Iñupiaq immigrant and naming him Bob Tuckfield. The story of Bob and his aunt's marriage to Joe offers a small diagrammatic picture of uprooted mid-1890s life in the northwest Arctic. Born in 1884 on the Kobuk River, Bob was named Qimiuraq in Iñupiaq. In the diseases and famines that claimed many in the lower Arctic, Qimiuraq lost his parents and his adoptive grandparents, and in 1893 the boy and his aunt Avagruaq (a shaman who would later marry Tuckfield) walked north to Point Hope in search of food and shelter. Stumbling up the beach from Cape Thompson, the two found shelter at Jabbertown, and within a few months they were living as a family with the whaling man from Swansea (Lowenstein 1973–89).

"Little Joe" Tuckfield (in Iñupiaq: Yuuguraq, meaning "Little Joe") had worked for Charles Brower in Barrow in the late 1880s and had traveled for Brower to the mouth of the Mackenzie River, where he discovered new bowhead feeding grounds (Bockstoce 1986:255). Tuckfield's find initiated the rush to Herschel Island, from where the final bowhead whale hunt was launched by steam whalers at the turn of the century. Like the footloose adventurers in Joseph Conrad's stories, Tuckfield lived comfortably in a world of traders-gone-Native and Natives who frequented the white "outposts of progress." On his schooner *Emily Schroeder*, which Koenig had bought from Peter Bayne after it was damaged in the 1893 storm, Tuckfield and his wife provided important refuge for travelers at their mooring site in the mouth of the Kuukpak River. One story, which evokes the social mix of travelers and residents at this intermediate point in local geography, describes how the Irishman Jim O'Hare perished just north of Tuckfield's home:

> One summer Yiiguuraq ("Little Jim") took off with Peter Pigaaluk, a coloured man, and a Noatak man who worked at Jabbertown. They sailed to Corwin to fetch coal and to get crowbills' eggs from Cape Lisburne. On the return, their boat was sighted by Tuckfield's wife Avagruaq and another Point Hope woman. It was a clear day with no wind, and the two women had a good view of the boat. After they had watched it for a while, they went back to their work and forgot about the boat. Then they remembered Jim's boat again. Perhaps they had landed for something? Travelers often used to stop at Tuckfield's to eat. But they never showed up. Some boats went up the coast to look for them next day. They found nothing. Capsized. Sunk… And people said the boat must have been swallowed by an *aqalugluk* (giant shellfish). This was an area where those shells were found… (Told by David Umigluk Frankson, in Lowenstein 1973–89)

An American world of schooner and coal mine, assimilated here with local folklore, provides a richly textured backdrop to the Point Hope, Noatak, African-American, and European individuals, while the story itself is seen through the eyes of an immigrant Iñupiaq shaman married to a Welshman. Two years later, a letter from Jim's brother in Ireland gives us a flavor of Jim's own dialect. Likewise, we can eavesdrop through Tuckfield's letters on his own Swansea idiom: "We are all well and happy"; "The old woman's pleased with the sewing machine": "The trader is leaving yere [sic]"; and "with love to hall [sic]." By 1907 there are "bycles" at Jabbertown, albeit "on the shelf…no pump and no tires, you had better send a set for Nuvuk's Bycles." In this audibly Celtic music of comradeship and practical affairs, some of Jabbertown's more endearing chatter survives.

Tuckfield's "old woman" at her sewing machine provides another kind of vignette. In addition to wood-burning stoves and metal kitchen equipment, the sewing machine was a significant addition to women's domestic technology. It was a superb new tool. With a mechanism of visible parts for which some replacements could be improvised, the machine complemented, but did not displace, the arduous process of skin-sewing, which still had to be done by hand. For making clothes from calico and cotton drill, the machine was

FIG. 19. *Beacon Hill grave of the shaman Talaaq, who emigrated from the Kobuk River to work at Koenig's trading station, ca. 1898. For several following decades, Point Hope people said the grave had power to change the wind's direction. Author's photograph, spring 1976.*

FIG. 20. *Beacon Hill grave of Jabbertown traders' children. Author's photograph, spring 1976.*

perfect. But it also gave women the opportunity to create for Europeans, and this brought women into the new economy. Unlike those early bicycles, which were only reasonably practical on flat parts of the beach, the sewing machine was there to stay, and it remains an important item in Point Hope women's working inventory.

Another of Tuckfield's and Koenig's friends was "Nuvuk Koenig," Koenig's brother-in-law through his wife Pausana. Before Koenig went south in 1907, Nuvuk had adopted the German surname and with Tuckfield took over the management of Koenig's station. Nuvuk had learned to read and write as an adult at the short-lived government school that was established at Jabbertown in 1904. Archdeacon Stuck, who made several visits to Point Hope and was resident for some months in 1918, called this an "abortive school at Jabbertown" on account of the "banditti of renagadoes" (a phrase he adopted from Vancouver's Northwest Coast journal of ca. 1792) that "infests newly-discovered shores" (Stuck 1920a:102, 109). "When the government school was built," wrote Stuck, "there was a school population, the offspring of Negroes, Portuguese, Hawaiians, Germans, Irish, English and I know not what other nationalities and Eskimo wives, whose fathers made a living by whaling" (Stuck 1920a:108). However "abortive" the school may have been (and there is otherwise no information about it, though the building itself was removed to Point Hope, where it survived into the mid-1970s), Stuck certainly accorded it some influence, perhaps on account of the "very decent folk" who married and settled at the Jabbertown station. These he exempts from "renagado" status and further differentiates them from the crewmen of the pelagic whalers, whose summer visits, he suggested, were followed by "riot and licence" (Stuck 1920a:109).

Nuvuk's highly accomplished letters certainly attest to the impact of the short-lived government school and express something of the mind of a twentieth-century Native who had converted to Christianity and who, like everyone in Point Hope, continued to struggle with diseases. Nuvuk was perhaps the most economically successful Point Hope man of his generation, and we may assume that Koenig had married into prominent *umialik* family. Here, in 1907, is Nuvuk writing to his nephew in Washington State:

> My dear Nephew Fred Koenig: Are you good? Me speak to you mama all time of Blessed Jesus. All people Eskimo Blessed Jesus come. And Bertha, Dredrich [Hachmann], Charly Marlin me give you best greeting. Point Hope men shot 4 white Bear...a Kobuk man shot other man dead. I go after fish come back quick. Me go to sled Kobuk. I will write to you letter. You write to letter come back. My baby boy... she is a little sick... Kakoon baby girl she is 3 day sick. (Koenig 1880–1914)

There are other Point Hope writings from this period, but Nuvuk's letter is the most fluent. By 1904 there was a sizable group of young adults who had attended school either at the mission or at Jabbertown. But while Driggs taught literacy so that people could read the Prayer Book, for most Natives writing became a tool of commerce, as is evidenced in this note from a twenty-five-year-old man to Koenig:

> I want 2 boxes cartridges
> one tin sugar not 3 of flour

8 tin matches
I have the stove and paper
next spring catch whale not
give to you stove
back again stove you. [signed] Ahviksinya[6] (Koenig 1880–1914)

At this stage of Point Hope history, literacy was one of the most helpful new cultural acquisitions, and Nuvuk's position as station manager was predicated on his ability to write to Koenig for supplies, note his expenses, and itemize deals. But given that he lived at Jabbertown and not in the Native village, there is no evidence that Nuvuk was also a Point Hope community leader. A clear line of development, which started with the "chief" Ataŋauraq, proceeded through Anaqulutuq (another "chief" imposed with no effect in 1889 by Commander Stockton; see Chapter 13), and culminated with the trader Nuvuk, may nevertheless be traced. Authoritative individuals, independent of the old ceremonial house structure, were starting to emerge, and it was the ability to use English and to operate in the world of commerce that made this evolution possible. (The next major stage, helped by the missionary Augustus Hoare, was the creation of Point Hope's first village council in 1920, a powerful body of men born in the 1880s who very effectively administered community life in a manner not unlike the old ceremonial house organizations.)

Most of Nuvuk and Tuckfield's letters are from 1907 to 1910. As the commercial hunt wound down, a sense of finality suffuses Tuckfield's correspondence. Supply ships from the south could still fail to arrive, and for two seasons of his stewardship, Tuckfield assumed that Koenig had died. The early days of Jabbertown had been difficult enough. Now there were fewer whales and the south-shore community was dwindling. Itinerant traders were plying the coast, but there was no love lost between the Jabbertown old-timers and the new irregulars. Tuckfield and Nuvuk quarreled with one of these, Louis Lane. Lane took legal action, and the U.S. Revenue Cutter Service and the new missionary Augustus Hoare had to intervene to apportion the baleen whose ownership they were disputing.

The Jabbertown hunt was pursued through the 1910s, and a note from Joe Tuckfield vividly itemizes the customary hits, misses, divisions, and losses:

> Nuvuk and myself got the only big ones... I think we will have a couple of thousand pound apiece. Lieb gets a quarter of mine for saving [the whale] and Nuvuk got to give a quarter of his to his crew... Jim Allen got nothing... George Marlin's gang got...about 50 lb bone... Nuvuk lost darting gun and got it back. (Koenig 1880–1914)

There were one or two good seasons ahead, but by 1907 the baleen market had collapsed, and it continued to be difficult for the whaler-traders to employ Natives. As demand for whalebone declined, ownership of Jabbertown buildings passed to Point Hope people, who lifted them onto boats and sleds and carried them to the village. As one indication that whaling that was purely for subsistence had returned and was in control of local people, Joe Tuckfield's adopted son, Qimiuraq (Bob), became one of Point Hope's leading *umialit.* That Bob Tuckfield and his wife from Kivalina became leading Point Hope whale hunters represented a further expression of the new mixed-origin population.

Jabbertown Social Life: Marriage and Subsistence Partnerships

One of the harshest critics of both ship- and shore-based traders was Archdeacon Hudson Stuck, who spent six months in Point Hope in 1918 helping the missionary William A. Thomas, who worked at Point Hope in the early twentieth century (see Chapter 26). The Jabbertown stations had vanished by then, but in an almost apocalyptic vision, Stuck evoked what he considered to be a continuing moral shambles. "One has the impression," wrote Stuck, as he contemplated the abandoned Jabbertown school ("where no one any longer jabbers") and the mission building from the far north side, "of an infected spot, from close contiguity with which even the agents of amelioration shrink" (Stuck 1920a:108).

Stuck was an idealist with a romantic interest in Native traditions, and his take on Point Hope's new demography was tinged with disquiet about a new Point Hope population that contained the "offspring of Negroes, Portuguese, Hawaiians, Germans, Irish, English and I know not what other nationalities" (Stuck 1920a:103). But Stuck also had a broader understanding of Northwest Coast history in which "traders, beach-combers and squaw-men" had exploited Natives and given nothing in return. Outraged at the corruption against which the churches had been impotent, Stuck continued: "Shall such unprincipled wastrels work unhindered? Shall drunkenness and lust and fraud and trickery and violence be the only teaching received from the white man's 'civilization'?" (Stuck 1920a:103–4).

Instances of those vices were of course relatively common in the 1890s, and some criminal behavior was confronted by Point Hope missionaries (see Chapter 20). Still, contact relations at Point Hope could be benign, especially through men like Klengenberg and Jim Allen, who were young enough to plunge themselves into hunting, love affairs, and games as though the whole process was simply cheerful play. Klengenberg's book is often historically naive and his later career was scandalous, but his descriptions of intercommunity life among equals at Point Hope are persuasive:

> I soon began to mix with the young fellows in the village, preferring them for their bright and hearty ways to the dull adults for whom I cooked at the trading post, and preferring them very much to the grumpy hellions of Blabbertown [sic]. Two of the young Eskimos had come forward and shown me how to cut up and skin and dispose of the first seal I killed. They were direct and friendly in their ways. If you played with them you would be accepted as one of themselves. (MacInnes 1932:60)

Klengenberg also witnessed some ceremonial house events, and poignantly evokes the optimism of Point Hope's traditionalists:

> I had some good times at that old lodge [*qalgi*], and it was there that I first began to estimate the superior social qualities of the Eskimos in comparison with the men at my own post. For one thing I noticed their readiness to share whatever of fun or food... and their happy-go-lucky ways. Night or day was all the same to them for a frolic... (MacInnes 1932:62)

Both Klengenberg and Allen also wrote about their courtship of the Point Hope women they would marry. Allen's account is significant in that it relates how a young widow,

Ellnou, made him a pair of boots, and proceeds to demonstrate how for both parties marriage represented a working partnership. Allen wanted "to get a wife" who would support his whaling and trading enterprise, while Ellnou, having consulted the missionary, demanded that Allen marry her "the white man's way." Thus economic ambition and mutual commitment—features of both Native and Euro-American tradition—converged, and the opportunistic cohabitation that defined most Jabbertown relationships was transcended. Driggs joined the couple in 1903, and they spent the next forty years as hunters and traders in Barrow and Wainright.

During the season of his courtship, Jim Allen also formed a hunting partnership with a Native man. Allen had learned to hunt on the north coast with Kayouraq, a Point Hope emigrant (Allen 1978:126). In 1902 they both were on the peninsula, and Allen asked if he could go out on the ice with him again. An enjoyable season of hunting and trapping followed, in which the two men agreed to a 50 percent share of any polar bears taken, while each kept the foxes that he himself trapped. At Jabbertown they could sell foxes to Koenig for $5 each and a bearskin for $40. This arrangement based on friendship and mutual profit between a white man and an Eskimo, which was also a feature of Charles Brower's story and which Dr. Driggs would repeat with his friend Anaqulutuq, was unusual. The ingredients of such partnerships represent an interesting cultural inflexion. On the one hand, they contain features of traditional Iñupiaq hunting partnerships in which products of the hunt were shared equally or according to rules of equipment ownership. On the other, there was a new element. While the Native partner had the skill, experience, and local equipment, the white man usually had firearms and economic resources. But the American (generally young and single) could not be too powerful or well established; otherwise, like Koenig, and later Brower, he would simply be an employer. These conditions generally rendered such partnerships quite short term.

The Sacred and the Profane—1

Following his vision of the infected spit, Stuck had a waking dream about the late Dr. Driggs, who was believed by the church authorities to have caught a similar moral infection (see Chapter 26). As we shall see, Driggs became a commercial whaler, and from 1898 onwards, Koenig and Hachmann, partly in return for scientific advice in their search for gold, helped Driggs with material for his own whale hunt. The friendship between the three men had started in 1890, and while Driggs detested the abuse of Native women to which Stuck referred, he saw nothing wrong with the commercial hunt and not only encouraged the Jabbertown effort, but in effect became a Jabbertowner himself.

Given that it had been established as a moral bulwark against the influence of whalers, the mission should, theoretically, have been Jabbertown's opposing institution. Driggs nonetheless records no active intervention until about 1894, and, on the contrary, used Jabbertown both as a personal resource and a place of recreation—a connection from which Native people, who now depended on the well-being of all the white men, indirectly benefited. Driggs's relationship with Jabbertown is documented in a number of surviving written, largely undated messages, mostly to Koenig. In addition to news about whaling and equipment, these hurriedly scribbled notes that were carried from the mission by

Native hunters evoke both a touching vulnerability and a variety of interests beyond business and survival:

> Thanks for the whale meat. I hope you capture several more with big bone. I would like to borrow 8 or 10 lbs of white sugar.
>
> March 15th /98 I send you the small screws that I have. If you wish any more, let me know. Oars, oar locks and sockets I expect up next summer. I will send the new [photo] plates the next time your sleds come down. Also I send you the carbolic acid. Tell John [Hachmann] that I congratulate him on the arrival of his daughter.
>
> I send you over a few books which you need not return. Good luck to you [at whaling]. There is a new moon.
>
> I send you the plates, printing frame and printing paper. The plates are in the boxes with the underwear. You sent four drawers and two shirts. Thanks. Any time you wish the developers, come down and I will mix them for you.
>
> April 26th 1900 Thanks for the tobacco, I believe this will be enough. I send you the eye water. Also I send you the four sacks of flour. I may be able to spare some more later later on.
>
> Tell Joe [Tuckfield] that that piece of rock is good indication [of gold]. He had better find the vein that it came from and put in a blast to see what is underneath. There is a trace in this piece of rock but not quite rich enough to work. It may pan out better underneath. The Native canoes went on the ice this morning.
>
> August 16th 1900 If you will let me know how many people there are around your place that you wish vaccinated, I will get the points from the *Bear* on its return. There has been considerable small pox at Nome and it may get up here. There are also letters here for Mr Lieb. His wife has been catching the *sigiriks* (squirrels) around here for the last three days. (Koenig 1880–1914)

In the whole of Driggs's correspondence with his Jabbertown friends, there is no reference to God or church, and one suspects that Koenig may have been the more committed Christian! He and Hachmann probably attended some of Driggs's services. Otherwise the sacred events on which the white men are recorded as converging are some rituals in the *qalgi* (1892), followed by Christmas dinner in Driggs's kitchen (see Chapter 20).

The Sacred and the Profane—2

White men occasionally visited the *qalgit*, and just as Native hunters attended church to bring them luck at whaling (see p. 271), the presence of economically powerful visitors may have had talismanic value. A more basic issue was the intrusion of strangers into sacred areas of Point Hope territory and culture. In contrast to the spaces consecrated by priest-centered religions, Point Hope's animistic spiritual geography was fluid. In the absence of monumental architecture and written texts, the existence of spirit places came and went seasonally or over limited periods of years. There were several kinds of such places, both on the ground and in the other world. The sun and moon deities were at their most powerful in autumn and winter, but absent during the summer. Other beings, such as the Itivyaaq whaling "family," were contingent only during activities such as whaling or ritual games (Asatchaq and Lowenstein 1992:54ff). There were powers that governed rules about eating certain foods in and out of season. Spirits of place might be present one day, but later less so. Within the village, the autumn rites made the *qalgi* buildings sacred. But at other times in the year, children and couples made use of the empty houses in which to play.

Point Hope's south beach was another place with a highly charged but variable spiritual identity. On the one hand it was simply a hunting area. On the other, it was peculiarly associated with myth. Several stories were told of young men who walked the south shore, where they encountered supernatural beings and returned to the village as shamans (Asatchaq and Lowenstein 1992:21). The south and north also had spiritualized gender: the north being male, while the female south contained "animal countries" from which whales and other migrants came. The sun and moon myths were likewise geographically located. The sun, which rose in January on the Cape Thompson horizon, was the moon spirit's sister, and her presence dominated the southern quarter on winter mornings. The ghosts of the historical past also inhabited the south side. The pre-Thule Eskimo Ipiutak people, whose houses, graves, and ritual artifacts were excavated in 1939–40, lived on, hunted from, and shamanized on the south shore. Over two millennia, these remains shifted to the opposite side of the peninsula as the north shore eroded and the south shore was built up. This subterranean migration of house mounds occasionally thrust the artifacts of mysterious predecessors to the surface. Apart from the site of Point Hope village, and on a smaller scale the area around Driggs's mission where the Itivyaaq spirits had their iglu, no other region of the territory contained such a convergence of spirits.

The Euro-American whalers and traders unloaded, camped, and built their houses among these shades. And just as from within a half-collapsed system of hunting taboo, Natives observed white men catch whales without resorting to ritual or observing taboo, so they witnessed the secularization of places whose imaginative identity was already periodic or uncertain. Adjustment to a non-animist worldview was, as we shall see in the history of the mission, relatively quick. But there were so many more problems of adjustment in this period of change that some traditions were simply let go in the modernizing process.

Jabbertown: The Name

The name "Jabbertown" evolved in the 1890s, but it has no agreed-upon origin or definition. That the place was dubbed "Jappa-town," after the Japanese who lived there, is certainly wrong because apart from Michael Healy's cook, no Japanese came ashore for long (W. Koenig 1962).[7] "Jabbertown" therefore, in the cuss-laden and derisory idiom of frontier self-description, evoked a cosmopolitan environment in which many languages and dialects were spoken, and to which the bluff American responded with a universalizing "no savvy." English was of course the lingua franca, and while Natives such as Ataŋauraq and Nuvuk quickly came to speak English, few Jabbertown men learned more than a few Iñupiaq nouns—for whale, ice, boat, caribou, harpoon, and so on. There were exceptions. Kelly was a fluent speaker (he prepared a seven-hundred-word vocabulary for his 1890 booklet), while Koenig and his circle had some knowledge of the language.

But English, in a variety of American dialects, with the added tones of African-American, Irish, Welsh, and Scottish voices, was just one of the imported languages. Portuguese, German, Hawaiian, and even Filipino Tagalog were also spoken. And then, in addition to this imported Babel, there were Iñupiaq dialects. These would have been familiar to Point Hope people, but the presence of settlers from Kivalina, Kotzebue, the Kobuk and Noatak Rivers, the Diomede Islands, and the Seward Peninsula was new in its scale, and a degree of hostility to outsider Iñupiaq dialects coexisted with the ongoing adaptation to English.

Jabber as Dispute and Struggle

Like "babble"—a word partly derived from the Hebrew *bavel* in allusion to Babylon, city of confusing jabber—the imitative word "jabber" means to "talk fast and indistinctly." Nineteenth-century writers such as Conan Doyle, Conrad, and Melville were humorously intolerant of the wilder or more "exotic" side of language. The "jabber" of Africans, Lascars, and Malays in Conrad's late-nineteenth-century stories is as superciliously evoked in the minds of British seamen as is Queequeg's mutterings on the *Pequod* or the chatter assailing characters like Watson (Sherlock Holmes's colleague) on the Indian Northwest Frontier.

"Jabber" thus implied the primitive, and describes a locution observed from the vantage of a "higher tongue." Thus, whether in British India or in the American Arctic, English was the language of good sense, progress, hygiene, economic power, and even Christianity. While Europeans speak languages, Natives jabber, both in their daily intercourse and even when talking to God.

But the name Jabbertown also expressed the grimmer aspect of station life. As we have seen, Christian Klengenberg also knew the place as "Blabbertown": "We called it Blabbertown, because the dozen white men there were always complaining, and being of contrary nationalities they did not understand each other very well, and would be saying the same thing over and over and louder and louder each time so as to make their meaning clear" (MacInnes 1932:56–57).

Somewhat later, in 1908, the explorer Ejnar Mikkelsen came through Point Hope and stayed in Jabbertown, "the place," he wrote, "where almost all the white whalers lived. The name of the place seemed very appropriate to us, as their chief subject of conversation was the doings and sayings of their neighbours" (Mikkelsen 1909:52). Thus while Klengenberg had interpreted "jabber" as backbiting, Mikkelsen understood it in terms of gossip.

Trading Jargon

Iñupiaq and English were the main languages of the peninsula, and these were crisscrossed, modified, sometimes enlarged, and at other times submerged by a trading jargon based on both languages, along with smatterings of Russian and Hawaiian. It is unclear who spoke the trading lingo, how it fit into Iñupiaq and English, and how often it was used. Since the jargon developed as a tool of barter, it operated mostly in conversations between Iñupiat and non-Natives. But a large number of new jargon-derived nouns to cater for imported items entered Iñupiaq. If there was a grammatical system to the dialect, it was rudimentary, and took two forms: a simple, pidgin-style English and a reduced Iñupiaq. Because English is an easier language to break down than the polysynthetic Iñupiaq, some sort of English was usually the vehicle of any complex utterance spoken by Euro-Americans. To denote local things such as types of ice and Eskimo equipment for which there was no English equivalent, Iñupiaq nouns and a few simple verb forms were used. Similarly, within a very short period, Iñupiaq people coined terms for many imported goods, and some these neologisms are discussed below.

Jabbertown and Point Hope were not linguistically unique. In the early twentieth century, Vilhalmur Stefansson recorded a trading jargon on Herschel Island and along the Mackenzie River (Stefansson 1909). Stefansson notes a three-way contact between Eskimos, white men, and Kutchin Indians, the Indians sometimes adopting the lingo for trade or when they interpreted for Iñupiat and white men. Stefansson does not mention Point Hope refugees at these outposts, but he presumably knew of them, and the jargon spoken in the eastern Arctic no doubt resembled the Point Hope–Jabbertown dialect.

Stefansson recorded both grammar and vocabulary. In the case of non-Natives, some individuals used simplified Eskimo structures to express subject-object needs ("He wants to go aboard ship," "I am hungry," "Give me whale meat"). But the majority of the dialect seems to have consisted of discrete nouns, pronouns, demonstratives, and a few verbs such as "want," "eat," "break," and "die." These dialect words were mainly deformations of Eskimo terms. But there was also a growing vocabulary of Iñupiaq neologisms that were generated through the correct use of Eskimo modifiers. For example, through the addition of *-hluk* or *-qluk* (strange, bad):

Jargon term	Literal meaning	English meaning
kam'mik-hluk	strange trousers	cloth trousers
oktcuk-hluk	strange or spoilt oil	kerosene
tuk-tu-qluk	strange deer meat	pork, bacon
tan-a-qluk[8]	strange alcoholic drink	molasses

Some categories might be represented, according to who was speaking, by different but mutually intelligible words. The Eskimo word *nuliaq* (wife), as adopted in trade jargon by a Kutchin Indian for use in conversation with both whites and Eskimos, corresponded to the Euro-American trading jargon *kuna*, probably from the Danish *kone*—a word whose consonants are found in the Eskimo language, which lacks the "w" and "f" of wife. Stefansson also recorded the way Euro-Americans learned some Eskimo suffixes then misapplied them, but then generated new words from the confusion: *nunamun*, "towards land" being translated as "on shore"; *nunamun*, grammatically the dative "towards land," but in jargon, the locative "on land," transforming to "tent."

Point Hope Jargon

"Jargon" is not quite the word to describe the lexicon for the new artifacts that entered Point Hope in the early twentieth century. Jabbertown came and went. But manufactured goods came to stay, and new words for many of these entered the language and were absorbed into twentieth-century Iñupiaq. Some of the new vocabulary came from English, when the English word fit Iñupiaq phonology. Words such as "gasoline" (*gasaulik*), "molasses" (*milaasiq*), and "tea" (*tii*) easily slipped into verbal compounds such as *tii-tuġ-niaq-tuŋa*: "I'm going to drink tea." Dr. Irving Rosse of the Revenue Cutter *Corwin* gives a good example of how Iñupiaq linguistic compounds evolved to describe Euro-American phenomena: "Canoe is *umiaq*; ship is *umiaq+pak* [big ship]; steamer, *umiaq+pak+igniġlik* [big ship with fire]" (Rosse 1883:32).

In the absence of precise phonological equivalents, a word such as "flour" could be spread into *palau*. There were also imitative words such as *qaqqaulaq* whose guttural "q" sounds convey the crackle of pilot bread being eaten. Some manufactured goods corresponded to Iñupiaq words and things. Different kinds of imported oil and fuel—kerosene, gasoline, motor oil, grease, gun oil—lubricated or burned in some of the same ways as did animal and fish oils. Iñupiaq coinages for each of these used a common noun stem (*uġruq*, "oil") plus a suffix to create a new lexical idea. Thus kerosene was "like oil," motor oil was "old oil," grease was "thick oil." New terms for stoves and stove parts (chimneys, ventilators) could likewise be conjured from words for traditional lamp, fire, and iglu features.

Food provided an even richer field for invention. Some rare items such as oranges (*asiaq*, "berry") might be somewhat vaguely assimilated into the language. But the colors, shapes, and functions of foods were often conveyed vividly and with a sense of fun. From Point Hope informants born in the 1870s, Rainey compiled a list of terms, many of which would have been coined in their own lifetimes:

Imported item	**Iñupiaq term**	**Literal rendering by Rainey/Ivrulik**
bananas	*usuuŋnaq*	like a penis
beans	*kumaurat*	caribou droppings
beans	*niliġuaq*	something that makes you fart
rice	*uraaq*	adaptation of "rice"
oatmeal	*siġri*	dandruff [Rainey gives ear wax]
sugar	*avu*	something you mix in
mustard	*iḷiḷgaam anaa*	baby shit
cheese	*tchi*	adaptation of cheese
liquor	*taaŋŋaq*	"tonic" (see previous page)
tobacco	*taugaaqiq*	adaptation of tobacco
coffee	*kuukpiaq*	genuine river (allusion perhaps to its diuretic property)

Whole new terminologies were also generated for culturally powerful and elaborately differentiated word groups such as pipe, pipe deposit, cigarette, chewing tobacco, and snuff. New coinages for household goods and other imports flowed into the language almost as quickly as the materials they signified: fork, spoon, bread pan, muffin pan, dish, basin, sugar bowl, coffee pot, flour sifter, scissors, camera, telescope, clock [small sun], wristwatch, looking glass, tea strainer, flashlight. Some of these words were shared by other Iñupiat, others were specific to Point Hope; a comprehensive list would be enormous. A number of deformed Polynesian words, such as *kow-kow*, "food," and *pani-pani*, "sexual intercourse," also entered the jargon; no doubt there were more.

The fact that the Iñupiaq language took in these new phenomena reflects both the wit of the people and the lively, accumulative nature of the language. Like any unwritten language with an elaborately differentiated vocabulary for all the features of an environment within which it must work and whose dynamics it must describe, Iñupiaq continued to express the relationship of hunters to arctic conditions. Species, places, and not least a minutely discriminated terminology for subsistence equipment and the scrupulously anatomized parts that went into composite tools represented just a fraction of the linguistic assemblage. It was a language of both tradition and uncertainty, with a subtle and elastic syntax that generated new forms with virtuosic efficiency. As changes within the tradition occurred, the language adapted. It grasped what came in and fashioned new terms from existing phonology so that new technologies were quickly put to work in the language. When the white man came, the language was already well exercised in the unforeseen. Terms for new domestic and subsistence equipment were needed if these goods were to be integrated effectively and put to use. Many of these words would enter twentieth-century Iñupiaq and survive for the duration of life of the things they described. And while some of the terminology assimilated by pre-contact Iñupiaq had fallen away by the mid-twentieth century, the process of language change of course continues both among Iñupiaq speakers and in the dialects of English that have since evolved.

Notes

1 This trend had started near the village in 1895 when the Episcopal Church requested missionaries Edson and Driggs to stake out land surrounding the Mission (see Chapter 23).

2 Kelly had been in the region since 1884 and based on five years' exploration had drawn a detailed map for Commander Stockton in 1889.

3 Records exist for Ward Vincent, Daniel Cahill, William Hill, Arthur James Allen, Jack Hadley, A. L. Thompson, Jim Clark, John W. Kelly, John Albright, Mr. Anderson, Captain "Little" George Washington Porter, Ned Herendeen, Frank C. Dean, B. F. Tilton, and Charles Sanborne.

4 "[F]or a negro is counted a white man—and so is a Chinese! a strange idea, but common in the islands" (Stevenson, *The Beach of Falesa*, 1893, Part 1).

5 Of the original six houses, only two remained by the turn of the century, and these then became groupings that held their ceremonials on or near the sites of the houses from early in the twentieth century (see Appendix B).

6 Aġviqsiiña, born ca. 1875, also named Frank. One of Rainey's informants in 1939–41.

7 The association with Japanese perhaps also arose on account of Native pronouncation of English, which transmutes "b" to the sound of "p." Thus Jabbertown became Jappertown in local English.

8 The root here is *tanaq*, which was already a mid-nineteenth-century loan word from "tonic," meaning liquor, as in "here's my tonic." Dr. Irving Rosse of the Revenue Cutter *Corwin* wrote that the word *tanuk* originated "with an old Eskimo employed by [Captain T.E.L.] Moore…in Plover Bay [1848]. Every day about noon that personage was in the habit of taking his appetiser, and usually said the Eskimo, 'Come Joe, let's take our tonic'" (Rosse 1883:32).

CHAPTER 9
Ataŋauraq: Shaman, Trader, Point Hope's "Chief"

> In his efforts to become absolute master of his people [Ataŋauraq] passed from tyranny to assassination. For trivial causes, such as losing something or failing to be prompt in paying tribute, he would sally forth on a shotgun expedition, and either hold the victim while one of his followers did the shooting, or shoot while some one else held the man; but usually he did the shooting himself, as he was the most powerful man of his tribe. He became the head chief through defeating the old chief in a rough-and-tumble fight.
>
> February 14, 1889, he was shot dead by two brothers, whom he had exiled, and who returned for the purpose of killing him. During his life he killed five men and one woman. He inspired such terror that every year people left for distant hunting grounds, till it seemed that a few more years of his reign would have seen Tigara depopulated.
>
> —J. W. Kelly, 1890

> The Innuits build their houses near together for mutual protection and assistance in hunting and capturing whales, walrus etc.; but although associated together in this way they have no hereditary chiefs or rulers. There is, however, one in each settlement who is regarded as a leader. This may be on account of his strong will or his success in hunting or fishing… The omalik at Point Hope is self-appointed and controls his band by force of arms. He is well armed, a good shot, and quick to act. He has two wives who take turn in keeping guard while he sleeps, to give warning of the approach of danger. It is said that he has killed several of the band, and the rest appear thoroughly cowed. Although a terror to his own people he is friendly to the whites, and seems anxious to possess their good will. Of course it is only a matter of time when some Innuit will kill him.
>
> —C. L. Hooper, 1884

Ataŋauraq's Reputation in the Twentieth Century

Asatchaq recorded the history of Ataŋauraq in March 1976, and his recitation came near the end of a long series of mythic and historical narratives. Asatchaq had started his series of recordings the previous December with Point Hope's two origin tales, and while the recitations that followed came in no particular order, Ataŋauraq's story, with its details of white traders and Revenue Cutter men, could only have come near the end. Asatchaq did not volunteer the Ataŋauraq material—I had to request it. But Asatchaq was not on principle withholding this often-explosive saga: the suitcase he kept in his cabin contained a short, unpublished typescript of a narrative about Ataŋauraq that a previous fieldworker had made (Asatchaq ca. 1970). Parts of that narrative are included here.

When Asatchaq started his recitation in December 1975, my own vague background knowledge had prompted me to request Tikiġaq's two creation myths (Asatchaq and Lowenstein 1992:5–17). By the time I'd been in Point Hope a few months, conversations in the village had already taught me something about Ataŋauraq, whose reputation was still alive almost ninety years after his murder. While the creation stories were no longer current, here was a nineteenth-century narrative that had the status of legend. In part, this reflected the prominence of Ataŋauraq's grave markers, which stood just west of the village, its whale-bone supports still prominent in the midpeninsula area. Memories of Ataŋauraq also continued to live because the "chief," with his extensive household, his many associates and enemies, had dominated the penultimate decade of the nineteenth century. Many people in twentieth-century Point Hope remained, after all, descendants of Ataŋauraq and his several wives. Finally, Ataŋauraq's reputation reflected the simple fact that he was a larger-than-life character, part hero and part villain, whose career lay at the meeting point of the traditional and the modern era, and whose own experience reflected many of the transitional issues with which the whole community struggled.

And while in 1975 Asatchaq, who was born in 1891, was one of the few elders who remembered the years that followed Ataŋauraq's death, younger people, who had little opportunity to hear traditional stories, knew quite a lot about Ataŋauraq or claimed a relationship with him. Rather like the legendary young man who broke taboo only to become the lunar deity (see Chapter 13), so Ataŋauraq's crimes became partly submerged within a generalized admiration for a memorable character.

Then, in 1980, the historian John Bockstoce discovered the portrait of Ataŋauraq reproduced here as Fig. 21. With a face that uncannily resembled that of a popular local man in his thirties, Ataŋauraq's reputation received a further boost. The merry and apparently self-confident figure wearing mattress-ticking trousers and skins cut in European fashion chimed well with eclectic late-twentieth-century arctic Alaskan dress ideas.

Taken in summer 1888, about six months before his murder, this is a revealing photo. Given his reported addiction to liquor, Ataŋauraq appears to be in good health. Poised on the stairway of the U.S. Revenue Cutter *Bear,* this portrait communicates several, often contradictory, messages and it is worth pausing to examine it alongside a photograph taken in 1881 (Fig. 21) and another photograph, not reproduced here, from 1887.

Two initial points should be made. First, the very existence of three photos attests to the prominence attributed to Ataŋauraq by white visitors to Point Hope, while his presence both within photo-technology and white men's environments offers a window onto the complex world in which the "chief" operated. Second, while the photos provide limited visual information, they illustrate aspects of Ataŋauraq's personal biography and suggest something of the history in which he was a player.

FIG. 21. *Ataŋauraq on board the Revenue Cutter* Corwin, *summer 1888. His stance and smile self-confidently directed to the camera somewhat belie E. W. Nelson's account of his shyness. Ataŋauraq is dressed in fine new clothes, not dissimilar to those of Koenig (p. 71). The Whaling Museum, New Bedford, Massachusetts, Kennedy Collection.*

All three photos show Ataŋauraq dressed largely in skin clothing but in Euro-American environments. The 1881 photo takes us quite deep into Native tradition while also superimposing an image of that apparent timelessness onto a backdrop of nineteenth-century America. The "stage" on which the five men are posed (the grouping suggests the photographer E. W. Nelson's arrangement) is the deck of the U.S. Revenue Cutter *Corwin*. In addition to the mast and ropes, the clearest background feature is an iron rail. In contrast to the ship's furniture, however, are the weather-beaten faces of the Eskimo men, their prominent labrets, and their clothing, of which four main styles are represented. As remarked by Burch (1981:18), the man to the far right wears a caribou skin with fur side in. His neighbor in the center wears a sea-mammal-intestine parka while the man to the right in the back row wears caribou skin, fur side out. The two men on the left have white cotton drilling over skin parkas. And while the man in gut skin is placed to suggest that he was group leader, the figure standing unobtrusively to the left is Ataŋauraq (Burch 1981:18), his build and features suggesting that the man next to him might be a brother or cousin.

FIG. 22. *Ataŋauraq (back row left) and a group of Point Hope men on the* Corwin, *1881. Ataŋauraq wears a white cotton parka cover and wolverine-skin parka hood ruff. The two men on the right wear caribou skin, fur side out and in, respectively. The man in the middle wears a waterproof parka of seal mammal intestines. Smithsonian Institution, National Anthropological Archives, E.W. Nelson Collection.*

While Ataŋauraq stands among what must be a group of high-status Point Hope men, he is not the focus of the composition and we might assume him to be a hunter and trader of no significance. But while this man might be a member of a tribal aristocracy, E. W. Nelson, who took this photo in July 1881, knew that Ataŋauraq had already become what the white men described as a "chief" and had taken upon himself the role of broker in trade between Point Hope and ship's personnel. Nelson also gives a vivid account of the somewhat difficult task of taking Ataŋauraq's portrait (no longer extant):

> I tried to procure his photograph, but he became very nervous and could not be made to stand quiet, until he was told that it would be very bad for him if he did not. As soon as I had taken his photograph he insisted on having me stand in the same position that he had taken while being photographed. Then he looked under the cloth covering the camera, and when he saw my image on the ground glass he appeared to be greatly pleased, seeming to think that he had thus counteracted any ill effect that might follow in his own case. (Nelson 1899:303)

This 1881 photo presents Ataŋauraq at a comparatively early stage in his dealings with white men, the cotton shirt representing the single mark of change, and Nelson's comments somewhat suspiciously approximating a stereotypical Euro-American view of Native "superstition." Apart from the cotton shirt, which would cover a caribou skin parka, Ataŋauraq appears in traditional guise. His hair is long, he wears labrets, and his parka hood is trimmed with wolverine skin whose claws, bound with bleached skin, dangle on his chest.

The 1887 photo is a portrait of Ataŋauraq with one of his wives, who stands behind him in front of a trader's building at Jabbertown. As observed by Burch (1981:16), the woman is probably dressed in a cloth parka, while both she and Ataŋauraq wear wolverine hood trimmings. Ataŋauraq's parka, however, is of reindeer skin, bought from Siberian traders. With his head tonsured and hair still growing to traditional length, Ataŋauraq smokes the pipe that perhaps featured in his fight with Bayne (see Chapter 10). Given that this photo was taken in 1887, when Bayne established his whaling station, we may assume the building was Bayne's. Facing the sun, the "chief's" puffy appearance suggests that he is in relatively poor health.

In the 1888 photo we see Ataŋauraq portrayed alone on the *Bear*, where he appears to stand comfortably in the white man's world and, as remarked by Burch, "in Yankee whaler costume" (Burch 1981:17). This summer wear, which was perhaps reserved for Ataŋauraq's visits to his friend Michael Healy, shows Ataŋauraq in his most Europeanized costume. The fur parka has been replaced by a jacket, trimmed with a variety of skins and adorned with ivory buttons and a wolverine ruff that apparently lacks a parka hood. Instead, atop his short-cut hair, Ataŋauraq wears a peaked Revenue Marine cap. His sealskin boots and clean new mattress-ticking trousers complete an assemblage that is coordinated to present the "chief" as an elegant, well-groomed, and aristocratic figure. But this photo also records a moment of some poignancy. He is a privileged ship's visitor and has perhaps been socializing with Healy (see "Ataŋauraq's Trade with White Men" below). The *Bear* would, however, soon weigh anchor and Ataŋauraq would be home again, fighting to maintain his supremacy in the village. Victim of the liquor he had recently learned to distill, and with at least four murders to his name, the "chief" would be dead by February 1889, and the majority of the village would be celebrating this outcome.

Ataŋauraq as *Umialik* and Shaman

Ataŋauraq's hegemony was a modern phenomenon, but as an *umialik* and shaman he lived within the tradition. Point Hope *umialit* were the elite hunters of the village whose economic success enabled them to attract crews to work their skinboats for the whale hunt. While *umialit* had no special authority within changing village hierarchies, an *umialik*'s charisma and generosity would sustain his influence while his fortunes held. *Umialit* who were also shamans enjoyed particularly high status. And while Ataŋauraq was a comparatively minor practitioner, the terror he inspired derived partly from his shamanistic status; he claimed, for example, that he could not be killed and that his own touch could be lethal. This (in addition to his physical strength, intelligence, business acumen, and ruthlessness) played a part in his acquisition of power.

Other aspects of local tradition informed Ataŋauraq's character. Many shamans and *umialit* in Point Hope history were preoccupied with acquiring social power, and Ataŋauraq had the same trickster-shaman flare that characterizes shamans in traditional stories. But unlike the over-reachers of tradition, Ataŋauraq lived in unstable times in which moderating social behavior had been weakened by a fall in the village population.

To organize a whale hunt meant providing food and equipment for up to eight crewmen over a six-week period on the sea ice, so an *umialik* needed a fair degree of economic security. For the months leading up to whaling in March or April, an *umialik* would therefore have to freeze down meat and accumulate other subsistence materials. To attract the best hunters, *umialit* must also present themselves as good bosses. To demonstrate their wealth, they and their wives must distribute gifts to the crew they wanted.

Ataŋauraq was born sometime in the 1850s into the family of an *umialik* who was said to "have lots of stuff," which suggests that his father was a high-status boat owner.[1] By the time Ataŋauraq was an adult, he was living on the western edge of the village in a group of five iglus with the collective name of *upsiksui* ("clump of willows"), named because the houses were, like willow bushes, close together and "hard to get through" (Kunuyaq 1940–41:1). The storage racks on the iglu tops were also closely packed, "so close they look like trees," according to Ataŋauraq's nephew Peter Kunuyaq. The other iglu owners in the group included Ataŋauraq's brother Tiguatchiaq and a successful shaman-*umialik* called Nuiraq. Collectively the group was known to some as the *upiksuamiut*, "the inhabitants of upsiksui," and to others as the *iñugailiŋmiut*, "the people who live together."[2]

Ataŋauraq and his family also belonged to Qaġmaqtuuq *qalgi*. One informant claimed that the family lived at close quarters because they were poor hunters (Rock 1940–41:3). But many people lived in similar groups to maintain hunting partnerships and facilitate *qalgi* organizaton.

The Feud That Began with Suuyuk's Murder

Ataŋauraq first appears in a story about a shaman called Suuyuk, who was murdered at Cape Lisburne in about 1875. Ataŋauraq's role here was marginal, but Suuyuk's story nonetheless underlies Ataŋauraq's fate, and his own murder would, in part, be a consequence of the Suuyuk feud.

This saga opens with Suuyuk's departure from Point Hope. Suuyuk is in a fury and leaves with the intention of cursing the whole village by scaring off the game.[3] The story behind this is unexplained. Fearing that Suuyuk will succeed, some men pursue him to Cape Lisburne and kill him. Within a few months the main assassin is murdered by Kunuyaq, the dead man's son, with his own cousin Ataŋauraq acting as an accessory. While Ataŋauraq would now initiate his trading career with white men, his cousin Kunuyaq remains a traditional hunter. This ends the first part of the Suuyuk revenge saga.

Ataŋauraq's Degeneration

Ataŋauraq next appears in stories of the 1880s, and while he cut a somewhat shadowy figure in the Suuyuk episode, he has now grown rich and powerful. He has also begun to fashion himself, as did some of his contemporaries, as a modern person by learning English and adopting white men's clothing and domestic habits. His degeneration was, however, explained by Point Hope storytellers in terms of traditional theory.

One local view was that Ataŋauraq had "turned bad" in conformity with a pattern exemplified in local stories. In these, a character known as *iñuqaġnaiḷaq* ("one who habitually murders people") is shown both dangerous and bereaved, the *iñuqaġnaiḷaq* killed in grief-stricken reaction to the loss of a child. Ataŋauraq was believed to have followed the same path. One twentieth-century informant put it thus: "One of his daughters died and he went crazy. He started beating his wives" (Lowenstein 1973–89). Significantly, the informant who offered this suggestion came from a family that Ataŋauraq had abused.

Even if this explanation is historically untrue, it demonstrates the frequently destabilizing dynamics of grieving.

Early Ataŋauraq Stories

The first story presents a supposed shamanistic attack by Ataŋauraq on one of his own kin. The victim is young Suuyuk (b. ca. 1870), grandson of the murdered shaman. Young Suuyuk is also Kunuyaq's son and Ataŋauraq's nephew.

Asatchaq (Jimmie Killigivuk) opens the story menacingly:

> This is about my father Kiḷigvak and his brother Suuyuk. Ataŋauraq wanted to kill them. He called them ducks: he said they were young ducks, easy to kill. He saw them when they were on the beach, and said that.
>
> One morning in the winter Suuyuk and Kiḷigvak got ready to go out on the ice. They were going to hunt seal. Then they separated. And when Kiḷigvak saw a seal, he could feel the inside of his body growing hot. Kiḷigvak had an amulet. It was a wolf's snout given him by Uyagaaluk [Ataŋauraq's brother]...he chewed it, then he rubbed himself with it...and the heat left him. At the same time, Suuyuk sat down on the ice by open water when he saw a seal. But it wasn't a seal: it only appeared to be one. Suuyuk shot at it. As he went to fetch it, he felt heat in his stomach, and...he began vomiting blood. (Asatchaq and Lowenstein 1992:157)

Suuyuk dies, but returns to life in the form of a polar bear. He goes to the village and calls to his father through the iglu skylight. The parents try, unsuccessfully, to bring him into their house, and the story ends with a final reference to Ataŋauraq: "Suuyuk's father [Kunuyaq] paid shamans to find out where his spirit was. He went to Ataŋauraq to locate his spirit. But Suuyuk's mind rejected Ataŋauraq's search, and the shaman could say nothing" (for complete story see Asatchaq and Lowenstein 1992:157–62).

The story took place around 1882, when Ataŋauraq was in his mid-thirties. But the inexplicable motive for the attack is compounded by Kunuyaq's paradoxical request that the alleged murderer should "find his spirit." To enlist the help of a family shaman would have been Kunuyaq's natural resort. But Ataŋauraq's attack on kinsmen is hard to understand, and nothing like it in traditional narrative has been recorded.

In the absence of a certain motive, two things are possible. First, that the attack was a symptom of both Ataŋauraq's instability and a destabilized community. As the storyteller said: "To Ataŋauraq, all people were alike, relatives or not. He'd kill anyone he pleased" (Asatchaq and Lowenstein 1992:179). On the other hand, given the hypothetical nature of any shamanistic diagnosis, Suuyuk's death might have had nothing to do with Ataŋauraq, and the connection between Ataŋauraq and the death was invented later. That "Ataŋauraq wanted to kill them" may therefore have represented an attempt to interpret Suuyuk's hemorrhage when it was safe to make such an accusation after Ataŋauraq was himself dead.

This enigmatic story is the only local narrative about Ataŋauraq the shaman as opposed to trader and strong man. Apart from its reference to firearms, the events, if not the kin relations, have a traditional character. The apparently motiveless assault similarly evokes a legendary atmosphere of menace and supposition.

Another of Ataŋauraq's claims helps explain the fear he provoked. This was the assertion that he was invulnerable and that the touch of his body could kill. "Ataŋauraq said that nobody could kill him, even if they slit his stomach and his intestines came out, he would not die" (Nasugluk 1940–41:8)—a boast that parallels Ataŋauraq's remark about his victims as "ducks." But some people also regarded Ataŋauraq as a mediocre shaman and hunter: "Ataŋauraq wanted to hunt beluga one summer. He wanted one of Piquk's charm songs. Piquk [a nonshaman] said there was no luck or magic involved, just hard work. Everyday he watched the weather, went in and out constantly, just getting a few seals at a time" (Asatchaq ca. 1970:15). Piquk's assertion expresses one aspect of local orthodoxy: that the supernatural and the pragmatic were interdependent.

Tense relations with this same Piquk are expressed in one of Ataŋauraq's most overt acts of predation. Piquk's grandson Asatchaq told this story:

> Ataŋauraq wanted my grandmother Kakianaaq, but she denied him. Ataŋauraq replied that he'd kill Piquk if she refused.
>
> So Piquk…told his daughter Niġuvana, "I am going out to meet Ataŋauraq. He can kill me if he wants to." Then he sent Kakianaaq to Ataŋauraq to tell him that he was waiting unarmed by the uprights of the storage rack… "He can kill me out here."
>
> Kakianaaq went and told Ataŋauraq. He said, "Let's just go and sleep in the new house." So they went through the entrance tunnel to a new room. She slept with Ataŋauraq. Then she sang him a sleep song. When he fell asleep, she went back to Piquk. They escaped to Aqalulik. (Lowenstein 1973–89)

In attacking the young Suuyuk, Ataŋauraq had cut across his own kin lines. This story shows him abusing a neighbor's family, and it demonstrates a calculated strategy: if Kakianaaq refused him, Ataŋauraq would kill Piquk and his children to preempt vengeance.

The family's exile was brief, but three years later, Piquk's daughter Niġuvana (unwillingly) entered Ataŋauraq's household: first as a helper and then as wife to one of Ataŋauraq's adopted sons. But the story illustrates one of the most serious consequences of Ataŋauraq's tyranny: the depopulation of Point Hope village. While Piquk returned, many of Ataŋauraq's enemies did not. And because they stayed away, their stories have been lost.

Ataŋauraq as a Trader

The next two stories are about Ataŋauraq's trading activities within Point Hope. The first shows his acquisition of local goods for trade with white men:

> Kunuyaq and family were hunting on the Kuukpak river. Ataŋauraq was there. And when Kunuyaq got a good caribou and had skinned it, he hung up the skin and meat and sent his wife to fetch it. The skin had vanished, so she just took the meat home.
>
> Next autumn, Ataŋauraq needed some rope. He sent a man from house to house with a caribou skin asking to trade it. When the man came to Kunuyaq's, and he had looked it over, Kunuyaq recognized the skin and took it. "This is my skin that was lost last winter. Go home," he said. "Let Ataŋauraq come and get the skin. He can pick it up himself." Ataŋauraq never fetched it. (Asatchaq and MacLean, ca. 1978)

The next incident also involves Piquk's family. This time we witness a younger man's defiance:

> Piquk, Kakianaaq [and their son Samaruna] went north to Utuqqaq for trade with other Eskimos. While they were at the Utuqqaq River they saw a four-masted ship off-shore, and they went out to trade. One man gave Samaruna a heavy, seamless woolen blanket and said that he wanted five tanned red fox skins in return. My uncle promised to bring him these next season.
>
> When they returned, Ataŋauraq heard that someone had a blanket. He sent to Samaruna and told him to give it up. Samaruna refused and the chief sent a second messanger. My uncle was angry, but eventually gave in.
>
> That winter Samaruna trapped red foxes. He picked the best ones to tan for the trading partner who had given him the blanket. But then Ataŋauraq heard that Samaruna had these beautiful skins, and sent a man to get them.
>
> Samaruna was angry. He had promised his white partner. So he sent a message back. If Ataŋauraq wanted the skin he could come and get it. If Ataŋauraq came, he'd kill him. Samaruna sat by his iglu behind his up-turned skinboat and waited. His house faced Ataŋauraq's. He sat with his .44 rifle in the bow of his skinboat. Ataŋauraq would have been an easy target; but hands were shaking. Still, if he'd come, Ataŋauraq would easily have beaten him. Samaruna was relieved that he didn't. (Asatchaq and Lowenstein 1992:173)

This incident can be dated with relative precision. Born in about 1870, Samaruna must have been around eighteen at the time, thus dating this event to ca. 1888. Samaruna was therefore roughly of the generation that Ataŋauraq was also assaulting: "Ataŋauraq tried to kill his [adopted] son Pauyuuraq.... Along the iglu wall some fox skins were hanging. Every time Ataŋauraq came at him with a knife, the boy dodged around the skins. All Ataŋauraq could do was cut off fox tails" (Asatchaq and MacLean ca. 1978).

Samaruna, of course, had a particular reason for hating Ataŋauraq. Ataŋauraq had raped his mother and threatened to kill the rest of the family. Also at this time, his sister Niğuvana had been trapped in Ataŋauraq's household, and young as he was, Samaruna single-handedly attempted to ambush Ataŋauraq. Waiting for the "chief" on the sea ice one day, he shot at him but missed. He escaped retribution when it turned out that he had aimed at the wrong man. This vignette coincides well with Samaruna's solitary defiance and the "shaking hands" of the fox-skins incident.

How Ataŋauraq survived for so long in a society where murder was relatively common and could be justified on grounds of revenge is unclear. Froelich Rainey expressed bewil-

FIG. 23. *Samaruna (born 1870) photographed ca. 1920. Samaruna's posture and the intensity of his gaze express an authority to which his nephew Asatchaq alluded frequently.*

derment at this in an interview he recorded with a married couple who had known Ataŋauraq:

> Rainey: How did Ataŋauraq become chief?
> Nasugluk: I don't know.
> Nasugruk: He took fox skins that didn't belong to him, and wives. He made friends this way.
> Rainey: Didn't that make enemies, weren't men mad?
> Nasugruk: A little mad sometimes, but things were different then, they changed wives all the time. But Ataŋauraq didn't exchange, he just took them.
> Rainey: Did you ever hear of other previous chiefs, or any stories about them?
> Nasugluk: No. Any good hunter who had lots of skins was a big bug [sic]. Anyone who owns an *umiaq* is an *umialik*, even if he's poor, also any man who may not have a boat but has lots of stuff and is generous with it, has grub all year round and gives it away, is *umialik*... There were two brothers who lived between Kotzebue and Kivalina on the north side of the point, it was all lined with whale bones...and the older one was powerful, carried half a whale. They were more like *umialit* than Ataŋauraq.[4] (Nasugluk 1940–41:13)

What baffled Rainey was the way in which Ataŋauraq's crimes might be interpreted retrospectively as acts of friendship. But this puzzle makes more sense if one views Nasugruk's assertion about how Ataŋauraq made friends both as a kind of shorthand for two things in one and also as a process. The two things may be looked at as the positive and the negative aspects, over a long period, of Ataŋauraq's trading activity, while the process might be described as the transformation from the socially useful to the predatory and extortionist. It would seem, in other words, that Nasugruk could be saying that Ataŋauraq's earlier period of eminence derived from his success as a "trading chief," and that this was generally welcomed because it brought white goods into the village. What started as communally beneficial led to Ataŋauraq's rise. Once he had become "chief," the act of trading that made him rich was interpreted retrospectively according to its later complexion, which included stealing from those with whom he had "made friends" as a result of the benign phase.

The consensus among survivors of the period was that Ataŋauraq had indeed been a good man who went wrong—either on account of the loss of a child (the traditional view) or because he had started drinking (a contact causation):

> Ataŋauraq was a good and generous man when he wasn't drinking. Hungry people could go to him, and the chief would give them rifles, ammunition, knives. Eventually he'd expect some return in the form of skins and meat. And people did pay him. The only thing he wouldn't part with was the molasses he used to brew liquor. (Asatchaq ca. 1970:17)

Initially, then, Ataŋauraq was viewed as a generous *umialik* making his surplus wealth available to co-villagers in the traditional way. But there was a new aspect to this activity. Any *umialik*, now and in previous years, would have anticipated some return from his crew, in particular in the form of support at the whale hunt. But the traditional expectation of a return would have been implicit and not expressed as a demand. With Ataŋauraq, the expectation was no longer understood as an implied obligation. Instead, Ataŋauraq offered goods within a system of enforced economic return that he had himself imposed.[5] In contrast to the competitive trading between near equals within Point Hope and beyond, Ataŋauraq used his super-*umialik* status to bind people to him as a central subsistence source. And whereas *umialit* would usually have acted within the sphere of an extended family and ceremonial group, Ataŋauraq now functioned as *umialik* to the whole community.

Ataŋauraq also instituted an internal shore-based Native equivalent of the relations that he enjoyed with the white captains (or "*umialit*") on the big ships. The goods that Ataŋauraq bought from the ships came into his possession and were then passed on, at a price, to fellow villagers. Ataŋauraq's trading activities, in other words, transformed gift exchange and traditional barter into the beginnings of a retail system. The "chief" was, in this respect, as we shall see later, Point Hope's first shopkeeper—one basis of which lay in adapting both *umialik* status and village trade to the new economic climate. Given Native experience of traditional inter-regional trade, it took very little time, and a minor shift of emphasis, to convert the visiting white man to a new class of trading partner.

Ataŋauraq's Trade with White Men

Occasional trade with early British and Russian ships, which yielded a few temporary and recreational goods, and regular commerce with Americans whalers were two very different things, for the Americans, from the 1850s onwards, became seasonal regulars. Following the introduction of steam, the most intense contact period at Point Hope was in July, when ships gathered offshore to receive supplies from the San Francisco tenders, or in early autumn when some of them were running south. Starting in 1879, the Revenue Cutter ships *Corwin* or *Bear* would meet the whaling vessels, check for contraband, take on board shipwrecked survivors, relieve the whalers of criminals and invalids, and, weather conditions permitting, put a boat into Point Hope and establish a brief official presence.

Contact with Point Hope during these weeks took two main forms. Skinboats would travel out to meet the ships, taking with them the skins, meat, ivory, and Native artifacts that crewmen liked to buy. Once gathered under a ship's bows, the Iñupiat would hold up samples of their trade and then be hoisted in limited numbers to the deck. Sometimes, to avoid crowding, the white men would be lowered to barter in dinghies. But there was also commerce on land. Ships' crews came ashore for wood and water, and, from the 1870s onward, to hire men for their boats and women to make clothes and as sexual companions.

Ataŋauraq was initially just one of the crowds who paddled out to bargain. But then in his early thirties he started to take over. Accumulating goods that white men wanted and requisitioning from co-villagers their own stock in trade, he became the agent through whom trade was conducted. A glimpse of Ataŋauraq's part in the process appears in the journal of Edward W. Nelson, who was on the *Corwin* in 1881. On July 20, Nelson wrote:

> Laid off [to south of Point Hope] ... The Natives came on board in an umiak. Among them being the Chief here who is such a thorough terror as he has shot four of them already and made the rest much afraid of him. During the trading when any of them received a fair price and haggled for more, he would quietly take the [American] goods and hand them to the man and the latter would take them and not say a word further. (Nelson nd)

It is worth noting two things here. First is the chief's assumption of leadership and of a superior knowledge of trading etiquette. Second is the implication that Ataŋauraq appears, in a peculiar way, to be on the side of the white man. When his fellow villagers, following a practice that had been noted by Cook, ask what Ataŋauraq judges to be too much, he puts them in their place. Thus Ataŋauraq detached himself from home practice, and in so doing developed a cross-cultural function. As the storyteller Asatchaq used to say, "Ataŋauraq was Tikiġaq's first white man."

Peter Kunuyaq also described the process:

> Yes, he traded. Also with the coast guard. First all the furs and whale bone [sold] for tobacco and ammunition and rifles…but when they got the big whale bone [baleen], they sold them mostly for rifles and ammunition. Ataŋauraq had a store… When Ataŋauraq first knew about the boats coming up here, he started buying whenever anyone shot an animal or got some [baleen]. When the whaling season was on, Ataŋauraq was busy all day. He had four men to take the stuff from people who came, and two men to store it on the racks. (Kunuyaq 1940–41:7)

This secretive, monopolistic activity, in which Ataŋauraq's storeroom was kept off-limits to a needy population was underlined by Kunuyaq's final comment: "He never let you into his iglu. He had fellows who told [Ataŋauraq inside] who was there. Then his fellows came out and said what [Native traders] would get. He traded rifles, tobacco, lead, powder, primers" (Kunuyaq 1940–41:7).

Kunuyaq's account given in English takes us as close as we ever get to how Ataŋauraq worked. Just as a traditional *umialik* needed his crew during the hunt, Ataŋauraq required helpers to manage a trading enterprise and also to spread terror. Kunuyaq's expression "he had four men" corresponds to another storyteller's use of the Iñupiaq *aŋutit* ("men"), but diverges from the same narrator's *savaktit* ("workers"). No names are given. In contrast to the narrative convention of naming the individuals in a story, Ataŋauraq's men are anonymous. Like shadows, they move between a super-individualistic boss and a subordinated community. The impression throughout is that these men are Ataŋauraq's thugs. Collectively, the seven of them, including Ataŋauraq, make up the equivalent of a skinboat crew, with its maximum complement of eight man.

In management terms, any *umialik* would have been used to organizing groups of roughly this size. And along with his eventual household of five wives, it is hard to imagine that he would have been capable of controlling a circle of many more people. Of interest here is the way in which Ataŋauraq has adapted one aspect of traditional skinboat crew structure to this new style of trade. The seven- or eight-man *umiaq* crew becomes a seven-man trading operation.[6]

But it is unlikely that the six helpers were in Ataŋauraq's whaling crew itself: "When the whaling season was on," said Peter Kunuyaq, "Ataŋauraq was busy all day. He had four men to take the stuff from people who came, and two men to store it on the racks" (Kunuyaq 1940–41:7). While Ataŋauraq, as traditional *umialik*, might sometimes retire to the village during the hunt, the crew itself could never be absent. Also the six men stacking baleen were not dealing exclusively with Ataŋauraq's whale products. They were stationed to accept materials from other *umialit* who wanted to barter for his goods. Given food shortages and each *umialik*'s obligation to feed his crew, Ataŋauraq's "store" may have been a useful provisioning service for the rest of the village crews. In monopolizing commerce in baleen, and providing fellow *umialit* with food and ammunition in exchange for baleen, Ataŋauraq thus helped support the accumulation of next winter's food. The process, which partly transformed the whale hunt into baleen trading, thus also helped support the subsistence effort.

Violence and Intimidation

Besides Ataŋauraq's physique and magical "invulnerability," it was the brutality of his entourage that made his dominance possible and enabled him to get away with crimes that would have otherwise been quickly avenged. Many of his offenses were conspicuously open: "he just took other people's things. For example he'd take a polar bear skin and in return just give a little brown sugar or some ammunition. Sometimes he'd lie at home smoking and the smoke would spoil the white fox skins that were hanging from the ceiling" (Asatchaq ca. 1970:7). And the following encounter with a group of villagers indicates the extent of their fear:

> Once four or five men found a polar bear in the village. They killed it and were skinning it when Ataŋauraq came by with his .44 rifle. "What are you going to do with that skin?" asked the *chief*. They told him that they didn't know yet.
>
> "If any one of you takes that skin home, I'll kill every one of you," said Ataŋauraq. "But if I take it home, everything will be all right with you."
>
> Even though there were a lot of them, they gave in and let the chief have the skin.
>
> "We wanted a bit of ammunition for our rifles anyway," they said. (Asatchaq ca. 1970:9)

Another story illustrates the narrow line that sometimes divided Ataŋauraq's generosity from his caprice. The "chief" had returned from the Noatak, where he had acquired fox skins from his partners. When two of these partners traveled to Point Hope they expected a large return. But the thought of visiting Ataŋauraq represented an ordeal. One of them said: "When we get to Ataŋauraq's place, it will be just like visiting white people." On arrival in Ataŋauraq's iglu, the man was too bashful to speak "because Ataŋauraq and his family were like white people... they ate *maktak*...but never said a word." The storyteller continued:

> After they had eaten, Ataŋauraq said to his partner, "There are rifles over there." In those days, they kept the guns standing on their butts in a corner of the iglu. Ataŋauraq told him to go and look at them and pick one out for himself.
>
> "Choose one," he said. But the partner didn't take the best of them...he took a small rifle, a .44.
>
> He had said that if Ataŋauraq offered him a gun, he'd take a long-distance rifle. He was not the only man who found it hard to speak in Ataŋauraq's presence. There were others too. Ataŋauraq was like a white man. (Asatchaq and Lowenstein 1992:167)

This story exposes a wealth of issues. Ataŋauraq's partner would have held high status in the Noatak region, otherwise the "chief" would not have been in partnership with him. As a stranger in Point Hope he is, however, at a social disadvantage. But the man's confusion is intensified by the host family's evolution into "white people," who were, for example, eating at a table. Additionally menacing were Ataŋauraq's charisma and a charming pretense at generosity underpinned by the threat of violence. These factors, in combination, subvert what should have been a partner relationship of equality into a game of humiliation. Generosity has been converted into a display of power.

Alcohol and Violence

Horrible as the murders surrounding the older Suuyuk were, the events they describe lay within a tradition of local feuding. In this, the search for justice rested on a process of revenge and retribution. Loss from within one group was "righted" by counterattack. This second murder satisfied the injured group by restoring an element of equilibrium, after which the feud would lose momentum. In contrast, Ataŋauraq's violence had become arbitrary and its causes often inexplicable. The bloodletting he unleashed was a contact phenomenon, and as suggested by most twentieth-century informants, alcohol was the inflamatory element.

Ataŋauraq's first two murders, unlike the train of events in Suuyuk's story, erupted out of nothing. Motiveless and chaotic, these were events that could be explained neither by bereavement theory (the death of a child) nor by the imperative of revenge. Asatchaq's account is as follows:

> Ataŋauraq invited his relatives Ilutuaq and his nephew Miliq to drink whiskey one winter. The uncle got drunk. But Miliq only pretended, and poured the whiskey into his parka. Ataŋauraq started fighting the uncle and killed him. Ataŋauraq went outside. Miliq followed, but could hardly walk. He had a broken leg. Miliq ran. He found himself in Ataŋauraq's brother's place. As soon as came out again, Ataŋauraq shot him. The wound never healed and Miliq died. (Asatchaq and MacLean ca. 1978)

The murderous "party" had been prearranged with the help of drinking cronies and the victims clearly invited for an attack. Different as this was from the event with Ataŋauraq's partner from Noatak, both incidents have the stamp of disordered new manners. Domestic hospitality had once implied trust, which bound host and guest to mutual protection. But to murder within the context of hospitality was to turn Point Hope values on their head.

The drinking party murders thus take place in a very new context. Alcohol provided the catalytic spark once the drinking had begun, just as alcohol was also the occasion for the "hospitality" itself. And if the attack was premeditated, then an additional element comes into play: namely that Ataŋauraq consciously perceived alcohol as a context, and perhaps a pretext, for violence. Alcohol, within the new historical dispensation, had become a "place" and intoxication the condition in which sadistic gratification might be organized.

It is interesting, however, that the outcome of that episode reeks of Ataŋauraq's own desolation. The storyteller continued:

> After this Ataŋauraq's character changed. Now he always went hunting alone, and other hunters stayed out of his way. The day after the murders, Ataŋauraq visited my grandfather Kunuyaq. Ataŋauraq's hair was cut short. My grandfather asked what had happened previous night. "I killed two men," replied Ataŋauraq. My grandfather asked why he hadn't told him before. He'd wanted to know. Then the chief went home. (Lowenstein 1973–89)

The focus remains on Ataŋauraq. His victims and their families are forgotten. The next morning we see Ataŋauraq alone and apparently traumatized. Why he cut his hair remains uncertain, but it harmonizes with the transitional moment in that it suggests an attempt to be another person from a different culture. It is not, however, just Ataŋauraq who has changed in character and appearance: the village now knows he has stepped beyond the limits of social convention. He has become stigmatized and his social environment has changed. The community at once avoids him and prepares for the endgame: "Now when

Ataŋauraq went out on the sea ice, men lay in wait for his return. He always wore white fox skin parkas on the sea ice" (Asatchaq and Lowenstein 1992:173).

Kunuyaq, who had killed Suuyuk's murderer, was a suitably experienced confidant. But Kunuyaq's action fell within a pattern of inherited practice. By the mid-1880s, at the height of Ataŋauraq's power, a history as recent as Suuyuk's had drifted almost to the realm of legend, while the surviving actors in that drama had moved into a new era.

Notes

1 Ataŋauraq's father and mother were Isumaq and Panigatchiq. He had two half brothers, Uyaġaaluk and Tigautchiaq, and a sister, possibly by adoption.

2 An extended group that Ernest S. Burch has called "compound" families (Burch, personal communication 2001). "These were the major organizational components of a traditional Northwest Alaskan Eskimo society…the local family formed *the* social unit… Dwellings were *not* scattered haphazardly about in a village site, but built in clusters (or neighborhoods), each of which housed the members of a single, local family… the [*qalgi*]…was in fact built by the members of a single local family in the vicinity of their dwellings" (Burch 1975:241, 246).

3 *Niġġutaiḷi*: shamanistically to scare off animals. For a similar nineteenth-century narrative see Asatchaq and Lowenstein 1992:187. For the Suuyuk story, see Asatchaq and Lowenstein 1992:152.

4 Here Nasugluk is probably referring to a myth about two giants who lived, according to another version, at Cape Thompson. See "Long unipkaaq" in Kaplan and Riordan 2007.

5 Another expression of this: "Whenever an orphan came for anything, Ataŋauraq would always give it to him. If he wanted a gun, he would give him the gun. Ataŋauraq would say: 'You can pay me back when you've become a good hunter.' Whenever anyone came for molasses, he would give them sugar instead. He would keep the molasses for distilling his liquor" (Asatchaq and MacLean 1978).

6 Through Niġuvana, we learn that of the eight *umialit* in Ataŋauraq's ceremonial house, three were Ataŋauraq and his two half brothers, Uyaġaaluk and Tigautchiaq. Given a pool of about thirty men, one may see how Ataŋauraq's trading party might both have emerged from whaling and ceremonial house association and taken its form from the organization of whale-boat crews.

CHAPTER 10
Ataŋauraq and the White Man

> I wish to again call the attention of the Department to the injustice that is being done the harmless people of northern Alaska by depriving them of breech-loading arms. No evil results can come of the repeal of the law...and a manifest act of humanity would be accomplished by so doing. They are a peaceful race. They have no tribal affiliations and no chiefs... They live apart in small villages...and it is not within the range of possibility that any combination can be effected for warlike purposes.
>
> —Michael Healy

Ataŋauraq and the Men of the *Corwin*: E. W. Nelson and John Muir

Edward William Nelson already had long experience in Alaska when he visited Point Hope on the steamer *Corwin* in the summer of 1881, so it was fitting that he was the first white man to record an encounter with Ataŋauraq. Introducing his *Eskimos About the Bering Strait* for publication in 1899, Nelson commented ruefully on "the length of time which has elapsed" since he left Alaska in 1881. And remarking on the absence of Alaskan research by anyone else, he added, "little has appeared on the customs of the Eskimo in the region visited by me" (Nelson 1899:21). While Russian ethnographers were active in Siberia, the only writer who had done comparable work to Nelson's was John Murdoch, who was in Barrow between 1881 and 1883. Of special interest here are Nelson's remarks on the fate of the Natives he had known before major contact. Formerly, wrote Nelson, referring to southeast Alaska, the Yukon, the Bering Sea, and the Arctic regions:

> [Most] Natives had not been so greatly modified by intercourse with white men... Since then the introduction of missionary schools and the gold-mining excitement have resulted in greatly changing the status of many of the people, and as a natural consequence their old customs and beliefs are rapidly falling into disuse or are becoming modified. (Nelson 1899:21)

Nelson had been stationed between 1877 and 1881 as a meteorologist for the U.S. Army at the St. Michael trading station, and it was from here that American goods traveled up the Yukon and along the coast. Besides his scientific duties Nelson also worked as an agent for the Smithsonian Institution, and by the time he reached Point Hope, he had made ethnographic trips through Alaska's southwest. His celebrated book contains a detailed study of the artifacts he collected. This, with descriptions of "the character and customs of the people," provides a major source of information about Eskimo Alaska.

Another celebrated writer on the *Corwin* that summer was the naturalist John Muir, who would write an invaluable account of the voyage. Before they reached the Arctic, Nelson and Muir had visited a number of places that, despite Nelson's disclaimer, had been significantly affected by white contact. On St. Lawrence Island, two weeks before they reached Point Hope, Muir wrote:

> Three years ago there were about fifteen hundred inhabitants on the island...subsisting on the seals, walruses, whales, and water birds.... Now there are only about five hundred people...nearly two thirds of the population having died of starvation during the winter of 1878–79. In seven of the villages not a single soul was left alive... In the one...only fifteen out of about two hundred survived.
>
> ...in passing along the north side of this island we stopped...at one of the smallest of the dead villages. Mr. Nelson went ashore and obtained a lot of [human] skulls and specimens of one sort and another for the Smithsonian Institution. Twenty-five skeletons were seen.

At another village Muir paints a yet grislier picture:

> [T]hose who had been the last to perish were found in bed, lying evenly side by side, beneath their rotting deerskins. A grinning skull might be seen looking out here and there, and a pile of skeletons in a corner... Thirty were found in one house, about half of them piled like firewood in the corner...
>
> Mr Nelson went into this Golgotha with hearty enthusiasm, gathering the fine white harvest of skulls spread before him, and throwing them in heaps like a boy gathering pumpkins. He brought nearly a hundred on board, which will be shipped with specimens of bone armour, weapons, utensils...

Muir concludes with a shrewd, if paternalistic, summary of the contact situation:

> About a hundred perished here, and unless some aid be extended by our government... in a few years...every soul of them will have vanished from the face of the earth; for, even where alcohol is left out of the count, the few articles of food, clothing, guns etc., furnished by traders, exert a degrading influence, making them less self-reliant, and less skillful as hunters. They seem easily susceptible of civilization, and well deserve the attention of our government. (Muir 1917:118–22)

Two weeks later, the *Corwin* was at Point Hope, where Nelson witnessed Ataŋauraq in action as a trader. Nelson had perhaps heard of Ataŋauraq because he resembled a trading chief on the Russian side. This was Gohorren, a Chukchi from Indian Point, Siberia, on whom Ataŋauraq may, in some degree, have modeled his career. As Muir commented, "these Eskimos, without newspapers or telegraphs, know all that is going on within hundreds of miles, and are keen questioners and alive to everything that goes on before them" (Muir 1917:142).

One of the first Americans to mention Gohorren was Herbert Aldrich, who suggested that his property would be worth $50,000 in San Francisco, observing that "he keeps a large supply of everything on hand, and almost controls the general trade of the Natives, particularly in whalebone" (Aldrich 1889:50). John Kelly's view coincided with that of Aldrich: "Kohorra, is the wealthiest Eskimo in the northwest. In addition to a large store of guns, whiskey, ammunition, and articles of trade, he possesses a large herd of tame reindeer.... Ataŋauraq aspired to be a chief like Kohorra...of whom he had heard" (Kelly 1890:9).[1]

As we have already seen, Ataŋauraq transmitted a good deal of village trade through his own private enterprise. But as Nelson described when he took a photograph of Ataŋauraq, the "chief" could also be shy, perhaps cannily so, in the presence of a white man:

> I secured several good photos of these Natives on board but when I took one of the Chief alone I had considerable trouble in getting him to remain quiet for he became extremely nervous when I covered my head in the (focussing) cloth and it was only when I told him that it would be bad for him if he moved that he (under the greater fear) remained quiet long enough to take a photo. As soon as it was accomplished he had me change place with him and he went through the performance of covering his head and peered through at me, and after having my inverted image shown him he came out with a grin of satisfaction. (Nelson nd)

While Nelson and Hooper knew that Ataŋauraq was a despot, neither man disliked him. Hooper wrote: "Although a horror to the Natives, [he] is friendly to strangers," and he was "much more cleanly in appearance than the others," suggesting that the "chief" cut his hair and wore clothes both to signify his acculturation and also perhaps to present himself as the chief he assumed that the white men expected him to be (Hooper 1884:41).

Nelson and Hooper landed on the Point later on the day of their arrival, and Hooper concludes his report with this vignette:

> The chief of the Point Hope Natives, or [*umialik*]...is quite a character.... He does not hold the position of a chief as an hereditary right or by influence of wealth, but by sheer force of arms. He is never without his Sharps rifle, and with it he overcomes all opposition. He is well built, rather quiet looking, not much given to talking, and more cleanly in appearance than the others. He is a good shot, and is quick to act. It is said that he has killed several of the tribe and has the rest thoroughly cowed. We were requested to take him away, but of course could not do so. Although a horror to the Natives, he is friendly to strangers. The colored deserter from the *Hidalgo* says that it was mainly due to this man that he received such kind treatment. Of course it is only a matter of time when some other Native will catch him at a disadvantage and kill him. In their present state of mind, however, they will not attempt it until quite sure it can be accomplished, for failure means death to the one who make the attempt. (Hooper 1884:41)

Hooper's paragraph raises two points of law. First, in connection with the Intercourse Act (the 1834 Act extended to Alaska in 1873), which prohibited the sale of repeating firearms to Natives, it is clear that Ataŋauraq was legally entitled to own his breach-loading single-shot Sharps rifle. Second, while the Revenue Service could temporarily suppress the sale and possession of liquor, it had no means of detaining offenders. And while, for example, the Alaska Commercial Company operated its own system of law on the Pribilofs, there were still no courts in Alaska. Alaska lay within U.S. jurisdiction, but the Revenue Service was the only agent of American law, and every community had to regulate itself as best it could.[2]

With regard to what American law would identify as criminal activity, what happened in Point Hope in 1881 could not be contained. Hooper knew what Ataŋauraq had done and that Point Hope people wanted Ataŋauraq deported. But where would the *Corwin* have taken him? What court in San Francisco or Seattle could accommodate the accused without English-speaking witnesses? Like every other Alaskan community, Point Hope therefore had to resolve its affairs according to local custom. As Hooper wryly concluded, it was "only a matter of time when some other Native will catch him at a disadvantage and kill him" (Hooper 1884:41). In so saying, Hooper acknowledged the *realpolitik* of Point Hope's situation and the limitations of U.S. authority.

Hell-Roarin' Mike Healy

On the *Corwin* that summer was a man so far unmentioned, though Hooper might also have described him, too, as "quite a character." This was Michael Healy, Hooper's second-in-command, who would take command of the *Corwin* in 1883 and then famously, the *Bear* until he was cashiered in 1895. Though never an Alaskan resident, "Hell-Roarin'" Mike Healy was a hard-drinking arctic tough who enduringly stamped his reputation on northwest Alaska, and, in his campaigns to bring order to both the lives of white whalers and to Native communities, became loved, feared, and respected by both constituencies. And with brilliant, picaresque energy, Healy pursued a career that, as we shall see, strangely paralleled that of Ataŋauraq. Before he succumbed to alcohol addiction, Healy probably did more than any other U.S. official to win improvements for Alaskan Natives.[3]

Michael Healy was known throughout his public career as a white man, and this is how he presented himself. In fact, Healy was of mixed ancestry, and this must have generated his sympathetic concern for the Natives of Alaska.

Healy's father was an Irish immigrant who acquired land in Georgia, and prospered enough to become owner of forty-nine slaves, one of whom, Eliza Clark, he married, sending eight of their children to be educated by Catholics in Massachusetts. Two of the older boys became priests, two sisters became nuns, and a third brother became a Jesuit and president of Georgetown University. Like most of his siblings, Michael was light-skinned and throughout his career passed as a white man. Born in 1839, Michael Healy did not take to an academic or church career. While his brother James was finishing his education in Paris, he sent for Michael, hoping to make something of him in a French school. But Michael absconded and by the time he fetched up in Boston in around 1861, he had, by his own account, sailed five of the seven seas, and was ready to enlist in the U.S. Marine Revenue Cutter Service.

Immediately on joining the service in 1865, Healy married Mary Jane Roach and was among the first in the Revenue Cutter Service to see duty in Alaska. Partly in an effort to regulate her husband's liquor habit, Mrs. Healy and their son Fred accompanied Healy on several of his arctic cruises—and both mother and son visited Point Hope in 1883. Looking back from midcareer over his Alaskan experience, Healy expressed pride in his long period of service: "Since…Alaska was ceded to the United States no officer of the Government has been on official duty in Alaskan waters more than I" (Healy 1887:18). In so saying, Healy was not merely vaunting his own profile; the welfare of Natives lay seriously within his interest, and he was proud to have helped ameliorate their often precarious subsistence.

Healy was, however, open with his opinions of Iñupiaq behavior:

> Every vessel cruising in the Arctic is visited by the Natives, and inflicted with their presence. They often remain twenty-four or even forty-eight hours at a time on board; and frequently depend almost entirely on the vessels visited for their food during their stay. They are the most persistent beggars I have ever seen. Tobacco is the principal object of their beggary, and they never pretend to eat their own food as long as they can beg that of the white man. (Healy 1887:16)

Healy's task, as Hooper's had been, was delicate. His brief was to guard U.S. commercial whaling while also protecting the Alaskan Natives from the whalers' abuses. Healy clearly understood the impact of the commercial whaling on the northern Eskimos, but the interests of commerce were never to be gainsaid, and neither he, nor anyone else, ever suggested establishing a quota to the whale or walrus hunts.

The Reindeer Herding Project

It was Healy, in collaboration with Sheldon Jackson (then U.S. General Agent for Education in Alaska), who came up with the inspired idea with which to solve the Native subsistence crisis. Healy's inspiration had emerged from a trip to the coast of Siberia. In the summer of 1890, accompanied by Jackson, Healy was "instructed to visit the coast of Siberia and distribute presents to the [Koriaks] around Cape Navarin in return for shelter and food furnished shipwrecked American whalers" (Jackson 1896b:5). The Siberians were, as Jackson reported "a hardy, active, and well-fed people, owning tens of thousands of domestic reindeer." The Alaskans, by contrast, were "on the point of starvation… [T]he whales and walrus that formerly had constituted the principal portion of their food have been destroyed or driven off by the whalers; and the [caribou] that once abounded in their country, have been killed off by the introduction of breech-loading fire-arms" (Jackson 1893:5).

To import Siberian reindeer to Alaska and create herds that would sustain otherwise destitute hunting societies looked like a perfect solution to the Native subsistence crisis. Jackson, with the confidence born of his relationship with the U.S. Bureau of Education, which would fund this ambitious project, was the ideal propagandist. This enterprise would, claimed Jackson, also be in the national interest. As he wrote in 1890:

> Northern and central Alaska is capable of supporting over 9,000,000 head of reindeer. To reclaim and make valuable land, otherwise worthless; to introduce large, permanent, and wealth-producing industries, when none previously existed; to take a barbarian people on the verge of starvation and lift them up to a comfortable self-support and civilization, is certainly work of national importance. (Jackson 1890–91)

Healy's visit to Siberia abruptly brought the plight of Native Alaskans to central government attention. While Natives on both sides of the Bering Strait had to be encouraged to support distressed white Americans, this, as demonstrated by enterprising and healthy Siberians, was a way of giving Alaska Natives a new means of self-help. If reindeer could be transported to the American side, then, as herders and reindeer owners, Eskimos would not only find permanent subsistence, but they would, as Jackson suggested, be "lifted" to a higher stage of civilization. The other, barely concealed, agenda behind this notion was that this process would facilitate missionization.

The characters of Healy and the pious, abstaining Jackson could not have been more different. But as their own writings demonstrate, theirs was a friendly and productive partnership, particularly in the early years of their association. It also helped that Healy was under orders from the Treasury Department to take Jackson on his summer cruises. There were, nonetheless, stresses. Jackson took advantage of his position, and in 1893 Healy complained, "the doctor is so persistent and unreasonable in his commands...I often have to be rude and choke him off... He is a hopper grass [sic]. I sit on him today and he jumps up tomorrow" (Healy 1893). But the two men collaborated sufficiently that within five years there was a growing herd of domestic reindeer on Alaska's Seward Peninsula, spreading to Kotzebue by 1900 and reaching Point Hope in 1908.[4]

Who Is the Greatest Man in America?

Given that commercial whalers hunted in both American and Siberian waters, the Revenue Cutter had an immense and complicated territory to cover. To board every ship in the region, check its liquor and firearms holdings; provide medical services; take on board the sick, destitute and criminal; and also to visit Native communities on the Alaskan coast was impossible. But in the latter respect, Healy was still able to make significant gestures, as when in autumn 1891, he visited King Island, Alaska, and found the Eskimos starving. As Jackson wrote in 1893:

> The appeal for food was so pressing that the captain detailed a lieutenant to make a thorough examination of the village... In a few hours we found the families, in their great distress had killed their sled dogs...[so] a purse was made up from the officers... and the captain steamed two hundred miles to the nearest trading post, and purchased all the provisions that could be obtained, which was taken back to the starving village. (Jackson 1893:12)

Jackson had nothing but praise for his colleague and he was not alone in his view. Healy was "more distinguished...in the waters of the far Northwest than any president of the United States... If you should ask in the Arctic Sea, Who is the greatest man in America? The instant answer would be, Why, Mike Healy" (*New York Sun*, January 28, 1894).

It was in Jackson's interest to maintain friendly relations with Healy. As we shall see, he would need the Revenue Cutter to transport the missionaries and schoolteachers he planned for the arctic communities. But the Healy who rescued shipwrecked whalers, provisioned King Island partly at his own expense, and patiently sat through reindeer herders' rituals before he could purchase animals for an Alaskan herd (Jackson 1893:7) was also a tormented alcoholic. The journal of Dr. James Taylor White, who sailed on the *Bear* in 1889, contains these withering comments:

> July 2 Wesley's birthday. A nice way to celebrate, steaming along in Behring Sea with a drunken commander.
>
> July 26 The Capt appears about crazy and is almost blind. He has drunk only four gallons of my whiskey besides his own and as much more from the whalers, beer and wine not counted. This makes four weeks continuous drunk. (White 1889)

Healy also appeared drunk among visiting Natives: "The Capt had had some sleep and with few fresh drinks in him he felt in good spirits. Without stopping to think about every body being asleep below he got the Natives to dancing and by giving them a little tobacco they kept it up" (White 1889).

A second observer expressed an equally dismissive view, albeit for other reasons. This was Isaac Stringer, the Anglican missionary on Herschel Island where five hundred American whalemen and a large population of Indians and Eskimos wintered during the 1890s. Game depletion, disease, and alcohol had also stricken the local and migrant Native populations, among whom there was a substantial north Alaskan presence. Stringer wrote:

> There are fifteen ships wintering at Herschel Island and some of the Captains appear to be more unscrupulous in the matter of giving whiskey than the ones that were here before. It is my opinion that the revenue cutter 'the Bear' is a sham and a fraud... Healy is supposed to inspect all the whaling ships and to allow only one case of rifles and a limited quantity of liquor for the men's use. And yet it seems that an unlimited supply of both rifles and liquor are brought to Herschel Island. (Stone 1981:113)

Healy's failure was of course inevitable, and the limits of his power were well illustrated by events at Cape Prince of Wales in 1893:

> Just prior to [co-teacher] Mr Lopp's departure...a good deal of aggressiveness was displayed by certain Eskimos, a small minority... A considerable quantity of whiskey was being received from the Siberian coast; and the school house had been broken into on several occasions... At the request of Mr Thornton, Captain Healy came ashore on July 10 1893, assembled the Natives, and talked to them; but it is not likely that the offenders thought that the Captain really intended to compel obedience. (Thornton 1931:xx)

The missionary Thornton felt peculiarly under threat in this village and may not have been an impartial observer. But his statistics about gun ownership in Wales illustrates clearly that Natives well understood the *Bear*'s limited influence. Among 135 adult men, there were 152 guns. Of these, seventy-two were illegally traded breech-loading repeaters—which it was impossible for Healy to confiscate (Thornton 1931:xx).

Trade, Partnership,and Collecting: Ataŋauraq, Healy, and Jackson

Healy appears in three short narratives about Ataŋauraq in which two main facts emerge. The first is that Healy and Ataŋauraq had a formal trading partnership. The second is that Healy indulged in drinking sessions with Ataŋauraq, although there is no evidence that Healy gave him alcohol for use in the village:

> Ataŋauraq had a partner. Healy, whom the Point Hope people called Mikkili. When Mikkili arrived each year, Ataŋauraq took him fox skins and baleen. When Ataŋauraq got on board, he went straight to his partner's cabin. Once, when they were inside, Mikkili ran out. Ataŋauraq followed him with a knife. Mikkili ran to the bow. Ataŋauraq followed. Mikkili went down and to his aft cabin... (Asatchaq and MacLean 1978)[5]

The storyteller's use of the word "partner" in this anecdote is significant. Trading partnerships represented a crucial element in Point Hope's relations with neighboring Iñupiat, particularly in the summer. Partnerships between *umialit* and other high-status individuals had special economic importance, and a man of Ataŋauraq's status would not enter such an affiliation without the promise of a good return. This would have been the case in the chief's dealings with Healy. "Where is my partner?" Healy is reported to have asked when he arrived in Point Hope after Ataŋauraq's death. Both parties took trade seriously. Healy looked to Ataŋauraq for baleen; Ataŋauraq ordered consignments of trade goods. In the year of his murder he was due to receive lumber for a house and an iron stove.

Trade in the second story is almost derailed by a drinking party:

> Ataŋauraq had two main trading partners among the white men. One was Healy. Once Ataŋauraq only had a few foxes, a polar bear skin and bundle of baleen. Healy couldn't give him much in return. They were drinking together and Ataŋauraq got angry when the Captain wouldn't give him much for his trade. So Ataŋauraq pulled a knife and chased Healy round the ship. Eventually they both grew tired. Healy gave in. He gave Ataŋauraq more than he'd first offered. Ataŋauraq got two boatloads of goods: among other things, plates, calico, flour, tools, knives, sugar, tea… (Asatchaq ca. 1970)

Alcohol aside, Healy was entitled to trade. Jackson was also a keen trader, and this grew into an interest in Native artifacts. The history of Jackson's collecting activity is complex (see Carlton 1992). By today's standards, his methods were careless and invasive, and while he never traded for profit, he did, in return for a very small outlay, assemble a large private collection.

Jackson's paradoxical aim was to preserve aspects of Native cultures whose intellectual and spiritual content he was committed to breaking down. But his real achievement was demonstrated by the collections he made for Princeton University, and then for his own Sheldon Jackson Museum in Sitka, whose first building opened in December 1889. In the first fourteen years of his museum's history, Jackson collected more than three thousand artifacts in addition to pieces he sent to Princeton and to the 1893 World's Columbian Exposition in Chicago.

FIG. 24. *Dr. Hamilton of the* Bear *holding an Eskimo drum and stick that he has collected, summer 1895. The whale bones recall Beechey's description of the Point Hope peninsula as a "forest of stakes." Candace Waugamann Collection, Archives, University of Alaska Fairbanks.*

Until 1889, Jackson's collecting had largely been confined to southeast Alaska, where in one notorious incident, he had ordered a Tlingit Indian to chop down his own clan totem pole. Jackson's journeys on the *Bear* in 1889 offered him diverse new opportunities. He employed two methods. Disembarking when he could, Jackson either did his business on the spot, or he contracted with local whites to collect for him during the coming year. To this end, Jackson would leave "lead, fabric, flour, beans, rice and matches" (Carlton 1992:186) for his agents to exchange for Native artifacts.

In 1891, Point Hope became one of Jackson's collecting sites where, independently of Driggs, he acquired masks from iglus, burial racks, and possibly ceremonial houses (D. Ray 1967:93). Jackson's heaviest collecting year was 1893, when he bagged a substantial proportion of the collection that would go to his Sitka museum. And then on August 1, 1894, almost certainly in Point Hope, Jackson's diary reveals that he "Bought a large number of masks at from 2 to 3 cents each" and that "Engineer D.H. Coffin [of the *Bear*] presented the Sitka Museum with an old bone adze, which he had picked up in the cemetery at Point Hope" (Carlton 1992:185). The fine collection of Point Hope masks still on display in the Sheldon Jackson Museum probably derived from this visit of 1894.

Healy, already committed to landing Jackson wherever there were schools and reindeer herds to inspect, was further tied down by Jackson's increasingly obsessive collecting. In 1892, the Revenue Cutter "ended up with the responsibility of building crates, packing, loading, and transferring five hundred some 'curios' bound for the Columbian Exposition from the *Bear*" (Carlton 1992:193). This caused one of their few disagreements. In January 1893 Healy wrote: "I do not wish to have anything more to do with a large collection of curios. It took two days and the carpenters and nearly the whole crew's services to get them properly boxed and placed on the docks" (Carlton 1992:214).

But in his dealings with Ataŋauraq, Healy had entered a new and different sphere of trade. The whaling chief and Revenue Cutter captain were partners, and at the center of their arrangement lay baleen and liquor. This was both irresponsible and foolish. In drinking with Ataŋauraq, and thus conniving at the turmoil for which he knew Ataŋauraq was responsible, Healy precisely reversed the object of his duty. There is, however, no record of how Ataŋauraq's drinking sessions with Healy affected his "day after" activities.

Ataŋauraq's Innovations: Peter Bayne and and Liquor-Still Construction

According to one informant, Ataŋauraq was the first Point Hope man to drink whiskey. Perhaps, in that Healy's bottle contained the commercially distilled real thing, this is true. And the story about whiskey connects with a cluster of innovations that Ataŋauraq effected—all of them having to do with Euro-American cooking.

The first innovation took place in spring 1885, when Ataŋauraq broke whale-hunting taboo by ordering his wives to light a fire on the sea ice to prepare tea. The other two firsts were the introduction of bread baking in Tikiġaq and the institution of a flour-soup recipe:

> In 1887 there was so many schooners anchored here they reached all the way to Jabbertown… When the boat crews came ashore for wood they would be helped by Point Hope boys, and gave them round white things [hard tack]… Sam didn't eat it right away, it tasted funny to him. (Rock 1940:5)

Charming as this picture initially might appear, the fact that children were rewarded with mere ship's biscuit for helping to deplete a local resource somewhat sours the idyll.

Sam Rock's first experience of hard tack at the age of ten marks the arrival of a product that would become important. Like many new foods, the biscuit acquired a local term, and became known as *qaqqulaaq,* a word created from *qaqquq-*, "to crack or crunch with the teeth" (MacLean 1980:47). The second half of the same testimony describes a recipe for flour soup attributed to Ataŋauraq (Ivrulik 1940–41:6). Rainey noted: "Ataŋauraq was the first to boil flour soup. They boil it and then put in molasses, sometimes seal grease, like gravy. (This approximated 'duff' which was served on board ships on Sundays and holidays.) Sam said then 'It is good' and when he was a man he would buy some. He thinks this is later than 1887" (Rock 1940:5–6).[6]

Unlike home-brewed liquor, the soup was of course a nourishing food, and unlike hard tack, it represented an amalgam of the traditional and the new. But "first" must not be taken literally. In a society with strict gender separation, the role of Ataŋauraq's household was perhaps to establish a process already in train.

Another innovation lay in bread baking, which followed Ataŋauraq's partnership with Healy:

> When the coast guard came, there was a Japanese cook on board. I saw that cook [in the late 1890s]. When the coast guard was leaving for Barrow, Ataŋauraq asked them

> to leave the cook here, to teach the women baking... Ataŋauraq kept him a week, and he taught some of the women to bake. (Asatchaq and Lowenstein 1992:169)

This was a second dietary development that was mediated through women, though how baking was done is unclear. Ataŋauraq had, perhaps, already acquired a stove. Otherwise one must picture a coal-oil tin or a large kettle, with seal-oil lamps or driftwood firing from underneath. In the absence of properly sealed ovens, it would take some fifty years for baking to become widespread in the village. But flour soup, pancakes, and deep-fried doughnuts quickly entered the nineteenth-century diet, and by 1900, boiled and fried flour-based foods had become staples. The use of carbohydrate with sea-mammal oils had, by this date, become transcontinental. A story about the spirit of a Siberian woman who, in about 1910, is said to have approached some people at Cape Thompson and begged for pancakes fried in seal oil provides a good example of this new culinary practice (Lowenstein 1973–89).

Cooking with flour also lay within a more sinister culinary project: the distillation of liquor. There are two stories about the introduction of home brew in Point Hope, and despite some differences, they are versions of the same events. The first story claims that Ataŋauraq learned to distill from Peter Bayne in 1887–88, the truth of this being more or less substantiated by Lieutenant E. P. Bertholf of the *Bear*, who spent February 1898 around Point Hope to prevent the "illicit distilling of spiritous liquor by the Natives" (Bertholf 1899:19). In the hunt for liquor stills, wrote Bertholf,

> I have made several trips to different villages and visited and searched all the Native houses from Cape Thompson to Point Hope, discovering and destroying six stills and about ten gallons of mash nearly fermented. On two of these trips I have been assisted by Capt. Peter Bayne and Mr Henry Koenig...who have whaling stations on this point, and I have been materially assisted by Mr [Rustam] Nelson acting as interpreter.

Nine years after Ataŋauraq's murder, Bertholf continued:

> Although whiskey was introduced into northern Alaska many years ago...the distilling of spirits was not...practiced by the Natives until about ten years ago, when white men who were engaged in the whaling business on Point Hope taught the old chief [Ataŋauraq] a process of obtaining alcohol from a mixture of flour, water, and sugar or molasses. The chief taught the process to his henchmen, and they gave it to the people of the villages they happened to visit, until there was not a single village on the coast from Point Barrow south that did not have one or more distilling apparatus going whenever the Natives could procure the necessary flour and molasses. During the past few years, however, the Point Barrow Natives have stopped making this liquor, having realised its bad effects, but from Point Hope all along the coast to Cape Prince of Wales, and even as far Unalaklik, the Natives brew and drink this "whiskey" as often as they can procure the necessary ingredients—molasses or sugar. (Bertholf 1899:20)

In addition to his expeditions on the *Bear*, Bertholf now made good use of time on land. What he did not know was that it had been the helpful Captain Bayne who had initiated Ataŋauraq with the knowledge of distilling technology.

Two more stories follow. The first describes a quarrel between Bayne and Ataŋauraq that fueled the white man's malicious gift of knowledge. The second gives an outline of the distillation method.

The first narrator is Charles Brower, who had spent part of 1884 with Ataŋauraq. Brower had returned on a second visit to Point Hope in 1887, and anticipated renewing his friendship with the "chief":

> Just before dark we arrived at Bayne's Place; he was not so glad to see me for some reason or other. I could not make it out, and finally asked him if he wanted me to move on. If I was not welcome I did not want to stay, but would go on to the village where I had many friends. That made it worse. For that was just what he did not want me to do. He thought I had come to trade with the Hope people, and was sore over it. He soon got over that when I asked him if he thought my sled looked as if I had loaded it up with trade goods. We had hardly enough grub on it to take us as far back as Kelly's house [five miles south of Jabbertown].
>
> Bayne was a funny man, he treated me fine while I was there, but he made some funny propositions to me when I visited him. He was running a whaling station for a company in San Francisco. There were eight white men in the party, some of them stayed on the Point many years before going outside, then they took their families along with them. The third day I stayed with him, I told him I was going to the village in the morning, which was six miles away. I wanted to see the old chief Ataŋauraq. I had lived in his house when I stayed at Point Hope two years before.
>
> As soon as I mentioned the chief, Bayne went off the handle, calling him all kinds of names, saying that he would have to kill him some of these days, telling me all kind of things that were happening this winter. He and the old chief could not agree about anything. It seemed Ataŋauraq was jealous of Bayne. Never before had there been anyone at the point who had a stock of merchandise, except himself. Now Bayne was getting most of all the trade, selling for less than the chief wanted to. This hurt him. Every time these two came together they had a scrap. Finally one day Bayne and he hooked. When Bayne got clear he pulled a gun on the chief. But instead of shooting, all he did was hit him over the head with it, not hurting him to any extent, only making him madder than he was, also breaking his revolver.
>
> Ataŋauraq was a powerful man. Grabbing Bayne, he threw him down and sat on him for a while. Bayne never forgave him for that. Now he wanted me to kill the old chief for him. The proposition was not a nice one to have sprung on one. I did not know what I was up against. I had to go to the village, and I did not want to be alone after what Bayne had asked me to do. So I told Kelly what was up. Kelly laughed for half an hour. Then he told me he would go along with me the next day. He also told me that Bayne had made the same proposition to him when he was there before, saying not to mind it, as all hands thought Bayne was crazy.
>
> The only thing Bayne wanted to get rid of the chief for was so he could run the point village all by himself. If I would accept his offer, helping to get rid of Ataŋauraq, then he and I could divide all the trade there was on Point Hope.
>
> Next morning Kelly and I went to the village. The chief and all his harem turned out to greet us, making me very welcome, same as they had always done, feasting us with the best there was in the place. I carried a message to Ataŋauraq from the head man at Point Barrow village, Ounalena. Ounalena wanted Ataŋauraq to meet him at Icy Cape the coming summer and they would have a wrestling match. Ounalena

> had heard he was a very strong man. He thought he was just as strong and maybe a better wrestler.
>
> I sure started something. Ataŋauraq just got up and declared himself, saying there was no one on the whole coast could throw him, nor anyone that could kick as high with both feet. Immediately he started to show his ability as a kicker. Standing in the centre of his igloo he kicked the window frame in the top of the house nearly eight feet with both feet at once. This was something I had never seen before, and I believed him. But when he asked me to carry a large rock all the way to Nuvuk [Point Barrow] just to show Ounalena what he could lift with one hand, it was too much, and I had to decline.
>
> I only stayed a couple of hours with him. But during that time most every one living in Tigara came in to see me, the house getting so full I was glad enough to leave.
>
> This was the last I saw of Ataŋauraq. Two years after he was killed in a feud that lasted the best part of two years, during which time thirteen people were killed, he being the last one to go. (Brower nd: vol. 2, part 1, 107–9)

While oral testimony proves overwhelmingly that Ataŋauraq was hated by the Point Hope community, there were also festive times when people converged happily on his otherwise guarded iglu. But the main interest here lies in the feud with Bayne, the outcome of which would become so dangerous to village well-being.

Ataŋauraq had, as we have seen, established himself as Point Hope's trade chief. Trade was channeled to the big ships through him, and his own store had become a further source of supplies. In summer 1887, Bayne had arrived with the intention of establishing his own trade monopoly, to discover that in Ataŋauraq he had a local competitor. First Bayne succeeded in undercutting Ataŋauraq's prices. But the *umialik* and the Jabbertown "captain" had another sphere of competition: the spring whale hunt. Ataŋauraq could do nothing to stop people buying Bayne's goods, but his opposition to Bayne's whaling effort did find local support. Rashly, Bayne proposed launching his boats southeast of the Point, which would have pre-empted Point Hope's effort by intercepting the whale migration. As self-appointed representative of the Native hunt, Ataŋauraq ordered Bayne to launch from the north, from where whales would be less accessible. It was largely this dispute—commercial whaling versus subsistence—that brought the men to blows. A Point Hope version of the story continues:

> Ataŋauraq sent a messenger to Bayne telling him not to take his whale boat from Jabbertown to the point. Instead, they should go whaling off the north side, where they'd be less likely to scare the whales approaching the local village skinboats.
>
> So Bayne took his crew of white men and Natives over to the north side. Bayne's crew contained a mix of whites, a few Point Hopers and men from the the Bering Sea region—Teller, Cape Prince of Wales and little Diomede Island. Whenever Bayne caught a whale he made sure he stayed as far as possible from the village, or anywhere else where'd meet Ataŋauraq. (Asatchaq ca. 1970)
>
> One day Ataŋauraq went to Bayne; he carried no weapon. But behind his back, Ataŋauraq was holding a long stemmed pipe. Bayne thought Ataŋauraq was carrying a pistol. And when Ataŋauraq got to Bayne, he wanted to talk to him. I don't know where Ataŋauraq learned English, but he talked to white people and understood them… So

> Bayne thought Ataŋauraq had a pistol. And he struck Ataŋauraq with his own pistol and cut his head... But Ataŋauraq did nothing. If he he'd had a pistol, perhaps he would have hit Bayne back or shot him. (Asatchaq and Lowenstein 1992:166)

Ataŋauraq was the one to bleed—the storyteller gives a description of the amulet that one of his wives held to the wound—but it was Bayne who, as described by Charles Brower, suffered the humiliation. Contemptuously refusing to hurt him, Ataŋauraq thus put Bayne in a position from which he had no motive for an attack, and from this no doubt came Bayne's attempt to suborn a third party. Once Kelly and Brower had turned down his suggestion, Bayne devised a subtler and ultimately more destructive route. He would, according to Asatchaq's oral testimony, attack him through liquor (Lowenstein 1973–89). Bayne well knew that Seward Peninsula Natives had already learned to distill their own alcohol; and since Bayne drew some of his employees from this region, it was a simple step to persuade someone to teach Ataŋauraq. And while this version makes no mention of Bayne, many Point Hope informants stated baldly, "Bayne taught him." The story goes as follows:[7]

> I heard from my mother how they brewed (*imiġai*). Ataŋauraq was still alive then, and a man from Teller came to Tikiġaq who had watched them brewing up in Teller. So when Ataŋauraq heard about this, he invited the man and told him to prepare the things he needed to make brew (*imiġakraq*). My mother watched too. There was flour and molasses stirred up with raisins and dried apple: these were added to the flour.
>
> But the man from Teller got the ingredients wrong. He didn't do it right. He knew the recipe, but he thought it would be wrong to pass it on. Now when the time came for the brew to be cooked, Ataŋauraq was the one to cook it. They used a five gallon can for the cooking part. And they put the barrel of a gun (*tuġlu*) to the five gallon can. Then another pot filled with cold water was put inside the first one. When the big can started to boil, the steam dripped through the gun barrel into the container. Attached to the end of the barrel was a strip of baleen so that the barrel would drip in one place, running down the barrel into the container. After it had dripped for some time, Ataŋauraq tried it. But it wasn't whisky.
>
> So then he sent for the man because he hadn't told him the right way. When he came in, Ataŋauraq said: "You didn't tell me the truth. You didn't tell me the right things to put in it." He told the the man to fix him the brew. "But if the brew doesn't make whisky this time, I will kill you. If the barrel doesn't drip whiskey, I will kill you right here."
>
> The man was afraid, and he made a proper brew. He added prunes (*asiaqpak*, "big berry") and some other things to the mixture (*akutaq*, "what has been mixed together"). My mother watched all this. The Teller man told the truth this time. Ataŋauraq *taamna* ("that Ataŋauraq"). This time it worked.
>
> While they were making their brew, two *qulliq*s (seal-oil lamps) were lit. I watched my father do the same thing when he made whisky. My father learned it from my mother. My mother saw it at Ataŋauraq's...
>
> There were many people who had *imiġun* (distilling equipment). When I was small a coast guard official came one winter [Lieutenant Bertholf in 1898, when Asatchaq was aged seven]. The stills were destroyed. I saw that happen. The pots were smashed, and what they had distilled was taken away. But first they had to be found—everyone

> had to say who owned one. Some were buried in snow drifts. People kept their whiskey outside, dug into the snow where it stayed cold.
>
> ...The basin sat in the middle and was sometimes filled with snow. When they do it right, it drips very well. And people tested it by putting a finger into the distilled liquor and then putting their finger into the lamp flame to see if it burned. They could tell when it was right because it would ignite straight away... Everyone who was able to had a still. (Lowenstein 1973–89)

As we have seen, alcohol had been coming in to Point Hope from the big ships since the 1860s. And given that the U.S. Army had taught Tlingit Indians to distill by the end of that decade, it was likely that the same knowledge would eventually reach north Alaska. The precise moment of entry can, however, be dated from Ataŋauraq's experiment, and despite the sporadic efforts of the Revenue Cutter, liquor stills would become commonplace by the late 1890s. As both Bertholf and Asatchaq recorded, home-brewed alcohol was never restricted to elite households. Through Ataŋauraq, as with his other innovations, the knowledge of distilling became widespread in the village. But it was also this knowledge that hastened the chief's own destruction.

Notes

1 John Bockstoce gives this vivid description: "At first [Gohorren] may have been staked by a whaler or trader, but soon he became a full trader in his own right. He amassed a great fortune in trade goods through both skill and deceit. At his zenith he owned three wooden store houses that were full of whiskey, firearms, ammunition, a wide variety of manufactured goods, as well as whalebone, furs, and ivory. With these large stocks of foreign goods he was able to trade throughout the winter, obtaining trade at favorable prices, and often offering credit so that many Indian Point Natives were in his debt more or less permanently... In 1891, for example, he had in his warehouses 200 sacks of flour, 80 boxes of tobacco, and ivory and whalebone worth between $5,000 and $8,000" (Bockstoce 1986:195–96).

2 As late as 1893, for example, miners at Circle City in the Interior were establishing their own courts in which they administered fines, banishments, and hanging. Nome, at the beginning of the gold rush in 1899, remained anarchic despite a small military and police presence, and the first federal justices of 1900 themselves flagrantly abused the law.

3 Healy's character may be recognized in sea-faring manners popularized in nineteenth-century literature. Joseph Conrad wrote of the paradoxical merchant skippers of the 1890s, as "turbulent and devoted, unruly and faithful" whose careers were fraught with "toil, privation, violence, debauchery." "My mind," wrote Healy in the mid-1890s, "is so occupied...with the care of the vessel, and my system so burdened by fatigue and exposure incident to the discharge of my duties...that I am wholly incapacitated for reading or noticing with proper care events that come under my observation" (Healy 1889:20). This was in response to a suggestion that captains should carry out scientific observations.

4 Healy purchased and transported Siberian animals to Alaska in 1892, and Jackson lobbied Congress and other influential East Coast figures to support central government funding. Two Point Hope men started training as reindeer herders in Teller in 1894. As noted by Neville Chance: "The Iñupiat herd at Point Barrow, which originated with 125 deer, expanded to 30,000 by 1935. Between 1918 and 1934 the Wainwright herd increased from 2300 to 22,000. By total count, the 1250 deer imported to Alaska from Siberia between 1892 and 1902 had

increased to over 600,000 by 1932" (Chance 2007). The history of the various Alaskan herds is beyond the scope of this study. But Point Hope's independent reindeer project got under way in 1908. It was run alongside the Kivalina herd by the Point Hope Mission, amalgamated with a rudimentary Point Hope store in 1920, separated from that operation a few years later and continued to exist until the late 1940s. As George Omnik told me in 1975, "then the reindeer ran away and joined the caribou." This statement encapsulated one of the major problems facing all the north Alaskan herding enterprises.

5 Compare the transaction onboard HMS *Plover* at Point Barrow in 1853 when "Erk-sin-ra… was brought into the Gun Room & took a glass of wine which he enjoyed very much—he had a cigar also which he equally relished" (Bockstoce 1988:291–92).

6 Charles Brower gives an account of introducing the same recipe in Point Barrow in 1886 or 1887 (Brower nd: vol. 1, part 2, 86).

7 Asatchaq also recorded the following: "Ataŋauraq wanted Bayne to teach him to distil liquor from molasses, and he was angry when Bayne refused. Eventually Ataŋauraq sent a man to Jabbertown to fetch the necessary ingredients—flour, molasses, dried fruit among them—and the man returned to Ataŋauraq, cooked up a mash, but failed to get a brew. Ataŋauraq set him to work on it again, threatening to kill him if it didn't produce liquor. Fortunately for him the process worked out" (Asatchaq ca. 1970). This version may be true. It appears more likely that it represented the surface reality stage-managed by Bayne to divert suspicion. As Bertholf asserted, it was a white man who taught Ataŋauraq to brew.

CHAPTER 11

Ataŋauraq and Charles Brower at Point Hope, 1884

Charles Brower spent several months in and around Point Hope in 1884 and knew Ataŋauraq well: not as a ship's captain, but as an approximate equal and sometimes as a subordinate. Brower wrote his story some forty years later, when he had settled in Barrow, and his narrative is unique in its historical detail. It provides the most vivid picture of turn-of-the-century Point Hope, and its main focus is Ataŋauraq.

Brower never enlisted on a whaling vessel. He traveled to the Arctic as an employee of the Pacific Steam Whaling Company to help supply its steamers with fuel. The PSWC also planned to establish a shore-based whaling station in Point Barrow. So en route to Barrow in 1883, Brower and three companions were set down just north of Cape Lisburne near open veins of coal that had been discovered by Beechey in 1826. Here, the four men erected a house and proceeded to mine the cliff face, carrying the coal across the beach to await passing ships. Like most Euro-Americans in the Arctic, the men also carried food, cloth, ammunition, and hardware to trade with Eskimos, and it was near Cape Lisburne that Brower made his first contacts with Natives.

A Point Hope man, Quwana, gave this vignette of life at the coal mine in about 1884:

> There was a house there, and…white men, including Amuguq (in Iñupiaq "wolf" after his name Henry Woolfe) and Charlie Brower. This was the first time that I traded with white people. My grandfather gave me a little bundle of baleen ends. These men at the coal mine camped on the north beach. Once, I saw a crowd of people there, and someone grabbed my little bundle and asked what I wanted. I said I wanted a pocket knife, and I got one with a white handle. It meant a lot to have that knife. I felt like an *umialik* (here "rich man"). That same summer I went up and helped pile coal on the shore, no bags. When the coast guard came, they gave the coal to Captain Healy, and he gave them grub and clothes. (Kuwana 1940:41)

Brower, aged twenty-one and already well traveled, had been warned by an anonymous source: "if at any time we saw a large party of Eskimo men coming, without their women and children, lock yourselves in and get all your guns loaded, for then they would mean trouble" (Brower nd: vol 1, part 1, 12).

This alarmist advice derived from fears arising from a small number of incidents. John W. Kelley, who stationed himself peacefully near Point Hope in the 1880s, wrote:

> [Point Hope men] have often attacked parties of whalemen who have been on shore after casks of water or driftwood for fuel. Four instances are mentioned: once when the Natives were driven back with clubs; another time when they surrounded a boat's crew, pricked the men's throats with knives, stole their tobacco, and cut the buttons from their clothes; again when they pursued the departing boats, caught hold of their painters, and tried to pull the boats ashore, which was frustrated by cutting the painter. Upon the last occasion, when they attacked a boat in force, the officer in charge stood them off with a whaleman's cutting-spade, inflicting many wounds, but it is not known whether any were killed. (Kelly 1890:10)

There are no further accounts of these incidents, but the fact of their occurrence cannot be contradicted. A second source of alarm derived from the so-called Gilley massacre of 1877, when thirteen Eskimo men were killed by a mostly Hawaiian whaling crew in the Bering Strait. There was a fear that, as a result of this, there would be region-wide revenge attacks. Brower heard the story from Gilley himself in 1886, Gilley's version being that he had refused liquor to some already intoxicated Cape Prince of Wales men, and the confrontation ended in a murderous brawl. The Gilley episode was still fresh in the minds of Bering Sea Eskimos and the desire for revenge on white men no doubt did extend along the coast—though not as far as Point Hope.

Brower himself was never alarmed by his encounters with Iñupiat. "Many [bad] things that were told us were all imagined, and hardly any had any foundation that I ever could find," he reported in his memoir (Brower nd: vol. 1, part 1, 12). And when two Eskimo families approached Brower's coaling station at Corwin and set up camp, Brower was positively charmed. One of the men, from Kivalina River, whom Brower nicknamed "Baby," became a lifelong friend—a relationship that was cemented by Brower's marriage to his sister. Also present at this meeting was the Point Hope man Anaqulutuq, a shaman whom, five years later, after the death of Ataŋauraq, Captain Stockton would appoint to be "head man."

Thus started Brower's life among Eskimos. But the first place he wanted to visit was Point Hope, where he and Woolfe hoped to trade with Ataŋauraq, whose enterprise had become a major institution that even offered credit. Woolfe, or Amaġuq, to call him by his unavoidable Eskimo name, vanishes from Brower's narrative at this point because the "chief" refused to trade with him.

So Brower set off with Point Hope's future "headman," Anaqulutuq. The white man's first taste of the "chief's" authority came when he entered the village and was taken to Ataŋauraq's iglu, where, in the interest of trade, he was compelled to stay. Ataŋauraq made a deep impression on the young American, and Brower's account is worth quoting at length for the light it sheds on the "chief" and his household. Brower's first remarks are forthright:

> At-tung-ow-rah was a thick set heavy man… He was thought by the Natives to be the strongest man on the coast. He was a surly brute and was considered to be a great *angatkuq* (shaman). Everyone in the village was afraid of him, as he was thought to have the power to kill any one by his devil driving, that he did not like. (Brower nd: vol. 2, part 1, 42)

Next Brower provides a sketch of Ataŋauraq's household: "Ataŋauraq had five wives, four of them he kept in the igloo where we stayed, but his first or head wife was in house by herself. This head woman ran all the others and the hard work was done by them." Ataŋauraq, Brower noted, slept in an alcove off the iglu entrance passage with two of his wives. "Sometimes he changed the older ones, but while he had a young woman, she was the favourite."

While one motive for Ataŋauraq's interest in Brower was trade, another was to impress the white man with evidence of his power. With this in mind, he took Brower on a tour of Point Hope's burial grounds:

> To show me what a bad man he was, Ataŋauraq one day asked me to take a walk to the graveyard with him. He wanted to show me where he had four men up on a rack that he had murdered at some time in the past. I went along with him to see the show. His private burying grounds did not appeal to me, but I did want to see the graveyard. (Brower nd: vol. 2, part 1, 43)

Brower next gives a detailed account of Point Hope's ancient and immense graveyard:

> We had seen the uprights as we came along the beach on our way down. I thought at the time they were wood. When I did get close enough to get a good look at them, they were all jaw bones of whales, standing upright in the ground. Mostly they were of the smaller ones, not over ten feet long. To these were lashed the lower jaw of a walrus. In the walrus jaw there was [sic] laid pieces of wood and on those were laid other pieces, making a platform, where most of the bodies had been placed. Some few bodies were on the ground, where they had been dragged on some old sled belonging to the person, or their family. The bodies were all wrapped in some old sleeping skins. When ever any one died all their personal things that were in daily use were left at the grave, as for instance, if it were a woman, her sewing gear, her *uluuraq* or woman's knife, and some times her cooking pots or her personal ornaments, beads, earrings were all buried. If a man, his hunting implements, sometimes his kayak, his pipe, or his *qiḷaun*: (drum). If a child, all his toys were left. Always whatever was left at the grave was broken so to be of no more use to any one alive, even the sled was broken at the grave.
>
> What interested me as much as anything, was the size of the graveyard. It extended for miles up toward the base of the point, and looked like a forest of small trees with the tops cut off. There were thousands of these whale jaw bones, and it seemed to me as if it must have taken a great many years to accumulate such a large number. Many had no bodies on them, and these were so old that the bodies had decayed and fallen down or been eaten by the ground squirrels that infested the grave yard, in the summer. (Brower nd: vol. 2, part 1, 43)

Brower's is the fullest surviving description of Point Hope's graveyard and this renders it an important document. Significantly, it echoes both Beechey's "forest of stakes" (see Chapter 1) and a Tikiġaq creation story (see Chapter 27) in which the tip of the peninsula is described as a forest whose trees are felled in anger by a bereaved father. Brower's account also serves as context for E. J. Knapp's description of Point Hope's "weird" burial ground, and most particularly the destruction of this ground by Augustus Hoare in 1908 (see Chapter 27).

Ataŋauraq then showed Brower racks holding his victims and his own family:

> One was quite a recent arrival but the others were old timers. He also showed me where his family were laid out to dry. His father, he told me, had been even a greater devil doctor than he and was buried here, with all his effects, among which were a lot of beads, and also two wolverine skins. These I did not see, I simply took his word for it. (Brower nd: vol. 2, part 1, 43–44)

Prior to this allusion to wealth as expressed in beads and fine skins, Brower had provided an impression of Ataŋauraq's caprice:

> [A]fter Woolfe left, Attungorah was peeved at me. I always thought it was because I would not give up my [trade goods]. At any rate the next morning he told me I would have to get out of the village. That was a sudden change that I was not prepared for. I was in his house at the time. I did not know what to do for a minute, finally I told him I would go out of his house, and that I would leave the village when I was ready. I got together what I had belonging to me in his igloo and took it all out side… On my way out I strapped my Colt revolver on myself, under my [parka]. Attungorah came over to me and again told me I had to leave at once or he would kill me. That was enough, if I was going to get bumped off, I knew that I could get him first, so I got my gun and laid it across my knees, then I made him understand that if he went into his house for a rifle, the minute he showed his head out through the opening I would shoot him.
>
> All his rifles were in the [entrance passage]... He was so surprised to think I was not going to run that he did not know what to do. It was a new experience for him, and to see me with the gun pointed at him, made him hesitate. He finally walked away, down to the other end of the village. Three other men went with him. I thought maybe he would get a rifle from some one and take a long shot at me, but he must have talked it over with them, for after a half hour or so he came back to where I was and told me that I was a good man, and wanted me to move back in his igloo, and stay as long as I wanted to. This I refused to do. I told him I would stay as long as I wanted to, but that I would never go to his house again. This seemed to hurt him a lot. (Brower nd: vol. 1, part 1, 44–45)

Brower spent the night in a neighbor's iglu and next morning took off on a hunting trip with Baby. It now became clear to Brower what this man and his sister had been doing at Cape Lisburne: they had been in flight from famine and illness on the Kivalina River. After another long journey, they all returned to Point Hope, and again Brower was compelled to lodge with Ataŋauraq. "Ataŋauraq had all five of his wives out and believe me they were jumping for him. It was not long before there was cooked meat of all kinds, and he even had them try and make slapjacks and tea for me" (Brower nd: vol. 1, part 1, 56).

It was now April and Ataŋauraq correctly assumed that if he invited Brower to join his whaling crew, the white man would contribute food and ammunition. Brower then witnessed Ataŋauraq in the act of shamanizing, or "devil driving" as Brower described it:

> Before going [out on the ice] the old devil drivers were to have a session that night… I asked Ataŋauraq if I could be present. At first he did not think that the other doctors would like to have me see what they were going to do, but when I asked them, they were agreeable, and seemed to feel as though it was an honor, at least they were pleased to have a white man see what they could do in the way of exorcising the evil spirits. They

informed me that in the evening three of them would go in the chief's house and do their stunt, that no one else would be allowed to be inside except myself, and I was not to talk or make any sound, as the devil, or *tuunġaq*,[1] as they said, would not like it.

When we went in the house all the openings were stopped, the gut window was covered with a deer skin and all the stone lamps were out except just a tiny flame on one lamp. The three [shamans] were stripped to the waist and sat around on the floor, each with his Native drum. I was told to get up on the sleeping platform, and to watch, and if I saw the *tuunġaq* not to be afraid as they were all powerful doctors and I would receive no injury.

The three of them started singing some of their medicine songs, slowly at first, but gradually growing faster and then faster and at the last shouting and making all kinds of outlandish noises, till they finally worked themselves in a trance, or at least that is what it looked like to me. As they worked themselves to this state, they rolled all around the floor, never letting go of their drums, but kept on beating them all in unison and keeping good time together.

The last performance was to froth at the mouth and then drop their drums and lay in all kinds of positions on the floor, as though they were dead, then first one would prophesy something would happen and then another would take his turn, until all three had several tries at it.

Some of the things that were prophesied were that the *umialik* would get three whales that spring, this was done by my old friend Anaqulutuq; another said the *tuunġaq* had told him there would be a good whaling season, and every one would have plenty to eat the next winter.

When it came to Ataŋauraq's turn, one of the things he said was that the devil told him that all my clothes were to be taken from me and that I was to have a new lot. The shirt I had been wearing was made of squirrel skins, and they were not allowed on the ice during the whaling, as the whales would not come near the ice if they knew that there was any such skins used. My sealskin pants and all my boots were pronounced taboo and were to be replaced by others.

It was warm in the igloo. I was wearing nothing but light clothing. All my skin clothes had been left in the hall where I was to sleep. All that the chief said that I could not use on the ice disappeared that night. I never saw any of them again. In the morning the chief's head wife gave me a whole new outfit of new deerskin clothing, made better than any that I had ever owned up to then.

If I said that I was not scared I would be lying, for a while I thought the whole lot had gone crazy. Although since then I have seen a lot of the devil driving I never have seen anything as weird as what happened that night...

When the lamps were all lighted Ataŋauraq and the shamans sat around for several hours, eating and telling me of the many different things that...I could not do on the ice, things that were forbidden by the doctors. Among them...were that I could not have any white man's food on the ice,that I would not be allowed to take any sleeping gear, not even to take change socks. No tents were allowed, and the only shelter we could have was a snow wall to use as a wind break.

All this seemed rather strenuous, but I wanted to see the Eskimos whaling so I would not give up. And in the morning was ready to start when the rest were. As I dressed in my new clothes I looked more like an Eskimo than I expected to. The Eskimo said I was no more *qavlunaaq*,[2] their name for a white man, but from then I was an Inuit... (Brower nd: vol. 1, part 1, 58ff)

There was no precedent for Brower's participation in the hunt, and Ataŋauraq did not offer him a share of the catch, which Native crew members would receive. But Brower's presence as a boat hand represented a dramatic reversal of roles. For thirty years, the commercial whalers had pursued their annual hunt. Now this white man—albeit in holiday spirit and studying a trade he would put to use in Barrow—lends his muscle to Ataŋauraq's whale crew as a subordinate. But this was also a difficult moment for the *umialik*. All Ataŋauraq's men had been ritually prepared for the hunt, and Brower's presence represented an imponderable element. To avoid the violation of taboo, Ataŋauraq must, at least partially, assimilate the white man. Brower mentions only one precaution, though there may have been others of which he was unaware. Whaling ritual demanded that everything used during the hunt should be new or freshly cleaned, so that whales would not be disgusted and thus repelled by human dirt. With this in mind, Ataŋauraq dresses Brower in new clothing, thereby assimilating him into the crew.

Brower's account of the hunt is one of the most telling in the memoir. Ataŋauraq seems to have conducted the hunt in accordance with precedent. But custom was changing and there was one dramatic aberration. Perhaps more than anything in local culture, whaling no longer existed in isolation. While the whale hunt lay at the heart of Point Hope's lifeways, it now also lay close to Euro-American commercial activity. Brower would soon be running his own outfit in Barrow. And the very fact that he was on hand at Point Hope was an expression of the changing milieu. For his part, Ataŋauraq's interest in having a white man in his crew suggests the flexible, improvisatory nature of the times. Once Ataŋauraq had invited Brower onto the hunt, a willingness to change in other ways would also seem likely. Brower's narrative, which is edited to include only passages that illustrate Ataŋauraq's behavior and changes he inaugurates in whaling practice, continues:

> Our boat was eight including myself. Taking the boat from the rack where it had been all winter it was lashed on the sled, then the crew lined up on both sides of it with harnesses over their shoulders, and hauled it out over the ice to the edge of the [open water], which was four miles from the land. The umialik did no work, walking along behind with both arms inside his [parka]. Seeing him not helping I asked him why he did make the other men do all the work. He said as he was a big *umialik*, he was not supposed to do anything,that the others were there for that purpose. Ataŋauraq was the only boat header that I ever saw who would not help. He never did even when we got caught [in the ice] and every man was needed.
>
> One of his wives walked along behind carrying a small bucket of water. At first I thought it was for drinking, but when I wanted a drink she would not give it to me, telling me it was for the purpose of pouring water on the bow of umiaq before launching it…
>
> All our whaling gear was pre-historic, nothing that we use [today in Point Barrow] was at that time used by the Eskimo. The harpoon was made of ivory with a cutting edge fitted in the end and about ten inches long. This was fitted in the end of the harpoon shaft. The wooden shaft, about seven foot long, was made of driftwood…
>
> The boat was placed on the edge of the ice, the bow overhanging about three feet. A small piece of ice was placed under the stern keeping the bottom of the boat away from the ice so that it would not freeze fast, then all hands cut blocks of snow which were built in a wall four feet high around the stern of the canoe, in a semi-circle. This

> was a shelter from the wind, and all the protection we had while whaling. The canoe sled was hauled behind the wall, and here we sat and watched for whales. Some one was on watch all the time, and when a whale was sighted whoever was on watch called the others if they happened to be asleep. I wondered what we were going to eat, the only food we took along with us being a piece of *maktak* [whale skin and blubber] and some frozen deer meat. This was all right for a while, but as a steady diet not so good. Every day some woman would come from the village and carry cooked meat of some kind on her back, so we managed not to starve. (Brower nd: vol. 1, part 1, 60ff)

After a two-day wait, Ataŋauraq's crew finally struck a whale:

> We were no time in getting the boat, to chase after our whale, but before we left the ice Ataŋauraq started to sing his whale song, as he told me after, it was a very powerful charm and had been handed down to him by his father, and was never known to fail. *Avataqsiun* (a song to make the whale float rise and buoy up a harpooned whale) is the name of these songs. Every man running a whaling chance had one and each was different. (Brower nd: vol. 1, part 1, 64)

The next scene contains a sudden and significant innovation:

> After the meat had been disposed of, the [baleen] was divided. Our boat took a fourth and the rest was divided share and share alike among the rest of the *umiat*... While we were working the whale, the old chief thought it would be all right if I sent in to the village and had my oil stove and some tea and hardbread sent out, that we had already killed a whale, and if we lit the stove and made tea back some distance in the rough ice the whales would not know it. This I also thought would be all right with me, so he sent one of his wives in for what we wanted. They were not gone long and that tea surely tasted fine. The old fraud, however, would let none of the others have any, making them believe that he alone had enough influence with the evil spirit [*tuunġaq*] to do these things that were forbidden. (Brower nd: vol. 1, part 1, 68)

This was a turning point in Point Hope whaling practice and it was perhaps inevitable that Ataŋauraq should have been at its center. To light a fire on the sea ice represented the taboo violation that moved directly towards the pragmatism and convenience of modern America. Given the "chief's" influence, the event must have contributed to further breaches in custom that would lead, rather quickly, to the erosion of hunting ritual. Like most of his contemporaries, Ataŋauraq had seen white men catch whales without ceremonial precaution, and Brower's record shows Ataŋauraq acting on this perception. And as with his cooking innovations, this incident would have a long-term impact on the role of women. Up to this point women were only permitted on the ice after a whale had been taken, and then very briefly, in order to haul the meat inland. The consequence of this moment reached far into the modern era. Women would subsequently spend long periods on the ice to cook; they would also join men on the *umiaq* benches.

Ataŋauraq soon regretted his taboo breach and, learning that crews to the north who had remained in taboo had taken three whales, he decided that the spirit that had told him he would catch three whales may have been displeased:

> At once Ataŋauraq was jealous of their luck, and we had to get everything in the boat quickly, and started up the flaw looking for another place to haul out. He told me he thought that maybe the devil [*tuunġaq*] was angry with him for letting me have the tea on the ice. At any rate we had not been in our new place more than four hours before we had another whale come right close to the canoe. He finished his spouting and went down just out of our reach. (Brower nd: vol. 1, part 1, 70)

Ataŋauraq now caught a second and bigger whale, and after the crew had towed it in, they returned to the village for fresh clothes:

> [W]e stayed ashore three days. The first night ashore, a woman died in the village. As soon as she was dead, the body was placed on a sled and taken about two miles up the sandspit. Ataŋauraq was all worked up over the death of this woman, and he told me that it was a bad thing to have her die just at that time, and to make it worse the woman was pregnant. (Brower nd: vol. 1, part 1, 71)

Having vacillated between the pragmatism of the tea episode and a suspicion that he had done wrong, Ataŋauraq now plunged into personal involvement with the latest taboo emergency. Ataŋauraq's relationship with the deceased is unknown. But the death of a pregnant woman during whaling might affect any crew that had a connection with her. As a leading *umialik* and shaman, Ataŋauraq took the woman's death as a matter for personal intervention:

> However, as he was a great shaman, [Ataŋauraq] thought he might be able to do something, if he went into a trance and heard what the devil had to say. This he promptly did. As before, he worked himself into a frenzy, and while in this trance said that the woman's body had to be opened and the child taken from her, the body of the child to be wrapped in sealskin and buried separate. If this was not done there would be no more whales taken that year. (Brower nd: vol. 1, part 1, 71)

Ataŋauraq ordered four women to operate on the cadaver and Brower went to observe. "Every superstition that these people had," he perceptively remarked, "seemed to me to have some bearing on their whaling." It came, therefore, as no surprise that Ataŋauraq now looked forward to the third whale, which had been prophesied at the start of the season. Late in the migration, Ataŋauraq accomplished this. And while all the other boats remained out on the ice, the "chief" said that "as he was a big *umialik*, and had enough meat and blubber for the next winter, he did not have to go" (Brower nd: vol. 1, part 1, 76).

Brower concludes: "There had been twelve whales taken that spring, but no one had taken more than one except us, so Ataŋauraq was feeling fine and strutted around the village, all his beads strung on him and arms hauled inside his [parka]" (Brower nd: vol. 1, part 1, 78)

The whaling celebration followed. But before this, Brower achieved the seemingly impossible: "I cleaned up managing to make some hot water and had a bath in one of the igloos." This was probably the first such event—and presumably he was provided with snow water and a basin in the iglu's subterranean space. The freshened-up young whaler now went

Fig. 25. *Kiḷigvak's house, built in 1913 with lumber from the trader John Backland, was the first Native-built frame dwelling, and the furthest northwest building on the continent. Asatchaq lived here until 1972. The whale-bone remains of Samaruna's iglu are in the foreground. Author's photograph 1978.*

off to explore the village, and his vivid observations complement the oscillations between tradition and the modern world that he had witnessed in Ataŋauraq's company:

> [The igloos] were all deserted now, everybody living in tents, scattered all over the sandspit. No other boats had come in, and for two days I bummed all over the village, walking all over the spit to have a look at the graveyard. It looked altogether different now to what it did in winter, and I could see where many bodies had never been placed on the racks. The ground was covered with bones. Skulls were laying everywhere, most of the other bones had been dragged away by dogs or other animals. Where ever there was a dry place it was occupied by a ground squirrel. I shot several with my shotgun and carried them to the village. But the women would not cook them for me, saying that these squirrels were not good to eat, as they fed on the dead bodies in the graveyard. Others were good, especially those living in the mountains. (Brower nd: vol 1, part 1, 76)

A final detail gives us a valuable insight into the acculturation process. Ataŋauraq, the most "Americanized" Point Hope *umialik*, still used stone-pointed harpoons. Privately owned firearms were another matter. Earlier we saw that the first firearms came into Native possession in northern Alaska between about 1860 and 1870, probably reaching Point Hope late in that decade. Brower doesn't give figures, but as in 1887, he watched seal hunters at work on the south shore. He writes:

> There were several Eskimo men at Tigara who did not own rifles, they were getting seals with their spears. I went from one place to another watching them. They were experts with their old *unaaq* as they called their spears. Those who had no guns would go off by themselves and after finding a place to their liking, would build a wall of ice about two feet high close to the water edge. This was used as a blind and at the same time to rest their spears and coiled line on. (Brower nd: vol. 1 part, 1, 76)

"Several Eskimo men" without rifles would suggest that by now the majority owned firearms or had one in the family. We may also assume based on the evidence presented earlier, that many of these rifles had been distributed in trade with Ataŋauraq.

These were the highlights of Brower's Point Hope narrative. On July 12, 1889, five months after Ataŋauraq's murder, James White, surgeon of the *Bear*, recorded a visit to Ataŋauraq's old iglu, which he found "pretty well filled with valuables. I saw some forty bags of flour, five Hotchkiss rifles, several bolts of drilling [woven material], a cook stove, an oil stove and some [kerosene] oil lamps" (White 1889).

Complementing Taylor's inventory, the storyteller Asatchaq described how Healy had, that summer, brought Ataŋauraq a stove and the lumber with which to build a house. These came too late. The chief was dead and the first frame house, owned by Asatchaq's father Kiḷigvak, would not be built until 1913. One of Ataŋauraq's widows told Healy not to bother with the delivery: "If they unload it here his relatives will leave me nothing" (Asatchaq and Lowenstein 1992:170). Healy took the stove, and perhaps also the lumber, to Brower, who was already in Barrow.

Of interest here is the presence, in Ataŋauraq's iglu, of two stoves. If in 1889 Healy sold Ataŋauraq's stove to Brower, Ataŋauraq perhaps had preempted that deal. Not everything had ended well between Brower and his *umialik*. Asatchaq described how the "chief" had "hit Brower in the face, and his face swelled up" while they were out whaling (Lowenstein 1973–89). If Brower's ultimate compensation would be Ataŋauraq's stove, it is not stretching credulity to assume that one of the stoves that White discovered in Ataŋauraq's iglu was the one that they'd made tea with, and perhaps appropriated from Brower, just two seasons before. Such were the vagaries of trade in the 1880s.

Notes

1 *Tuunġaq*, "shaman's helping spirit" or generalized spiritual force.

2 *Qavlunaaq*, lit. "eyebrow ridge." A term used for white men by Canadian Inuit. Point Barrow Eskimos called white men *tanik*; in Point Hope, *naluaġmiu*. Writing in 1940, Brower may be using a term brought by returnees from the Mackenzie region. Brower's use of the word "Inuit" is also unusual. Alaskan Eskimos call themselves Iñupiat.

CHAPTER 12
How Ataŋauraq Was Assassinated

> We also learned that the Esquimo chief, living at the village near by, was killed by some of his own people. He was too tyrannical, oppressed his people and kept them poor, taking every thing away from them and keeping it. He had already killed one man and two girls of the tribe besides his own wife. Shortly after this his son was killed by the father of his wife whom he had killed.
>
> —James Taylor White, nd

> February 14, 1889, he was shot dead by two brothers, whom he had exiled, and who returned for the purpose of killing him. During his life he killed five men and one woman. He inspired such terror that every year people left for distant hunting grounds, till it seemed that a few more years of his reign would have seen Tigara depopulated.
>
> —John Kelly, 1890

The Death of Ataŋauraq

On the morning of February 14, 1889, a crowd gathered near the western end of Point Hope and two armed men approached Ataŋauraq's iglu. Kayuktuq, the son of Quliuq, in whose murder Ataŋauraq had been an accomplice, climbed to the roof, while his brother, Aniqsuaq, walked through the entrance passage. Aniqsuaq was a shaman and he sang a magical song to keep the iglu occupants asleep. Lifting the cover of the entrance hole, Aniqsuaq pulled himself into the chamber, moved aside the sleeping women, and shot Ataŋauraq in the head. On the iglu roof, Kayuktuq had lifted the gut window and was poised to fire downwards.

Exactly why this action took place on this day is impossible to know. But three main factors converged on the event. First, Ataŋauraq's health had been declining since autumn 1888. The women in Ataŋauraq's circle who suffered most from his drinking were keenly aware of this, and were desperate to be rid of him. Concurrent with this, the two sons of Quliuq returned to accomplish a revenge killing.

Four of Ataŋauraq's young contemporaries described the events that attended the murder; extra detail was provided by Asatchaq, who was born two years after the assassination.

Nasugluk's Account

Nasugluk, born in about 1875, had been briefly adopted by Ataŋauraq, and it is through his eyes that we first view the "chief" in autumn 1888. He provides a disturbing picture. While other men hunted or met in their *qalgis*, Ataŋauraq withdrew to the iglu passage alcove to brew liquor. Nasugluk said: "Ataŋauraq got worse and worse, cruel, made whiskey on the lamp in the entrance passage cooking place" (Nasugluk 1940:8).

Apparently isolated and socially estranged—the cooking place was a women's work area—Ataŋauraq's violence fell increasingly on women and children:

> Nasugluk and his cousin were sitting cross-legged on one side of the room and Ataŋauraq on the other. The women were out fetching meat and Ataŋauraq took a big double-edged knife—like the first kind that white men brought—and lunged at the two boys. They whispered to each other that one would grab an arm and the other a leg, but before Ataŋauraq got to them he must have taken pity on their faces. He turned back and hung up the knife on the wall. He was maybe just trying to scare them. (Nasugluk 1940:8)

This was not Ataŋauraq's first assault on children. He had already lunged at a boy called Pauyuuraq in an arbitary and unprovoked knife attack through some fox skins that were hanging in the iglu (Lowenstein 1973–89). But Nasugluk's account reveals something slightly new. While both assaults seem to have arisen from nothing, this later one peters out, perhaps for the reason the informant suggested. More serious was Ataŋauraq's attack on one of his wives. Of this, Nasugluk said: "One night Ataŋauraq killed his wife Qinaruna with an *ulu* and a blow on the head. But he did his shaman work and she came alive again."

This assault, one of many, hastened the revenge prophesied by Captain Hooper (see epigraph to Chapter 9). But the wife beatings were also habitual. Asatchaq commented: "When Ataŋauraq drank, he beat his wives. Ataŋauraq had adopted my mother Niġuvana. When Ataŋauraq beat his wives, they would run to Niġuvana. He never harmed any woman who hid behind [her]" (Lowenstein 1973–89).

Kuwana's Account

Some of the background to Ataŋauraq's attack on his wife was outlined by Kuwana (born ca. 1875), and his narrative creates a link between this assault and the decision to kill Ataŋauraq. It was February 12, 1888, and Kuwana had walked out onto the ice with his uncle in search of bearded seal:

> [W]e were far out on the ice and could see the village very little… As I kneeled at a breathing hole, there was splash. It blew off the ice and I saw the seal's whiskers. I bit off my mitten, aimed down and shot it. Then I tried to find it with my spear. The other hunters ran up, scolding. The seal had escaped. Next day, my uncle took me out again. The wind had shifted. A boy called Aglaq caught a seal and we brought it home…took the gut off the skylight and let it in through the hole. We let it thaw in the house. *They start after Ataŋauraq to kill him* [emphasis added]. While we had been out, the chief's

> wife had escaped and Ataŋauraq had hired someone to take him after her. They caught her by Cape Lisburne and brought her home. Then the chief began drinking. He told his wife if she did this again he would cut the tendon out of her ankles. Then he got mad. He grabbed an *ulu* and hit her on the head. When she fainted he tried to revive her. (Kuwana 1940:29)

Up to this point, only two men are recorded to have opposed Ataŋauraq. Now for the first time, we observe the revulsion of a woman, although she is helpless. This nonetheless was the start of the process, fueled by women, which brought Ataŋauraq down.

Two separate groups now planned an assassination and tension was rising:

> All the men decided to kill Ataŋauraq. And though they didn't tell him, Ataŋauraq knew this. So when he went hunting, he took Nasugluk [aged thirteen] with him. One day when they came back from the ice…they afterwards found out that four men had been lying in wait with bows and arrows. But these men were Nasugluk's relatives. They were afraid that if they hit Ataŋauraq, he would kill Nasugluk. (Nasugluk 1940:8)

It was not just the men who were preparing an attack. Nasugruk, who was then a girl of about thirteen, told Rainey:

> When they came home from the Kuukpak river, Ataŋauraq's sisters and wives all pleaded with them to kill him. Everyone was trying to get them to do something, because they wanted their children to live…They urged Aniqsuaq and Kayuktuq to kill him because Ataŋauraq had said *he would kill their children and then themselves.* The old women said they should kill him first. (Nasugruk 1940:8)

Starting in late 1888, the endgame was thus generated from the particular fears of women. While Ataŋauraq's two assassins were living outside the village at the time, they were evidently within reach, and that autumn the older women urged them to act. For some months nothing happened. But the attack on Ataŋauraq's wife hardened his avengers' resolution, and two of his co-wives decided to take action. Nasugruk said:

> Two of his wives were preparing to kill Ataŋauraq: the one who had been wounded with the knife and another. This second one had a baby, so she took it to someone else, and the two women got dressed in men's clothes. However, when they heard that some men were planning to kill him too, they put their own clothes on again and went home to bed. (Nasugruk 1940:9)

The next details are confusing. According to Kuwana, Ataŋauraq knew the attack was coming that evening, and "didn't want to sleep. He'd heard about it"—though another informant suggested that given a warning, Ataŋauraq would have gone on a killing spree:

> There was only one wife who didn't know what the men in the village were planning. The other wives were all friends of Ataŋauraq's enemies. This one wife was the most beloved to Ataŋauraq, and she would have told him if she had known, and he would have killed many people. (Nasugruk 1940:9)

Fig. 26. *Kuwana, born 1870, photographed ca. 1920, perhaps during the whaling festival. He wears caribou-skin trousers, sealskin boots, and a white cotton shirt. Photograph from lost original in author's possession.*

If Ataŋauraq knew of that evening's plan, the passivity of his response is difficult to explain, although this down-to-earth interpretation was offered by Asatchaq: "On the evening they planned to kill him, Ataŋauraq was drunk; he was sleeping off his liquor" (Asatchaq and Lowenstein 1992:173).

Nasugluk stayed up on the night of February 14 and waited for action:

> I looked towards Ataŋauraq's house and saw two men going inside with rifles, and lots more outside. One man was singing a song to make Ataŋauraq sleep. My uncle told me to go inside our house. But before I went in, I heard a shot, and then other guns. A man came jumping out of Ataŋauraq's iglu, as if flying.
>
> The two men who went inside Ataŋauraq's house had found him sleeping between his wives. Aniqsuaq took his rifle, straightened Ataŋauraq's head so as not to hit his wives, and shot him. (Nasugluk 1940:9)

Among those who crowded to Ataŋauraq's skylight the next morning were Nasugluk and Kuwana. Kuwana described how he had woken to see the two assassins talking with his uncle:

> Just before they shot him, I fell asleep. Then someone violently grabbed my hair. It was my mother. When I woke up, the men who had done the murder came to our iglu and talked. My uncle in his hurry started putting his clothes on inside out. It was nearly daylight. When I went out, the three men were out there: my uncle and the brothers. (Kuwana 1940:30)

Kuwana went to view the scene. While his description of the murderers had been cursory and blurred, what follows becomes circumstantially detailed:

> My mother took me to Ataŋauraq's. The skylight was covered with an old piece of calico. They had moved him already and covered him with a blanket.
>
> I heard a strange noise and peered through the skylight. I thought Ataŋauraq was going to come back to life. My mother told me to look inside again. The strange noise went on. The chief's older brother [Tiguatchiaq] was in there; he was weeping. Ataŋauraq wasn't going come to life. His head was smashed. (Kuwana 1940:30)

Nasugluk provided extra detail with dramatic, semi-shamanistic emphasis:

> Someone said: "Our men have killed someone. They have killed Ataŋauraq." It was daylight already. We went to Ataŋauraq's...and my aunt urged me up to the iglu window. The gut window was loose. I peered into the iglu. There lay Ataŋauraq and his wives beside him. All were asleep, except his first wife. She was leaning on her elbow, just going to have a smoke...Ataŋauraq's little daughter was standing naked in front of the lamp with her fingers in the corners of her mouth.
>
> As they watched from the skylight, some people came in. These were Ataŋauraq's sisters. All but one of them was crying. This was Pasigvauna. She went over to the woman who was sleeping in the corner, kicked her in the head and told her that she hadn't taken care of her husband. But the woman didn't wake. Then Pasigvauna grabbed her hair. That didn't work. The third time she tried, the woman woke up. Like Ataŋauraq, she had been put to sleep with a song. This woman was pregnant.
>
> That same day, they took his body to the graveyard. Two days later [Aniqsuaq and Kayuktuq] went out to his grave and cut his neck and ankles through from the front—right through the bone to the back skin. (Nasugluk 1940:9)

This ritual mutilation was presumably a mark of contempt, but also no doubt served to preempt revenge from Ataŋauraq's spirit.

The Aftermath of Ataŋauraq's Murder

The blast in Ataŋauraq's iglu stunned the household. Recoiling from the impact, the assassin emerged "flying," as if hurled up by shock waves. An isolated minority—one wife, the chief's siblings, three adult children, and that sad little daughter—mourned his passing. The majority of the village woke happier that morning and the rest of the winter seemed to pass quietly. But as village historians somewhat wearily attest, the aftermath of Ataŋauraq's drama was yet to be played out. Two stories about Ataŋauraq's murderers come first, and these are followed by an account of an adopted son, whom Ataŋauraq had appointed as his "heir." Last is the story of a murder at whaling.

Kayuqtuq's Story and the Fate of Aniqsuaq

Soon after the murder, Kayuqtuq left Point Hope both to escape the dead man's brothers and also to avoid the Revenue Cutter's summer visit. He was right to be worried. When the *Bear* arrived and Healy missed his "partner" Ataŋauraq, he apparently steamed in pursuit of one of the assassins. An extraordinary scene at Cape Lisburne followed. Kayuktuq and his family were camping on the beach. And when Healy sent a boat in: "Kayuktuq went into his tent and loaded a shotgun, handed it to his wife and pushed her out with it. She stood there with the gun raised as the white men landed. But when they saw her standing there, they went away again" (Lowenstein 1973–89).

It is an inexplicable incident. Did Healy intend to arrest the killer? How would he have identified Kayuqtuq? And with what witnesses would he send him down for trial? And if Healy could thus travel through Lisburne's stormy waters for Ataŋauraq's killer, why had he not previously detained Ataŋauraq for worse?

No more is recorded of Kayuktuq. But it is interesting to reflect on the nature of historical change here. The entire Ataŋauraq saga had started with the murder of the shaman Suuyuk at Cape Lisburne in about 1875. Now in the same location, less than twenty years later, spears have been replaced by firearms, and instead of a skinboat crew, it is uniformed Americans who are approaching. As we have seen, virtually every stage of Ataŋauraq's history is infused with these images of transformation.

The drama now continues with the story of Kayuktuq's brother and the manner of his death in the autumn of the same year. Kuwana reported:

> Aniqsuaq stayed in our iglu that autumn and one day brought in a musket barrel. He had put a stove in the house. Now he loaded wood into the stove, opened the top and shoved in the barrel. [His wife] had brought a bucket of water. She was going to heat it for some washing.
>
> Then Aniqsuaq took the barrel off the stove, put up the bucket and pushed the barrel into the stove, so it pointed outwards. Then the barrel exploded. A bullet hit him in the shoulder. He turned away and put his hand to his chest. Then Aniqsuaq's uncle got up, and the gun went off again, just missing his head.
>
> Aniqsuaq started to rub his chest. He was a shaman, so he called his spirit. He rubbed his chest and called his spirit. And he said there was no bullet there. The hole had gone too. He called his wife to fetch his medicine from the entrance passage—this was some old grass covered with soot. He tested with his hand to see if there was blood. No blood had come through. The bullet had just missed the breast. His wife brought his medicine and he rubbed his chest. The hole came back. When he looked at his hand, he saw blood…
>
> Aniqsuaq lay wounded. His uncle took a piece of wood that he'd saved for a bow and arrow, made it into a spoon and cleaned the wound. He got paper and lead shot out. They stopped cooking in the iglu. They ate raw meat…
>
> When Aniqsuaq was almost well, his mother went out one morning. But in the passage her hand fell on something. She went back to inspect it. When they saw it in the lamp light, they found a dead lemming. All dirty with oil. And the wound started bleeding, and next day he died. The day before, a man had come to the skylight to ask Aniqsuaq if he was getting better. Aniqsuaq lied. He called out he was dying. He thought that man would go after him. Someone had put that lemming in the tunnel. The lemming was dangerous. (Kuwana 1940:36ff)

The first issue to emerge here concerns Aniqsuaq's position among the factions in the village. While the chief's death signified a closure, it also represented only one step in a process that had started with the death of Suuyuk. The moment the two brothers killed Ataŋauraq, the feud was reignited—despite the growth of Jabbertown, with its extra population of southern Iñupiat, which had diverted local energies from internal conflict.

The manner of Aniqsuaq's death, for which the lemming could not exclusively be blamed, helped to abort future violence. The accidental shooting had, after all, been a neutral incident. "When people talked about it, they said that gun had been timed just right," said Asatchaq in another account (Lowenstein 1973–89), a comment suggesting that things "happened" to converge on an outcome desirable to the enemy faction. But the manner of the killing has a character that paralleled that of Ataŋauraq. When the barrel exploded, it was as though the entire contraption of distillery and iglu had, as it were, been sprung for its latest victim. Just as Ataŋauraq had died where he had sat and watched his still, so his murderer had been destroyed in secluded devotion to alcohol production.

The shamanistic element to the story thus takes place within a context of material transition. Just as stove and bucket suggest new domestic goods that would become common by the 1920s, so the wounded man's amulet situates him in old Point Hope culture. Until Dr. Driggs arrived the following year, there was no alternative to local medicine beyond contact with the *Bear*'s physician. The wound itself is one point of convergence, in which lead shot and cartridge paper are probed by amulet and a driftwood spoon. The other, more lethal, convergence lay in the combination of alcohol and firearms. There had been, and would be many more, victims.

Violence at the Time of Whaling

Ataŋauraq was killed in February. His murderer died in the autumn. Between these events came two more fatal shootings, neither of which could be explained by Point Hope informants.

In the first incident, a man from Buckland shot, apparently at random, a Native from Cape Prince of Wales as he was leaving an iglu to join a whale boat. Escaping to Jabbertown, the murderer was shot dead by two white whaler-traders (Kuwana 1940:31). The second incident, involving two Point Hope men, occurred on the ice after a whale had been taken:

> Kakianaaq's crew were sitting behind their upturned skinboat. One of them suddenly shot another. He had been eating *maktak*. And Iḷaqpak shot him in the mouth. He would also have killed another man, but he escaped.
>
> No-one understood the reason. Then someone shouted: "Better shoot everybody now!" They all grabbed their rifles and waited to see who'd start that war. The men stayed in their crews. That's why the fighting never started. Because the crews weren't made up of families. They were all mixed up. (Kuwana 1940:31–32)

Conflict, generally over the ownership or apportioning of a whale, did sometimes break out during the traditional period (Asatchaq and Lowenstein 1992:185). Again, as in the episode with the two Seward Peninsula men, violence appears to have erupted inexplicably. "No-one understood," said Kuwana, who witnessed the event. The changing demography of the village, to which the story also alludes, does, however, seem to have played a role in the second incident. Reduced, but still dominated by local people, the Point Hope community now hosted increased numbers of Native visitors, and this mixed population had changed the composition of the local skinboat crew, which, suggested Kuwana, was one reason that the violence did not spread. "The crews," he remarked, "were not made up of families. They were all mixed up." The implication here being that *umialit* no longer had enough local families on which to draw, but must enlist anyone they could. The shooting took place within one crew, while other boats gathered in its aftermath. Had crews been formed on a traditional basis from within ceremonial houses, they would, suggests Kuwana, have continued the fight out of group loyalty. As it was, there were no solid groups to be either for or against. Demographic changes in the community population appear thus to have exerted a new kind of social control.

Ataŋauraq's Heir and His Destruction

That spring saw further violence, although this was partially contained by the intervention of its perpetrator's associates. The aggressor was Alauraq, one of Ataŋauraq's adopted sons, who is said to have been the chief's heir. The significance of this legacy is unclear, but presumably the young man inherited Ataŋauraq's skinboat and part of his trading operation. There are differing accounts of Alauraq's status in the village. According to one informant, his mother was Kunuyaq's wife, while the whaler-trader John Kelly maintained that Alauraq was adopted from another village:

> The Tigaras exchange arms and ammunition, obtained from whale ships, with the Noatak and Kivalina [people] for furs. Whale oil and seal skins are bartered for deer skins and children. A child costs originally about one seal-skin bag of oil, or a suit of old clothes. If not resold they are brought up in the family purchasing them to do menial work as soon as able. As they grow up they are made to believe that their parents were bad people and they would have died if they had not been taken in by the present owners. When grown up they are free. Not knowing any other dialect or having any other home, they stay where they are and become recognized Tigaras. [Ataŋauraq] had a Noatak boy, whose fidelity he afterwards rewarded by creating him chief, next in power to himself. (Kelly 1890:10)

According to Nasugluk, who was a young man at the time, Alauraq's brief career went wrong during the 1889 whaling season after he had assaulted one of his wives:

> When Nasugluk joined crew, Alauraq asked after his [injured] wife. Nasugluk said, "Where there is no sickness, people are fine." One night later, Alauraq took a whale. They sent a boy to the village to tell his wife. When the boy returned, he sat down on the sled looking sad. Alauraq's wife was dead.
>
> Alauraq sat on the sled for a while with his head bowed, then asked Nasugluk for his rifle. Nasugluk wouldn't give it to him; nor would Anniq. When Alauraq got home, they were binding the dead woman. "This is not a good a place," said one of the women. Alauraq's tears ran down. His other wife got up and left. Her father and brother had come from Kivalina to fetch her. (Adapted from Nasugluk 1940:12ff)

All but abandoned by a horrified community, Alauraq is transported from the village by dogsled in pursuit of his wife:

> When he got to the south shore, he left his sled and ran on and caught up with her. She, her father and her brother, waited with their rifles. As Alauraq came up, the father shook his gun at him. But he kept coming till the son shot him from behind some driftwood roots.
>
> They covered his face with a mitten, put snow blocks round him and some men carried him back to a burial rack. That was the end of Alauraq. (Adapted from Nasugluk 1940:12ff)

Asatchaq gave a shorter version of this story, and added one detail: "Alauraq was copying his father Ataŋauraq, drinking. Kunuyaq, his real father, told him to be a good man, not like Ataŋauraq. 'If you take my advice, you'll live to be an old man. Otherwise, you'll end like Ataŋauraq'" (Lowenstein 1973–89).

Ataŋauraq: Conclusion

Alauraq was a young man without *umialik* charisma or a ceremonial house entourage, and he vanishes abruptly. His story suggests two complementary things. The first, perhaps reflecting the whole Ataŋauraq saga, is the pragmatic message that the influence of alcohol was so destabilizing that a drinker could devastate his social and domestic life, disrupt subsistence, and finally destroy himself. In contrast, the narrative presents examples of decent, normative conduct, which the majority of people in Point Hope continued to follow. Repeatedly through his story, we observe people in Alauraq's circle who continue their struggle to pursue an ordinary existence, and in episodes not recounted here, Alauraq's associates conspire to moderate his conduct by confiscating his pistol, refusing to give him a rifle, and helping him get away from the community so as to avoid further trouble.

Accompanying the recital of Alauraq's crimes are also expressions of community anger and sorrow. Nasugluk's remark "Where there is no sickness, people are fine" directs shame on the young leader, while Nasugruk (later to marry Nasugluk) cries out at him bitingly, "This is not a good place to stay." With the knowledge that Alauraq's wife has died from his assault, and dumbstruck with grief at the news, a boy "sat on the end of the sled looking sad." Equally striking is the way "Nasugluk and Siuluk didn't feel like cutting up the whale when one of their crew [the *umialik*'s wife] had died." Traumatized by Ataŋauraq's rule and other contact ills, these are people who are simply trying to put life back in order and pursue subsistence. Alauraq's attack on his wife at this crucial juncture of the year was also an assault on a community whose well-being was dependent on the hunt. The contrast between the urgency of the hunt and the alcohol abuse that threatens it deepens the impression of malaise. In sober moments, Alauraq is shown vividly to understand what he has done. Twice the storyteller describes him with "bowed head" and once "his tears ran down." Here we are reminded how Ataŋauraq cut his hair after he had confessed to his first murders.

But in contrast to the picture we are given of Ataŋauraq, who brooked little opposition, the people around Alauraq speak out. Their actions and his disgrace highlight vividly the discord between destructive individual behavior and communal welfare which both the whaling season and the continuing subsistence crisis made urgent. This tension between individual freedom and the demands of social obligation ran deep in Point Hope society, a conflict frequently expressed in shaman stories. The whole Ataŋauraq saga had, for example, begun with Suuyuk's egotistic declaration that he was "going to starve the village." And while the community had dealt summarily with Suuyuk's threat, Ataŋauraq's later ascendancy had, for a decade, subdued the village. With the death of Alauraq, the community had reasserted itself. While epidemic diseases would continue to reap their miserable harvest, the history of Alauraq's containment suggests ways in which the village was reintegrating with its traditional focus on subsistence and community sharing.

CHAPTER 13
Disease in Alaska

Who knows what kinds of sickness they may bring us?

—Kodiak elder, 1763 (Quoted by Fortuine 1992)

First Imported Epidemics in Alaska

The Europeans who first arrived in Alaska in the mid-eighteenth century followed a route similar to that of the Ice Age migrants from the Old World twenty thousand years previously. But while the Paleo-Indians who entered what would become America had left most of their sicknesses behind them, the Russian expedition, dispatched by Peter the Great to chart where Siberia ended and America began, brought Old World infections with them. How far the Aleuts who met Vitus Bering in 1741 on this, his second journey, were affected by disease is unknown, though a small number of Natives may have contracted tuberculosis and venereal disease from Bering's crewmen. For once, in fact, it was the Europeans' suffering that dominates the story of this voyage. Debilitated by scurvy and hunger, crazed by their own quarrels, and disoriented for much of their journey, they had little time for contact with Native Alaskans. At the climax of this voyage, Bering and the majority of his crew died of scurvy and exposure in and around the Aleutian and Commander Islands in 1742. And apart from the observations of Georg Steller, the young German scientist whose plant and bird studies provided evidence that Alaska and Europe were indeed separate continents, the immediate result of Bering's journey lay in sightings of sea otters and the abundance on Bering Island of "Steller's Sea Cow" (the manatee), whose clumsy tons and trusting nature made them vulnerable to the Russians. The rush for otter furs and the promise of abundant food led to the arrival of Russian hunters (*promyshlenniki*) in the 1750s, and this brought Alaska Natives face to face with Europeans for the first time. It was not, however, until the fur hunters began moving east along the Aleutians over the next two decades that the process of disease transmission set in with serious effect.

The killings, forced labor, and epidemics that had disfigured the Caribbean and Mexico following contact in the sixteenth century were reproduced, albeit on a smaller scale, by the Russians for forty years of unregulated activity. Just as the Spanish had enslaved the Caribbeans to work gold and silver mines, so the Russians dragooned Aleuts to catch sea otters and cohabited with Native women while the men toiled from their *badairs* (kayaks). The Aleuts resisted sporadically until 1764, when their final serious attempt at opposition was crushed. The early 1780s brought a measure of relief from *promyshlenniki* atrocities, when Catherine II established a trading company that would supposedly establish friendly relations with the Natives of Alaska. But a partially state-controlled semi-regulation also brought more Russian settlers to Kodiak, the Yukon, and southeast Alaska, and this involved the further spread of diseases through southern Alaska and into the Seward Peninsula.

The disease history of Kodiak Island, starting in 1784, provides a useful but depressing indicator for much of what happened throughout Alaska during the next hundred and twenty years. This vast rugged home to the Koniag Eskimo lies to the north of the Aleutian chain in the Gulf of Alaska. As outlined in Chapter 3, it was here, in 1784, that the Russians established their first trading settlement. Open warfare with the Koniag was followed by an uneasy period of equilibrium during which the Russians operated the skin trade, opened a school, and commandeered flotillas of men to pursue the sea otter hunt and to help extend Russian trade into southeast Alaska (Fitzhugh and Crowell 1988:75). Only the arrival of Orthodox priests, who took over the school in 1794 and, from there, expanded their ministry into southern Alaska, somewhat ameliorated the abuses of the traders.

The relatively dense Native and European populations on Kodiak were in frequent contact, and this created a fertile environment for the spread of infections. When the Russians landed in 1784, the Koniag were found to be "tall, healthy and well-fleshed," although venereal disease had been introduced by the *promyshlenniki* (Fortuine 1989:28). At least nine epidemics struck Kodiak Island in the nineteenth century, but as Fortuine has noted, it was always the Natives who sickened and died, while "the Russians and other Europeans seemed to be spared, a circumstance which led many Natives to believe that the White man had unleashed these terrible powers expressly for their destruction" (Fortuine 1989:199).

As if to probe the heart of Native spiritual and healing culture and to expose their impotence in the face of epidemic, a respiratory disease of 1804 killed many Koniag shamans and spared the rest of the population—an uncanny and inexplicable occurrence. The next recorded sickness arrived in 1819. According to the Russian manager of the trading company, this "began with a fever and a heavy cold, a cough, shortness of breath, choking, and three days later death followed" (Fortuine 1989:201)—a description that fit diseases in many other Native Alaskan communities in the nineteenth century.

The Kodiak epidemic of 1819 has further historical significance in that it highlighted the charitable work of Father Herman, an Orthodox missionary who lived just off the Kodiak coast. Born in Russian Lithuania in 1760, Herman spent forty years at his Koniag mission and was canonized by the Orthodox Church in 1970, not least for his frequent political intercessions on the behalf of Natives. His devotion during the 1819 epidemic was memorably described by the Russian manager on Kodiak. More than a hundred people had taken refuge in a ceremonial house:

> Some were lying already dying, their bodies growing cold next to those who were still alive; others were already dead: the groans, shrieks, that were heard tore at my heart. I saw mothers already dead, on whose cold breasts hungry children crawled... Father

> Herman nursed them ceaselessly, tirelessly and at great personal risk visited the sick, not sparing himself in his role as priest—counseling those suffering to be patient, pray, repent, and prepare themselves for death. (Quoted by Fortuine 1989:201)

Herman built an orphanage for the children who survived the epidemic. But their generation would fall to repeated disease visitations. A "contagious rotting fever" swept the Aleutians and Kodiak four times between 1806 and 1824. In 1827, 158 Koniag died of an unidentified sickness. Measles hit Kodiak in 1860 and 1874 with a possible death toll during the latter epidemic of 515 people (Fortuine 1989:207). This was followed rapidly by typhus, mumps, and scarlet fever, which killed an unrecorded number of weakened and distraught measles survivors. Influenza struck in 1881, causing "many deaths," while "black measles," a disease resembling both measles and flu, arrived the following year from southeast Alaska—where, as on Kodiak, it spared most of the Europeans (statistics selected from Fortuine 1989:Ch. 11).

Native Health in Alaska to the Early Nineteenth Century

The Europeans who first landed in Alaska in the eighteenth century noted the Natives' vigor, good health, and self-confidence—this in significant contrast to the deteriorated condition of mainland Indians during the same period. The Aleuts, wrote Georg Steller, were "strong and stocky, well-proportioned and fleshy on arms and legs" (Steller 1988:103). Lieutenant Chirikov, also of the Bering expedition, found Alaskan Natives to be "healthier and tougher" than their Siberian neighbors. The Koniag Eskimos, according to Shelikhov on Kodiak Island, were "tall, healthy and well-fleshed," while two of Cook's officers on HMS *Discovery* in 1778 had found the Chugach Eskimos to be "fat and jolly, as though they lived well," and the Kuskokwim Yupik to be "tall, well-made and wild, fierce-looking people" (Fortuine 1992:27, 28).

Further north, arctic and subarctic people appeared still to be healthy into the 1820s. In 1816, Otto von Kotzebue described both the St. Lawrence Island and Kotzebue Sound Eskimo as "robust and healthy" (Fortuine 1989:29). And as we have seen, Beechey wrote similarly of Point Hopers in 1826 as "robust…very honest, extremely good natured, and friendly" (Beechey 1831:I, 262). Still innocent of disease and firearms, the people Beechey met at Cape Thompson "had had a very limited intercourse with Europeans." All this would change once regular contact started in the 1860s and intensified after 1880.

The arctic record goes relatively silent during the first three decades of commercial whaling. However, between 1852 and 1854, when HMS *Plover*, commanded by Rochfort Maguire, was stationed off Point Barrow, there was a quiet episode of British activity during the search for Sir John Franklin. Both Maguire and the ship's physician, John Simpson, made scrupulous observations and came to the help of Point Barrow people during a period of famine. And while hunger claimed many Native lives during their stay, neither Simpson nor Maguire mention diseases apart from the occurrence of dysentery (not endemic to the area) and the possibility of their men transmitting gonorrhea. Like Cook, Maguire was concerned that his men not pass on venereal diseases, and during an outbreak of gonorrhea on board the *Plover* in December 1852, Maguire forbade any further intercourse with Native women. "I feared," wrote Maguire, "the contagion spreading amongst the Native population, where the effects of it were to be the most dreaded and indeed not to be thought of except with a degree of horror" (Bockstoce 1988:131). Since the outbreak had originated in one identifiable crew member, the infection had presumably spread among the men as a result of sexual contacts that had already taken place. Maguire, however, subverts this conclusion in the same entry: "strange to say we have never heard of a case of [gonorrhea] amongst [the Natives]." There is no further mention of venereal disease in the rest of his journal, and the diagnosis must remain open.

John Simpson, in his *Observations on the Western Esquimaux*, also provides a first, short ethnography of north Alaska. But of significance here is the picture given in Maguire's journal of Simpson's activities and the complaints to which he attends, as he walks on what appear to have been regular rounds through the Native village. All the cases described by Maguire, except those of dysentery, whose causes were ambiguous, were of a routine local nature: fractures, gashes, and pulmonary sickness following exposure on the ice. Of particular interest are the three-way interactions between patients, their shamans, and the physician. Sometimes Simpson was welcomed, sometimes he was sought on board ship, at other times his assistance was refused. "Doctor Simpson," wrote Maguire on November 22, 1853, "went to the village as usual, having a companion or two from the crew, one of whom carrys the small Kettle containing his patient's dinner—who seems to do well, but he has something seriously wrong in the right side of his chest, but the pain is more circumscribed than before" (Bockstoce 1988:312). The following day "Doctor Simpson went to the village as usual and put a blister on his patient's chest, which he seemed glad of, as he expects it to suck away the internal disease." (This process suggests a convergence of shamanistic theory, in which tribal doctors sucked out harmful spiritual matter, and the action of poultices, which drew on the internal system.) Four days later, however, "Doctor Simpson on his way to the Village to visit his patient & give him his usual dayly meal was met by [a woman] who came to tell him that he was not to go to the sick man's hut today, as they were going to perform some miracle or ceremony of their own for his cure" (Bockstoce 1988:314.) Two weeks later: "Doctor Simpson's patient who was so long adrift on the ice, is afflicted again with the pain in his chest, and [a shaman] is engaged with him beating it away with a drum" (Bockstoce 1988:317).

The Barrow shamans both competed with Simpson and allied themselves variably with and against him. But despite the doctor's scientific curiosity, he showed no interest in what the shamans did, and Simpson would sometimes simply take over from Eskimo practitioners, while graciously allowing them to observe: "Another of Doctor Simpson's patients," wrote Maguire, "who has received a deep cut on his shin was down and dressed on the lower deck. As this was a case beyond [the shaman's] skill he was content to look on approvingly" (Bockstoce 1988:318). And on another occasion: "Doctor Simpson visited the Village to see the man with the broken leg. He found it tied up in three short sticks, but unfortunately with a bend in it, which Doctor S promised to remedy the next day. Doctor Simpson visited the Village to set the broken leg...[and] felt that [the patient] appreciated the great comfort & relief from the change from Eskimo splints to his" (Bockstoce 1988:363). Simpson offered his help for free. But since Eskimo shamans seldom worked without payment, Simpson, too, was generally offered a fee—because as Simpson and Maguire both divined, "it is necessary to have the good will & wishes of the person consulted or the cure will not be effectual, and the fee is given as much to ensure this as to secure the advantage of his skill" (Bockstoce 1988:388). An observation on the therapeutic relationship which anticipated Freud! That the shamans accepted Simpson's interventions was a credit both to the doctor's skill and to good nature on both sides. But even in this benign encounter between two very different societies, we witness a transformation among the resident majority into passive spectators when confronted by the application of combined surgical and technical skill. Simpson was, of course, even within his own culture an impressive physician. But to the Eskimos, who had no other European physicians with whom to compare him, he was representative and typical. Throughout the contact experience in the Iñupiaq region, Natives showed themselves to be open and curious about new ideas and artifacts. On one level this led to a creative cultural balance, especially in the period between 1920 and 1960. The technical ascendancy of Euro-Americans, on the other hand, sometimes had its demoralizing aspect. This point was expressed bluntly by David Umigluk Frankson, a Point Hope man born in 1900: "Before whites came, everyone respected shamans. When missionaries and whalers came, they had big property, like ships that were powerful—whaling guns, whaling outfits, food and clothing. Right there is their power" (Lowenstein 1973–89).

This process of relegation is one of the most distressing aspects of contact history, and paradoxically it was often the same white men who transmitted disease as exhibited admired skills. The next part of this chapter sketches the background to Point Hope healing and the beliefs that underpinned these. In this way we can explore how the late-nineteenth-century epidemics undermined local views and traditions that had explained and supported life for so many centuries.

Point Hope Native Medicine

Point Barrow and Point Hope shared similar disease theories and roughly similar shamanistic practices. As Maguire learned, disease and even accidents were caused by "bad spirits" (*tuunġat*). But perhaps because it was geographically closer to Siberia and Indian Alaska, Point Hope had developed the more complex shamanistic and ceremonial system. Thus, while Native medicine at Point Barrow was, in its resemblance to the central Canadian Arctic and Greenland, more purely Eskimo, Point Hope's ritual and belief systems also contained elements that had, over the centuries, been borrowed from other societies. Here, to show something of Point Hope's medico-religious life as it existed before contact, and to suggest which medical issues could and could not be coped with in both traditional and nineteenth-century contact history, it will be helpful to outline something of old Tikiġaq's beliefs and to describe how the shamans worked. The extent to which disease undermined Point Hope's medical and shamanistic traditions will thus, I hope, become clear.

Births and deaths in traditional Point Hope were attended by rites, which exerted a continuous pressure on social life, and these exigencies helped to shape the affective milieu. Childbearing and death were spiritually loaded and metaphysically vulnerable events (*qiruk*) (Rainey 1947:273), and the consequences of any oversight in the ritual handling of either could be dangerous. In the first place, neither event was permitted, so far as was possible, in the domestic iglu or ceremonial house (*qalgi*). Infant mortality was relatively common. Obstetric complications, the difficulties of neonatal care within taboo constraints, enforced perinatal isolation, and female infanticide meant that birth was frequently followed by a death. These factors alone rendered birth an ambivalent occasion. The death of a child created in the bereaved a polluted condition that was peculiarly dangerous (*qiruk*) to others and placed them in a taboo state. Added to all this were year-round taboos associated with vaginal blood, a substance that in most contexts, especially those of hunting, food handling, and sewing, was both powerful and threatening.

Preparations for a birth involved the construction of a small snow hut or sod house where mother and baby must live for up to three months. Here they would be visited by their elders, and it was through these latter that names, amulets, and ceremonial affiliations were given. These donations linked the baby to a ceremonial house, a wide range of living and dead individuals who would be its guardians and namesakes; connected it through amulets with animal spirits; and ordained which foods would be taboo. All these lifelong associations were established during the confinement. Only after the isolated but socializing aftermath of birth was complete could mother and child return to the iglu. By this time, the baby had become symbolically integrated into the community. As with shamans who were also initiated in isolation, complete personhood was achieved during a period of separation (Rainey 1947:273).

The rituals connected to death were less extensive, but as with a newborn child, the dying or deceased could not be kept at home. Of course, a proportion of deaths occurred away from the iglu, and perhaps more than most Native people, Eskimos were subject to accidents and exposure. The sea, darkness, snow, and windswept tundra often claimed the most able hunters. But while death rites were less prolonged than birth rites, the ghosts inhabiting the Point contributed to a supernatural environment that was threatening, even if familiar.

In traditional Point Hope, where multigenerational family life was conducted in iglus and tents, taboo demanded that the dying and everything that belonged to the deceased be moved into the open. The mere fact of death would otherwise infect the living, the souls of the dead would find no easy release, and the bereaved would be troubled by unhappy spirits. The face of the dead individual was hidden with an animal skin and the body was transported, with its possessions, to the east of the village, tied up in caribou skins and used boat skins, and laid on a driftwood-log platform supported by whale bones. With knees bent and feet apart, the legs were oriented to the east. This pointed to a place beyond the Kuukpak River, to a "land where it is always good weather" (Rock 1940:8). Articles belonging to the deceased were then laid on the rack. Bows, harpoons, and, later, guns were broken; women's knives and sewing equipment were similarly set out. When death happened in the winter, snow sarcophagi were made, and the platform was erected in the spring so as to "give the dead good weather." In June, after the whale hunt, *umialit* would visit these graves and give their dead relatives and associates small pieces of *maktak* and whale meat. Then the family built a fire by the rack, and since death and misfortune were, in themselves, thought to be contagious, the sled that had carried the body to the burial ground was burned (Rainey 1947:271). At this point people were allowed to weep openly. All this was followed by a *qalgi* feast of cooked meat and then dancing. These rituals were largely abandoned during epidemics when there were too many dead for the survivors to cope with, and both social and ritual life were disrupted.

Point Hope people believed that human beings had three principal souls. There was a life soul called the *iñuusiq*, which was possibly the component that traveled to the land of the dead. Second was the *iḷitqussiq*, a personality soul that could be reborn. Following death, there was a waiting period before the spiritual rebirth was complete. During this time, the *iḷitqussiq* hovered near the body. This was also the soul that had gone away or been captured by a *tuunġaq* (disaffiliated spirit or shaman's helping spirit) during sickness. For men and women this period of "sleep" (*siñiktaq*, literally "dream") lasted four and five days, respectively. At the conclusion of this period, the bereaved knocked on the upright posts of the iglu interior as a signal for the soul's departure. Some souls who had loved too much or who were addicted to life haunted the village. These were unwelcome because they added to a spiritual environment that was already crowded and often difficult to deal with (Rainey 1947:271).

The third human soul component was name (*atiq*), and this was the soul that provided social continuity (Lowenstein 1973–89). When newborn babies were given their names in the birth hut, the soul of a dead namesake entered the child, and this constituted an ancestral rebirth or the return of a recently deceased elder. Through *atiq*, the child thus *became* the person whose name it inherited.

While some of these beliefs and customs were shared by other Eskimo societies, one unique to Point Hope was a system of amulet affiliation, by which most people in the community belonged to a group associated with either "things to do with childbearing" (*iġñiruaqtuqtuq*) or "things to do with death" (*quŋuqtuqtuq*). Those who belonged to the birth group enlivened their personal amulets, such as animal skins or parts of birds, through contact with pregnant women, placentas, or umbilical cords. The death group did the same with human relics and grave goods. Human remains were important to death-amulet people, and the bodies that lay within range of the village were a component in their therapeutic inventory. To give one example cited by Rainey, corpses on racks built over the slough that ran through the center of the peninsula would decompose into the water, and those who belonged to the *quŋuq* group could, in the summer and autumn, take healing drafts of an elixir that had been thus impregnated with mortality (Rainey 1947:273). In other circumstances, people would rub their amulet on a human bone or a burial wrapping. Members of each amulet group were, however, dangerously allergic to objects in the opposite category. A large section of the community—what proportion remains unknown—was thus divided more or less equally into two opposing taboo areas, both of which, and in combination, connected them to mutually antagonistic life principles. The tension between the two parties was usually resolved in the home through a one-category membership per iglu, though little is known about this (Rainey 1947:272).

The two amulet classes were one among a larger set of symbolic dualities that informed Point Hope thought, another important set being the sun and moon spirits who were

Point Hope's main deities (see Lowenstein 1993:68 and passim). According to the myth, these spirits had been a Point Hope brother and sister. The brother, Aliŋnaq, had raped his sister, who then cut off her breasts and offered them to him in the *qalgi* macerated in her waste pot. Appalled by his sister's response, Aliŋnaq rose in a spiraling motion through the skylight until he reached the moon, while the sister flew to the sun, where she displayed her wounds. These are the red streaks that mark the sun when it emerges in January. While the angry-looking sun appeared to threaten the village as it struggled above the sea ice, its spirit was also a source of strength, particularly to women, who raised their children from their iglu tops towards the winter sunrise (Lowenstein 1993:129).

While the moon spirit was of a type with the trickster creators found in most Native American religious systems, the transformation of an incestuous rapist to lunar deity lent a somewhat sinister potency to Aliŋnaq, which gave him power over the uncertainties of human life. In the moon spirit's keeping were the conditions of human welfare and, not least, the souls of the animals that Point Hope men hunted. The success of these hunters depended partly on whether people observed the sort of taboos that Aliŋnaq had himself violated, and whether the hunters' wives were able to raise a pot of ritual water through a hole in the sky so that Aliŋnaq could drop animal souls into them. These were symbolic and invisible representations of the animals that men would bring home. But in keeping with the changeable character of trickster spirits, Aliŋnaq also, somewhat unpredictably, punished. The pot-bearing women who shouted to Aliŋnaq to give their animals to their husbands also begged for life and health from him; those who had broken taboo, Aliŋnaq struck down with accidents or afflicted with bad luck in hunting. Bad times, when animals became scarce, were thus the result of an uneasy interaction between an irritable deity and imperfect humans who could never predict when their actions might cause offense. Most aspects of this moon lore were shared by Eskimos of both the Iñupiaq and Yupik regions, so while the following narrative was recorded in the Yupik area, the widespread nature of the myth gives it representative status:

> Nearly all epidemic diseases are supposed to come from the moon, but occasionally they descend from the sun. An eclipse of the moon is said to foretell an epidemic... Among the inhabitants along the lower Yukon it is believed that a subtle essence or unclean influence descends to the earth during an eclipse, and if any of it is caught in utensils of any kind it will produce sickness. As a result, immediately on the commencement of an eclipse, every woman turns bottom side up all her pots, wooden buckets, and dishes. (Nelson 1899:431)

Here the rite of catching souls and longevity in women's pots is reversed; the "womb pot" that begged life being also susceptible to sickness from the same source. In another Yupik story, two shamans travel to the moon in an attempt to discover the origin of an epidemic:

> [T]here was an eclipse of the moon, and soon after throat disease caused the death of about a dozen people. Two shamans, father and son, started to visit the man in the moon to find out why the disease had been sent and to learn how to stop it. The pair... returned and reported that when they had climbed nearly to the moon the old man became tired and stopped for a while, but the young man was very much frightened,

> but told the reason for his approach. He was then told that the disease would kill several other people before it would stop. (Nelson 1899:430–31)

In this story the spirit ambiguously predicts that the epidemic will "continue until it stops" without entirely admitting responsibility for its onset. What comes through is that the community *believed* the moon spirit to have been responsible. The same myth, the same shamanistic relations with the moon, existed at Point Hope, and while there are no stories about lunar intervention in that village, Nelson's report suggests that such an interpretation was current during the epidemic period.

The healing arts were practiced in Point Hope by a variety of therapeutic specialists. Some of these pursued healing alone, others cured—and cursed—as part of a broad shamanistic repertory. At one end of the spectrum were the male and female healers and female midwives who massaged pregnant women and performed a variety of perinatal interventions. People with internal, particularly intestinal, pain might also go to these nonshamans for massage and organ manipulation, while others went to them to have headache-generating "old or bad blood" (*augluk*) removed through an incision at the top of the head. Niġuvana and her brother Samaruna, born ca. 1875, were two of the last nonshamans with a range of these skills. Niġuvana practiced traditional midwifery and massage, set broken bones, and reset dislocations. She drained swellings in the joints and, with an incision above the bridge of the nose, treated snow-blindness. Her son Asatchaq said that she could also remove cataracts with her fingers. Using flint and metal blades to make small incisions through which to draw blood, Niġuvana treated backache, stiff knees, and headache. Niġuvana was also a clairvoyant (*tautukti*) whose dreams took her into the spirit realm. Asatchaq shared his mother's visionary dream life. Each composed songs, some of them in the course of dreaming, which they used in communal dances and for healing.

Another semi-shamanistic group were diviners (singular *ivrukuq*) who diagnosed illness in nontrance states using a wooden rod and a sealskin strap (Rainey 1947:275). A patient's head that was heavy when lifted in the apparatus indicated a sickness caused by invasive spirits (*tuunġaq*). Accidents and illness were invariably believed to have been caused by such spirits. It might be perfectly clear that a man had broken his ankle because he had walked on a loose rock. But the same accident would also be attributed to a spirit sent by a shaman, some taboo violation, or the anger of an animal whose soul had been offended. Illnesses that had no traceable cause would also be attributed to the supernatural: a spirit had invaded the body, or the victim's soul (*iḷitqussiq*) had been abducted. Shamans might deal with any of the problems mentioned, but often it was only a fully initiated shaman—one who had successfully enlisted the help of spirits in an initiatory ordeal—who could deal with a disorder that had been generated by spirits. In cases such as these, shamans summoned their own spirit helpers to drive out bad spirits or eject harmful spiritual matter from the patient's body. To this end, shamans drummed and danced until a trance was achieved. While a shaman's body lay in sleep, his spirit went on a journey to fetch a helping animal soul or some other spirit that he could control (Rainey 1947:274). In the course of the voyage, he might have to fight the harming shaman or some other rival. If the patient's soul had been captured by a spirit, he would try to release it so that it could make its way home. Success and failure in this was signaled by either the patient's recovery or death.

In addition to spirit journeys, a shaman's repertory included healing, harming, and hunting songs, massage, and the extraction of invasive substances. Saliva contained healing power, and shamans used this in massage and also for the transmission of songs. Shaman and patient might also chew an amulet whose therapeutic properties were activated by saliva, and this was then applied to the patient's body. Beyond the use of amulets consisting of animal parts, rocks, crystals, or meteoric pyrites, coastal people had few resources that they could use for medicines beyond oils (beluga and seal oil for constipation), root poultices, and powdered driftwood, which was packed onto sprains and swellings. Subarctic and inland Iñupiat had a larger repertory because they could adapt plant products such as artemisia, willow bark, and roots for medicinal preparations. Thus, while these Iñupiat had access to a pharmacopeia resembling that of the interior Indian societies, coastal people depended on more purely shamanistic animal- and spirit-based healing systems similar to those of the central Canadian Inuit. Herein, with regard to Point Hope, lay an inevitable discord. While ritual life in Point Hope had, as mentioned earlier, been enriched over the centuries by ideas and practices from the Indian Northwest Coast and the Alaskan interior, hands-on shamanistic healing remained of a piece with the high arctic tradition. But no Native society, whether it lay within or above the boreal zone, had a medicinal repertory that could protect it from contact diseases. While many indigenous healing systems have survived to this day, this was one of the factors that would lead to the demise of Iñupiaq shamanism.

CHAPTER 14
Sheldon Jackson Takes on Alaska

The Jacksonian Imperative

The drive to get Protestant missionaries into Point Hope and other parts of Native Alaska was spearheaded initially by the Reverend Dr. Sheldon Jackson, a single-minded visionary who by the early twentieth century had brought mission-led education to thousands of Native Alaskans. Jackson's career, a mixture of high-minded philanthropy, religious zeal, and propagandistic intensity, is central to late-nineteenth-century Alaskan history. And while Jackson's initiatives may have been benign in their intent, their major outcome nonetheless was to undermine Native traditions and erode local languages in the vast areas in which Protestant missions operated. To illustrate the background to these, it will be helpful to examine the career of the remarkable man who conceived them.

Born in 1834 to a Presbyterian family near Albany, New York, Jackson developed an early passion for collecting, which in adulthood he expressed by establishing mission stations and churches, and, not least, by acquiring Native Alaskan artifacts—most significantly for the museum in Sitka which still bears his name. Five feet tall, a victim of chronic dyspepsia and an irritable temper, Jackson first had his missionary enthusiasm aroused at Princeton's Theological Seminary and where yet more intense evangelical inspiration came from lectures by Alexander Duff, a missionary who had worked in India since the 1830s. Duff, who championed a Christianzed European curriculum in the schools he had created in Bengal, was the architect of what he called "replacement theory": a process that demanded "pulling down" non-European cultures and then "building up" a new Christian, English-language-based culture (Paton 1923:62).

After three years at Princeton, Jackson was ordained as a Presbyterian minister, and his first job took him to Spencer Academy, a mission school on a Choktaw reservation. The idealistic Jackson was responsible for the oldest "half-tamed" boys, but he found the disciplinary regime inhumane and distasteful (Stewart 1908:34). Poor health soon came to his rescue, and in 1859 he was posted to Minnesota, where he spent the next decade setting up churches for the Home Mission Society. It was during these years that Jackson had his closest and most rewarding contacts with Native Americans.

The next stage of Jackson's career came in 1870 when, despite opposition from the Board of Missions, he assumed control of an immense missionary area in Iowa, Nebraska, Dakota, Idaho, Montana, Wyoming, and Utah. Now that his work had taken him west,

Fig. 27. *Reverend Sheldon Jackson, D.D. Frontispiece engraving to Jackson's* Alaska *(1880). While many other images in the volume are products of a zealous imagination, this portrait gives a vivid impression of the Doctor of Divinity's unbending determination. Scanned from the original.*

Jackson set up an office in Denver, Colorado, and it was at this point that he became aware of the two constituencies that he felt the church should be serving. There were the white settlers streaming across the plains, and there were also the western Indians who in the early 1860s were being persecuted by the U.S. Army in atrocities such as the Sand Creek massacre of 1864 (Prucha 1984:150). By 1872, and with the support of the Presbyterian Home Missions Society, Jackson was establishing the Indian mission schools that he believed might provide some compensation for U.S. Army abuses (Mitchell 1997:64).

Jackson's ideas about evangelical strategy diverged, however, from the Presbyterian mainstream, whose policy it was to send their preachers into the field and just hope for the best. Education, Jackson believed, represented an essential first stage. Only after this "civilizing" experience in childhood, which included the acquisition of English, could conversion follow. "They won't come to hear preachers: send us a teacher" was his pragmatically straight-forward cry (Mitchell 1997:63)—a point of view that John Driggs, Point Hope's missionary teacher, would put into practice twenty years later.

Jackson's First Journey to Alaska

Sheldon Jackson's first trip to Alaska happened more or less by accident. Jackson had been in Oregon in summer 1877, but his projects there had been stalled by the revolt of Chief Joseph's Nez Perces Indians. In Portland, Jackson was introduced to Amanda F. McFarland, who had been recruited to travel north to help a group of Christianized Tsimsian Indians in southeast Alaska. Identifying a new opportunity, Jackson volunteered to accompany Mrs. McFarland, and so it was that in August 1877 the diminutive and irascible Jackson together with the voluminous and benign Amanda McFarland arrived at Fort Wrangell on the steamship *California*. Having "supervised" Mrs. McFarland's installation at a school being run by Clah, one of the Tsimsian leaders, Jackson returned south. He had been in Wrangell for less than a week. But one taste of the north and his appetite had grown prodigious.

Fort Wrangell was a jumping-off point for gold prospectors headed for the Stikine River and a center for Indian, mixed-race, and white fur traders. Rainy and ramshackle, Wrangell's bars, gambling, and bawdy houses sported the kind of moral turbulence that so often excited missionary horror. Here, "prayerless" specimens of a supposedly civilized America encountered Native Americans and infected them with alcohol and diseases. But as fast as white Americans streamed through, the Tlingit Indian populations in the region spiraled downward. These were precisely the ills that pelagic whalers were already inflicting on the coastal Iñupiat.

A letter to the local U.S. Army chief in 1876 from one J. S. Brown, an infantryman, outlined the situation. Calling for a school and mission to be established, he wrote: "Since the advent of traders and miners among [the Indians], lewdness and debauchery have held high carnival and the decimation of their numbers is the result" (Stewart 1908:295).

The same writer described how ten Christianized Stikine River Indians had arrived in Wrangell in 1876 to seek a "new life." Here they had worked as wood cutters, while an army captain had provided them with a room for worship and where they opened a school. In this way Native people themselves had prepared the field for a missionary outpost. It would be the first American station in Alaska, and Jackson somewhat rosily painted his first moments in Wrangell as though the station had already taken root:

> About twenty pupils were in attendance, mostly young Indian women... Soon a thoughtful Indian man of about thirty years of age came in... It was Clah, the teacher. The familiar hymn, "What a friend we have in Jesus," was sung in English; a prayer followed in the Chinook jargon, closing with the repetition in concert of the Lord's Prayer in English. After lessons...the school arose, sung the long-metre doxology, and recited in concert the benediction. Then the teacher said, "Good afternoon, my pupils," to which came the kindly response, "Good afternoon, teacher." (Jackson 1880:140)

While it is hard to credit its accuracy, Jackson's somewhat sentimentalized story does provide a recognizable picture of Alaskan Native devotion: orthodox liturgy, nineteenth-century hymn singing, and prayers in a Native language converging, as they would at many Native missions—including, in the twentieth century, St. Thomas Church Point Hope. Over the next two decades Jackson would travel through much of Alaska, and from 1889 onwards would travel in the summer with the Revenue Cutter *Bear.* This took him on many visits to Iñupiaq villages such as Cape Prince of Wales, Point Hope, and Barrow, whose missionaries he would be instrumental in appointing.

Early Propaganda

Shortage of experience and absence of reliable information did nothing to stop Jackson from writing immediately about the Great Land, and in 1880 he published a book. Lavishly illustrated with miscellaneous engravings, the volume was titled *Alaska*, with the misleading subtitle "And Missions on the North Pacific Coast." Of these Jackson knew only the station at Wrangell and the British Columbian Metlakatla community, which would move to Alaska only in 1887.

Jackson's text was a mixture of hearsay and wishful thinking, much of it quoted without acknowledgment. The book set out to broadcast two opportunities of immense proportion. First, Jackson represented Alaska as a barbarous territory where Native communities cried out for Christian evangelists. Second was the promise of economic reward for this enterprise of the spirit.

On the first count, suggested Jackson, the Natives were ready for missionaries. To underpin this optimism, Jackson claimed that Alaska Natives were not "American Indians." Here he echoed a commissioner of Indian affairs who in 1871 had claimed Native Alaskans "were not members of 'Indian tribes' as the term was used in the Appropriations Act" of July 1870 (Mitchell 1997:71). Eskimos might, according to Jackson, be "low in the scale of humanity" and Tlingit religion "a feeble polytheism," but Native Alaskans were physically and intellectually superior to American Indians, and might, given missionary support, be brought into the Protestant fold.

Jackson's book also delivered a flood of visual misinformation. An "Eskimo Village, Alaska" depicts imaginary engravings of Canadian snow houses, while Natives are represented as degenerate primitives who "leaped and howled around the [cremation] pyre like demons, holding long poles in their hands, which they thrust into the seething, blistering corpse, with dismal cries of '*Wu-we-wu!*'" Another picture shows a "Shaman and Sick Man" in a Plains Indian tipi that has been transposed to an unidentified location. The patient's eyes are turned fearfully at a shaman, who gazes into the distance with malign stupidity (Jackson 1880:92, 110).

A contrasting picture depicts what Jackson wished to accomplish in Alaska. Wrangell harbor, the site of Mrs. McFarland's school, is ringed with European houses, while in the background stand an (anticipated) church and industrial school. Indian canoes, stripped of ceremonial decoration, fringe the beach, where a man in naval uniform strolls with his wife. If, this imaginative engraving implies, the Natives could similarly be "cleaned up," then this region might become as orderly and prosperous as any God-fearing mainland U.S. "section."

A second area of Jackson's promotional activity was directed to Alaska's economic potential. Here was a land whose resources ached to be released onto the American market. Unifying the themes of religion and commerce, Jackson wrote quite presciently that "It would be much cheaper to spend a few thousand dollars in now educating [Alaskan Natives] to citizenship, than a few years hence spend millions to fight them, when the encroachment of the white shall drive them to desperation" (Mitchell 1997:58).

Jackson proceeds to catalogue the furs, fish, cheese, lumber, minerals, and petroleum resources ready for export from Alaska. Even turnips, radishes, and berries, for the profit of both Alaska and the mainland, could be shipped to San Francisco. Alaska was not just a "great country or continent"; it would become America's productive new empire.

Jackson also adduced examples of the "civilizing" process already in train. In Wrangell, Mrs. McFarland, who in 1878 had been joined by a minister, stayed on to open a home, "where she could gather such girls as were in danger of being sold, and train them to be the future Christian teachers, wives, and mothers" (Jackson 1880:217). In this connection, it was also important to the evangelical process that the moral order should be reflected in personal appearance. One engraving in Jackson's volume shows Clah, the teacher, in citizen's dress and with his hair cut short (Jackson 1880:159). But such standardizing, post-Christianizing transformation also had its drawbacks. As Jackson observed of some Indian students: "[The girls] began to comb their hair, to dress more neatly, and keep their persons more cleanly... They became more attractive; and...white men were the more anxious to buy them for base purposes" (Jackson 1880:217). In 1894, Jackson would note a similar sartorial and hygienic transformation at the Anvik mission, and in 1895 Elijah Edson in Point Hope would call urgently for supplies of soap and towels while, with complete justice, deploring the exploitation of Native women by Euro-American traders.

A vivid demonstration of the uneasy coexistence of Native and evangelical beliefs is expressed in Jackson's description of a meeting in 1879 in which Christian converts gave an entertainment representing Native lifeways that the converts had renounced. One participant was Toy-a-att, a Christianized chief, who performed a war dance, then "reappeared in the form of a wolf, and, with a mask, rolling eyes and snapping teeth, gave the dance of the 'invocation of the spirits for success in hunting.'" Following this, Toy-a-att launched into a speech of apology:

> When I was a young man I danced vigorously; now, since I have become a Christian, I have almost forgotten how. When I was young I was a great fighter; now I have learned from Christianity to fight no more. Christianity has changed us. Formerly we thought the [Raven] made us, and made these mountains, and the water; now we know God made it... My brothers, I thank you. You come into this house to see how we used to do. We were foolish. Now we know better now God show [sic] his kindness to us; now he send his ministers to teach us the new way.... Now we forsake the old way. We not like as we used to do—fight, shoot, wound, trouble, all the time.... I fight no more. I give up my spear. All peace, all love now. I have a Saviour. He died on the cross to save me.... My heart is very happy now. I am in bay where no wind; no wind now to upset my canoe and trouble me. I am in a safe harbour. (Jackson 1880:107)

Jackson prints other such speeches. And while the passage quoted may only approximate the original speech and is biased towards Christianity, the text as published—in mixed Northwest Coast oratory and Christian-revivalist idiom—express some of the conflicting energies that Native Alaskan people experienced during this period. Jackson's transcription is of course both artificial and approximate. Nonetheless, two tendencies emerge. On the one hand are the currents of local tradition to which many communities continued to adhere and which remain vivid despite Toy-a-att's apparent rejection of these. On the other is the voice of the convert that Jackson wanted to hear: although within it, behind it, and in Toy-a-att's attire ("his red cloth shirt, covered with mythological emblems worked in white pearl buttons and beads, his embroidered and painted deer-skin blanket thrown over one shoulder") the life of the old ways continues, even as he speaks and dances. As we shall see in Chapter 23, about millenarian Alaska, these conflicting currents take the form of tensions between Iñupiaq shamans and the modernizing visionary Maniiḷaq. In nineteenth-century Point Hope, the dialogue is quieter and represented in terms of a gradual conversion process resulting in the rise of the church, almost literally, over the ruins of the ceremonial houses. As late as the 1970s, however, in Iñupiaq churches of many denominations, the thought and imagery from Toy-a-att's oratory—guilt for past aberration, relief at having found spiritual security—was still quite routinely being expressed in public testimonials. Whatever Toy-a-att may actually have thought or indeed said in that meeting of 1879, and however his cultural identity continued to evolve, Jackson captured an idiom of archetypal significance that anticipates what was to come in the Christianization of the arctic villages.

CHAPTER 15
The Search for Order

Ideal Christian Natives: Metlakatla, Unalaska

Besides their religious and educational fervor, a dominant aspect of Jackson's Alaskan writings lay in a passionate desire for order. And in the absence of territorial government and a resident system of justice, only church and school could provide structures, the finances, and the personnel that might bring this about.

In the decade before Jackson's arrival, there were no Protestant churches in the territory. About fifty Orthodox churches were scattered throughout southern Alaska, but until the arrival of Mrs. McFarland in 1878, there was no American missionary. On the upper Yukon and in British Columbia, a small number of Methodist, Presbyterian, and Anglican missions were active, the most remarkable of which was the Anglican mission at Metlakatla, founded in British Columbia in 1862 by the lay preacher William Duncan. In 1887, Duncan had been forced out of British Columbia, and led his community of Tsimsian Indians to Annette Island, southeast Alaska. This, by 1890, had become a "model" Europeanized village with a unique community of Christianized Natives. Writing to Jackson in 1891, an American visitor described Duncan's church and the cultural transformations that went with it:

> Metlakatla is the realization of the missionaries' dream of aboriginal restoration. The church...can seat 1,200 persons. It has a belfry and spire, vestibule, gallery across the front...organ and choir, Brussels carpet in the aisles, stained glass window...and it is wholly Native handiwork. The dwelling houses are neat and attractive. They have enclosed flower gardens and macadamized sidewalks.... The women weave cloth for garments and people dress tastefully in modern garb. (Jackson 1891–92:876)

"Aboriginal restoration" meant, above all, conversion and moral purity. But mastery of English, European-style education, compliant manners, short-cut hair, and "citizen's dress" signified a repudiation of shamanism and a new regulation of sexual behavior. The "restoration" process similarly involved replacing Native dwellings with frame houses like the church and school buildings. In Point Hope, this latter would be Driggs's dream—in part to provide, in the face of epidemics, a hygienic family environment. Metlakatla did not, in the form Duncan molded it, long survive the death of its charismatic elder, but it was the first and most successfully controlled of Alaska's early missionary experiments.

Wherever Jackson traveled, he noted the degree to which Native society was being brought within an orderly dispensation. At a boarding school on Unalaska, for example, there were

> numerous children…of [absent] white men and Native mothers... Left to the care of poor, ignorant, and often drunken mothers…in destitution and want…it is not wonderful that this central home seems to them the very gate of heaven. Once under its Christian influences they rapidly become an independent, well-behaved set of children. Those who have been inmates of the home for three years or more not only read, write and speak, but seem to do their thinking in English. This is one of the bright spots in the midst of the general darkness and heathenism of western Alaska. (Jackson 1891–92:1430)

Jackson worked fast, and by 1890 he had found an ally in Michael Healy, commander of the Revenue Cutter *Bear* who, at least temporarily, appeared to offer muscle to the secular arm of the ordering process: "Ten Eskimo police [at Cape Prince of Wales] were appointed by Captain Healy...to assist the teacher and take charge of the drunken Natives who might be inclined to be disorderly. These Native police worked with great efficiency and were found exceedingly useful in preserving order" (Jackson 1891–92:875).

Another concerned onlooker was Commander Stockton, who was patrolling north Alaskan waters in summer 1889. Stockton's surprising remarks on the character of some twelve hundred American whaling men that he observed at Point Tanget are worth recording: "I can not speak too highly of the skill, seamanship, courage, and endurance of the whaling masters," he wrote, in contradiction to the generally held view of churchmen and Revenue Marine officers. "They are a fine body of American seamen" (Stockton 1890:181). And Stockton painted another picture of commercial good order at Point Barrow, where Euro-American whalers appeared to work in harmony with Natives:

> The scene on shore was one of abnormal activity for this region, the erection of the house of refuge, the hasty landing and transportation of stores (in which the whalers assisted), the movements of the Eskimos about their village (which was dotted with the white summer tents of the residents and the visiting inland Eskimos), and the clustering and trading about the Whaling Company's station (Ray's old meteorological base), gave a life and movement which was as shortlived as the season... (Stockton 1890:181)

What Stockton would report from Point Hope would, nonetheless, belie this positive appraisal of contact relations.

John Chapman on the Yukon

Apart from Metlakatla, the process of reordering was nowhere more in evidence than at Anvik, on the Yukon River. Christ Church Mission was set up by the Reverend John W. Chapman, whose purpose on arrival in summer 1887 was to evangelize and educate the Ingalik Athapascans who hunted and fished in the lower Yukon valley and who formally sold Chapman a site on the river. At first Chapman worked alone, but with the help of local people with whom he quickly established good relations, he completed a schoolhouse, which opened in December 1889.

Deep in the subarctic interior, Anvik was far more remote and much colder than Wrangell. And unlike cosmopolitan Wrangell, the Native community was more or less homogeneous. There was, nevertheless, a transient Euro-American population, and the mission became a routine stopover for travelers (J. Chapman 1891–92:66–69).

While Chapman's report for his first year's work describes a tidy and efficient project, he was clear-headed about what was a difficult and often unstable situation. Like most missionaries, Chapman performed several roles, and for the first seven years worked as teacher, physician, cook, housekeeper, gardener, and builder. In 1892 he wrote that "conditions of life here are trying to a degree that can hardly be appreciated at home'" (J. Chapman 1891–92:68). Severe winter cold and mosquitoes in the summer, isolation in a forest landscape, and the enormous difficulty of the language represented constant challenges. Then there were epidemic diseases, mainly flu and measles, that moved upriver from coastal settlements. And, wrote Chapman, "it is...difficult to treat the Indians... because they do not obey directions'" (J. Chapman 1891–92:68). The river, too, was vast and threatening. In May 1890, the mission was flooded and was forced to move. "This is vexatious, but seems unavoidable," wrote the imperturbable Chapman. But in spite of daily and exhausting logistical challenges, he also found time to become fluent enough in Athabascan to publish a volume of local folklore (J. Chapman 1914). With the exception of mosquitoes, Chapman's difficulties represented problems that Driggs would also face in the very different environment of Point Hope. And in the 1890s, their very partial evangelical impact would be roughly the same.

While Chapman's early writings are realistic about the limits of what he could achieve, his reports were, on the whole, optimistic. "There has been no great and signal movement, and yet there are cheering signs that the unseen work of the Holy Ghost is being performed," he wrote in 1892. By then he had a small, semi-nomadic congregation and had performed twenty-two marriages. The foundations had been laid for a church building and "the rough and untidy appearance" of the settlement had improved. In an effort to establish a village community around the mission, Chapman had planted a garden and distributed seeds to the Indians—"principally turnip." "There is no good reason to doubt," wrote Chapman, "that they will soon begin to look upon the product of their gardens as an essential part of their living" (J. Chapman 1891–92:68). Since the caribou herds, here as in the Arctic, had all but vanished, Chapman thus promoted both a useful supplementary diet and the pursuit of horticulture, which had long been advocated by writers on the "civilizing" of nomadic hunters.

Chapman himself was also relatively mobile. Once the ice broke in spring, his mission boat took him down river to St. Michaels and up the Yukon in search of pupils and to assess evangelical prospects. Here Chapman found that Russian and Roman Catholic missionaries had reached some Ingalik communities before him:

> It appeared...at various villages, that they have been made acquainted with the chief tenets of the doctrine of our blessed Lord... Of the doctrine of Atonement they seem to have received no idea. They and their children have been baptized, but there appears to have been no such thing as a renunciation of the heathen practices to which they are wedded, and which appear to involve something like the worship of images of animals, as well as a system of sorcery. (J. Chapman 1891–92:68)

This is one of the best statements of the period about the ways in which Native Alaskans adopted new ideas without entirely abandoning traditional thought. As Driggs and Edson would find at Point Hope in the mid-1890s, conversions were almost always partial and ambiguous. Bring a "fallen" white man to religion, and it was possible to rake up from him a consenting view on his fallibility. Native American people never had anything for which they felt they must atone, and the doctrine of original sin was perhaps the most alien of all imported ideas. Indians and Eskimos had been happy and confident in their lifeways. The new God and rituals might be useful. But of all the Christian doctrines, those of sin and atonement were perhaps the most undermining, for while Native people must often make temporary reparation for breaking a taboo, there was nothing in either Ingalik or Point Hope religion to suggest that human beings were inherently sinful.

In 1894, when Chapman was established, Sheldon Jackson, in his new role as General Agent for Education in Alaska, traveled up river and approvingly described both a new church and language teaching at the mission school. Jackson wrote: "A few of them are beginning to talk considerable English… A number of the psalms and hymns have been memorized by the school." And he was impressed by the girls who

> take great interest in housework. Margaret, 10 years of age, makes excellent bread, biscuits, boiled rice, cleans fish or grouse, and keeps everything about the kitchen neat and bright... During the year Dr Glenton [temporary medical assistant] has done much to mitigate the suffering of the people from disease and to teach the younger portion of them more attention to sanitary laws... (Jackson 1894–95)

Scenes such as this are far less exaggerated than Jackson's account of instantaneous piety at Wrangell in 1877. But the picture is still part of a propaganda effort that was directed to sources of funding on the mainland. Children like Margaret of course existed, but what Jackson describes here represents only a tiny and partial detail of life within a small corner of a society, most of which was still beyond mission influence. Jackson's focus, in other words, is on the mission itself and *its* ability to adapt to the Alaskan interior—not the other way around. In accounts such as this, we learn nothing about Ingalik society as a whole or of contact issues that the mission created. But through Chapman and Jackson we do witness the first stages of what the missionaries wanted to achieve, and may observe the extent to which they worked to enclose, within a missionary precinct, the energies of Native people. And this, at Episcopalian headquarters, provided grounds for optimism that Dr. Driggs might run an equally tight ship at Point Hope—which, as we shall see, he did in some departments and in others didn't.

Sheldon Jackson Goes to Washington

After his second trip to Wrangell in 1879, Jackson relocated to Washington, D.C. where he campaigned for the new cause that he had adopted: Christian education for Natives in a U.S. territory that was still without civil government. *Alaska*, the book he published in 1880, and which he distributed to likely congressional supporters, contributed to his argument and to his own (self-proclaimed) credentials. Inflated as Jackson's 1880 writings appear today, his book brought together two big themes that both stirred the imagination and stimulated political action: the spiritual needs of Native Alaskans and the advantages of exploiting Alaska's resources.

To enthusiastic audiences in Washington, Jackson now argued that to educate Native Alaskans—and thus to bring them peacefully into the mainstream of American society—would be both prudent and beneficial. The background to his thought lay in a government statistic of 1870 that the Indian wars (thus far) had cost the U.S.Treasury "$1 million per dead Indian" (quoted by Mitchell 1997:79). "It would be much cheaper," wrote Jackson in 1880, "to spend a few thousand dollars in now educating [Native Alaskans] to citizenship, than a few years hence millions to fight them, when the encroachment of the whites shall drive them to desperation" (quoted by Mitchell 1997:58). Furthermore, said Jackson disingenuously—still innocent perhaps of the epidemics that had already destabilized so many Alaskan societies—the Natives of Alaska were a healthy and self-sufficient people:

> It is not necessary that the United States should feed or clothe them, or make treaties with them. This enables our Indian policy to take a new departure; and treat them as American citizens. All that is necessary to be done is to afford them government and teachers, which they cannot procure for themselves. (Quoted by Mitchell 1997:82)

These views argued a relatively unthreatening extension to Indian policy and set up a new horizon of U.S. territorial responsibility. Dragging Alaska to a point where it loomed more distinctly in the northwest offered Congress the opportunity both to act with enlightened future self-interest and also to respond with decency towards people who might later threaten them. Hurrying between lectures and fund-raising events, writing letters and articles in astonishing profusion, Jackson quickly earned celebrity in government and church circles. Besides his church supporters, Jackson had two particularly helpful allies. The first was the Presbyterian elder Benjamin Harrison, who in 1889 would become U.S. President. The second was Senator Orville Platt, who, with Harrison's help, introduced the Organic Act of May 1884. It was this act of Congress that took a civil administration to Alaska, and in Section B of the Act, thanks to Jackson's lobbying, provided funds for Alaskan schools. In April 1885, Commissioner of Education John Eaton appointed Jackson as First General Agent of Education in Alaska. Jackson had arrived, and education in Alaska was delivered into his custody.

Jackson's so-far-unpaid Alaskan labors were now finally recognized by the Presbyterian Board, and in 1885 he was made supervisor of the Alaskan Presbyterian missions. That year, in the course of six months, Jackson transferred his headquarters from Wrangell to Sitka and consolidated the new Presbyterian school there into the Sheldon Jackson Institution: an industrial or vocational school for both local and boarding students that theoretically would serve the whole territory.

The Comity Agreement: The Division of Alaska

With a governor, a judge, a district attorney, a marshal, and a town clerk installed in Sitka, and with $25,000 of federal money allocated to schools, the territory was divided into four administrative districts: the Sitkan (southeast), Aleutian, Arctic, and Interior. In the first three years of his new administration, Jackson concentrated on settlements along the southeastern trade routes. But while he made a genuine attempt to provide schools for children in both Native and white settlements in the Southeast, this limited provision fell a long way short of his proclamation in 1880 that the churches of the United States must obey the call of "God's providence…to enter and possess the land."

Jackson had, in fact, been possessed of far greater evangelical ambitions when in 1880 he had convened a group of East Coast church leaders and devised the First Comity Agreement for Alaska. Comity agreements, in which groups of church administrators met to share out territory, were a well-tried method of cooperation between Protestant missionary societies. The aim of Jackson's meeting was to divide up as much of Alaska as a select group of church officials could think of: the aim being to cover maximum ground, avoid duplication, and to exclude the Catholics and Russian Orthodox. The territory was thus by 1880 theoretically carved up into the following missionary spheres:

Southeast Alaska and Barrow	Presbyterian
Yukon River	Episcopalian
Kodiak and Cook Inlet	Baptist
Aleutians	Methodist
Kuskokwim River and Nushagak	Moravian
Cape Prince of Wales	Congregationalist

(Jackson 1890b)

This, Jackson believed, represented an ideal division of the Protestant missionary force. But since only a small part of Alaskan geography was known at the time, only a fraction of the territory was considered. Anyway, none of the church societies had the funds to put more than a handful of missionaries into this remote field.

Beyond Jackson's immediate sphere of influence in the Southeast, the first outlying missions came little by little. In a newspaper article of 1890, Jackson summarized the position:

> Another advance is being made in Alaska… In 1877 Presbyterian Missions commenced in south east Alaska. The Moravians followed in 1885 in the valley of the Kuskokwim, and in 1886 the Nushagak. In 1886 the Episcopalians and Roman Catholics commenced in the valley of the Yukon. In 1885 the Friends established themselves at Douglass Island (southeast), and the General Government commenced a series of day schools. In 1887 the Swedish Lutherans opened missions at Unalakleet and Yakutat; and in 1889 the Methodists entered Unalaska. (Jackson 1890b)

Missionaries to the Arctic

At this stage, the Arctic was beyond evangelical reach. But this was changed in summer 1889 following a cruise to Point Barrow by the U.S. Navy steamer *Thetis*. In his newspaper article of 1890, Jackson outlined how Commander Stockton had called his attention to the "degraded condition of the Esquimaux in Northern Alaska." Jackson accordingly reported to W. T. Harris, U.S. Commissioner of Education, who entered heartily into the plan of establishing schools among them. The region being so remote and barren, Dr. Harris preferred placing the schools under the supervision of well-known missionary organizations (Jackson 1890b).

Stockton's correspondence has been lost. But a sketch from his biography outlines what happened:

> [Stockton] stopped at a town named Tigara...and there found a colony of Eskimos, living in a most degraded state...brought about, it was said, by their extreme isolation and the contaminating influence of white men who came there in their whaling ships. The impression made upon Commander Stockton's mind...was so great that he felt compelled to communicate the facts to the Board of Missions of the Protestant Episcopal Church, with the hope of...establishing a Church mission at Tigara to remedy if possible the appalling conditions prevailing at that time. To this end he wrote a strong letter setting forth what he discovered and urging the immediate and pressing need of establishing a mission.... The Board was so deeply impressed...that the question of establishing a mission at Point Hope met with a unanimous response, and at their meeting held in June 1890, Dr John B. Driggs of Wilmington, Delaware, was duly appointed... (Benjamin 1925:18)

The Moravians, who were already active in southwest Alaska, had been offered this part of the territory, but they were already overcommitted. Jackson described how he visited the missionary societies in New York City, with the result that the Presbyterians took Point Barrow, the Congregationalists would support a mission to Cape Prince of Wales, while Point Hope would receive a man from the Episcopal Domestic and Foreign Missionary Society (Jackson 1890b).

Initial funds from the Organic Act of 1884 were limited to $25,000, and Jackson's work underlines the importance of private donations in the funding of missionary societies. The schools within Jackson's new demesne would be so-called contract schools. This meant that government money for building and materials had to be matched by salaries and subsistence from the church societies that were sending their people. For the mission stations to remain open, Jackson must therefore energetically continue to raise funds. Among the Episcopalians, fund-raisers such as the Women's Auxiliary would remain vital.

Jackson himself undertook to advertise the north Alaskan job openings. On March 13, 1890, he issued a notice to the religious journals: "Please publish this Call. The time is so short I need the help of all papers friendly to the effort for Civilizing and Christianizing the Eskimos of Alaska" (press cutting in Jackson 1890–1900). At the same time he published an article with this headline:

> CHRISTIAN TEACHERS WANTED FOR MISSION SCHOOLS AMONG THE ESKIMOS OF ARCTIC ALASKA. These Schools will be Government "Contract Schools" under the management of the Board of Home Missions of the Presbyterian Church.

This half truth was followed somewhat casually by a brief reference to the Episcopal Missionary Society's search for an appointment at Point Hope.

All Jackson wanted at this stage was "two or three gentlemen teachers" for the posts at Cape Prince of Wales, Point Hope, and Barrow. Twenty-four persons volunteered, and Thomas Lopp and Harrison Thornton were chosen for Cape Prince of Wales, "Professor" Leander Stevenson for Barrow, and Dr. John Driggs for Point Hope. Marcus Cherry, a carpenter, was also chosen to work alongside Chapman at Anvik as his assistant; after one year, he retreated to a sanatorium for the insane in California.

The contract school arrangement of course violated the principle of church and state separation. This was well understood, and its illegality was waived because Alaska was so far away and too expensive to otherwise administer. Thus institutional Alaska, a world away from the constitutional niceties of Washington, D.C., had finally got going.

With contracts in place, the missionaries were set to travel. They would convene in San Francisco in early June 1890, and, with materials for school buildings and provisions for a year, would steam north on the schooner *Jeanie* to Port Clarence, Alaska, where they would transfer to whaling ships, which would take them to their different destinations.

By arrangement with the Treasury Department, Jackson would be traveling (to Driggs's disgust!) for most of the journey in a comfortable state room on the *Bear* and would be on hand to help each missionary settle at his new station. This would be the first of Jackson's annual tours in the company of hard-drinking Michael Healy—an unlikely colleague for the abstemious Jackson. But the collaboration between the two men would continue through the mid-1890s, and on return from his first arctic voyage Jackson wrote:

> As enviable as is [Healy's] record in the relief of shipwrecked sailors, his record is equally brilliant in protecting the thousands of half-civilized Natives from the rapacity of the white man, and the demoralization that comes from the white man's rum...even some whalers were not above trafficking in the accursed stuff. The temptations were very great when a bottle of whiskey would purchase $500 worth of furs, and the profits were a thousand fold. At that time the demoralization of intemperance was threatening the extinction of the Native race. Through the vigilance, firmness and tact of Captain Healy, this trade has been almost broken up on the American side of the coast: so that no part of the United States is today so free from intemperance as Arctic Alaska. (Jackson 1891)

Jackson's remarks on "intemperance" were unrealistic. But the following year saw the beginning of the two men's most famous venture: the introduction of Siberian reindeer to Alaska's Bering Sea region, a daring and altruistic attempt at alleviating the Iñupiaq subsistence crisis through the introduction of a ready supply of meat and skins. The history of this project remains outside the scope of the present volume, but the initiative ultimately foundered because Jackson had assumed that subsistence hunters could (indeed *should*) be converted into herders. As to Point Hope, while its first two trainee herders traveled to Teller in 1894, it was not (to Driggs's dismay) until 1908 that one of these men brought his stock to the Tikiġaq region. The Point Hope herd was managed until 1920 in association with Kivalina's; but prey to parasites and wolves, and on account of the difficulty of managing the grazing, the herd dwindled, much of it having joined the caribou. Point Hope's herders preferred, in the end, to be village-based subsistence hunters (see Chapters 10 and 16).

CHAPTER 16

John B. Driggs: Medical Missionary to Point Hope

Landfall and Exposure

Dr. Driggs arrived at Point Hope in July 1890, and his first letter to his employers was scribbled on the south beach where he slept rough for almost eight weeks. Of the many letters he tried to write during this period of exposure, this stained and battered page was the first to survive:

> I arrived here on July 11th and have camped on the beach ever since, my house did not arrive until the night of the 12th, the erection of the house was started on the 14th and finished to day, my freight is s-even miles from here and will be brought arround just as soon as the s-ea becomes smooth enough... (Driggs, Letter of July 21, 1890)[1]

Shortly before he struggled ashore, Driggs had trained a telescope onto the south beach from the whaling ship *Narwhal* to see "a number of Natives, and, what seemed best of all, a white man was reported."

Driggs already knew of the station's existence, but from missionary writings about wild white westerners, he had expected to meet only what he called "the worst elements." "From all accounts," he had written from San Francisco, "they are a very hard set I am going among; especially the whites" (Driggs, Letter of June 5, 1890). But confronted now with the prospect of isolation in a Native village, this sighting of Jabbertown calmed his anxiety. The view was soon made concrete. Once Driggs had been rowed ashore from the *Narwhal,* he was met by Koenig, Hachmann, Marlin, and some Point Hope hunters. Among these was the shaman Anaqulutuq, Point Hope's new "chief," whom Commander Stockton had appointed the previous summer. All these encounters would lead to mutually supportive long-term friendships. But Driggs's first full-length letter focused on logistics:

> I had upon landing my things carried up the beach to the grass, and laying down a part of the 1,000 foot of boards that were for the mission, I placed my boxes on them forming a small inclosure just large enough for me to crawl into and sleep nights. Then over all I threw the sail that Capt McGregor had so generously offered to loan me. That night I did not sleep much, for every little while a corner of the canvas was pulled up and an Eskimo would take a good long look. (Driggs, Letter of Sept. 25, 1890)

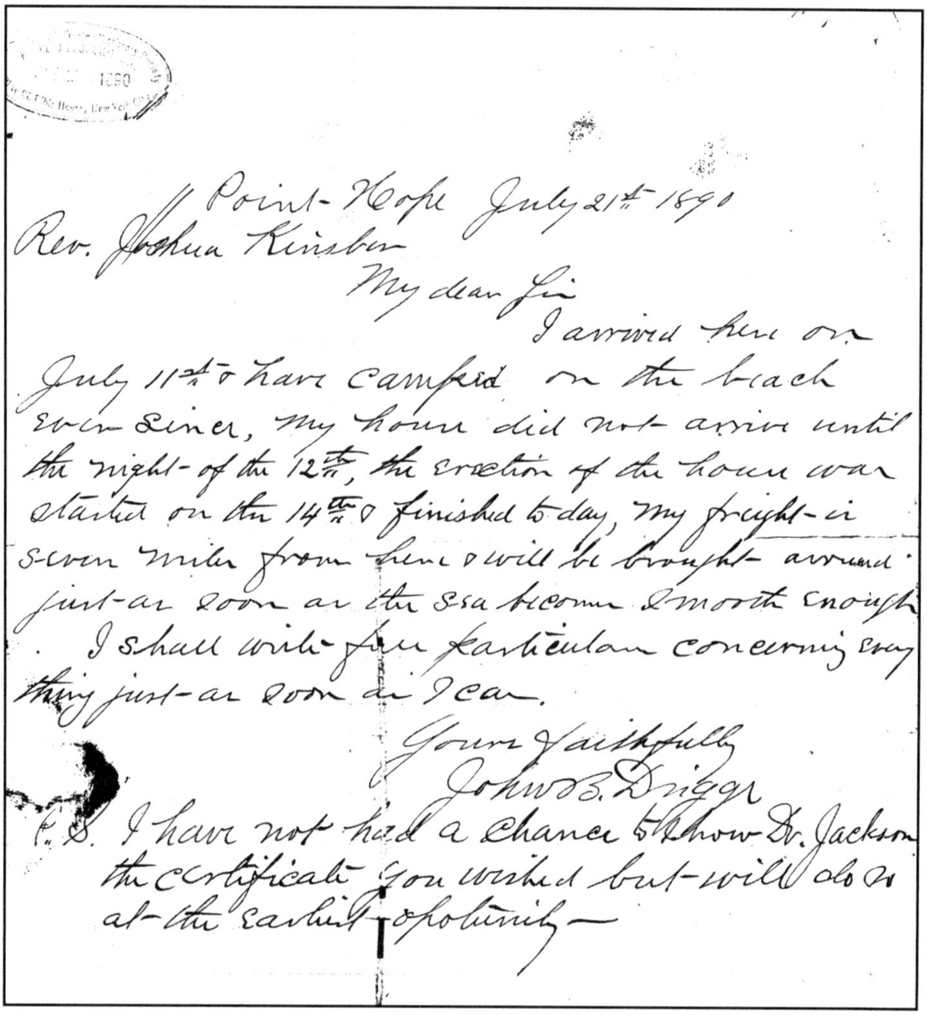

Point-Hope July 21st 1890
Rev. Joshua Kimber
My dear Sir
I arrived here on July 11th & have camped on the beach ever since, my house did not arrive until the night of the 12th, the erection of the house was started on the 14th & finished to day, my freight is seven miles from here & will be brought around just as soon as the sea becomes smooth enough. I shall write you particulars concerning every thing just as soon as I can.
Yours faithfully
John B. Driggs
P.S. I have not had a chance to show Dr. Jackson the certificate you wished but will do so at the earliest opportunity—

FIG. 28. *Driggs's first surviving letter from Point Hope expresses the bare facts of his exposure with typical understatement, but the hyphenated "s" of "seven" perhaps expresses haste or anxiety. Archives of the Episcopal Church, Austin, Texas.*

Driggs's coal arrived on the whaling ship *Orca* four days after his own landfall, and eight miles from his camp on the beach. With the help of four Native men and Hachmann, Driggs dragged it above the beachhead, where it lay for the next month until the bluffs collapsed in a freak summer storm and his coal was washed away. The remarkable Anaqulutuq, cruising tirelessly through the inshore water, salvaged all but six of the doctor's bags of fuel.

Driggs had instructions from Jackson, who was at this moment traveling north on the *Bear,* to plant the mission in the village, or at least on its outskirts. Point Hope was about six miles from where the doctor was encamped. But Driggs was too anxious and exhausted to make a decision. Besides, he knew nothing of the place, and no one had arrived to build the house where he would live, teach, and open his surgery.

During the second night in his camp, Driggs was woken by an Eskimo carrying a note informing him that Captain Haverside[2] of the supply ship *Oscar and Hattie* was anchored off the point. Next morning, guided by an unidentified man whom he described as a "sound Native," Driggs walked seven miles to the north beach. On this third morning since his arrival, there were at least half a dozen men in the village, and Anaqulutuq recruited enough hands to paddle Driggs out to the *Oscar and Hattie* in a skinboat. This was the first of several off-shore trips Driggs made that summer. Anaqulutuq's skinboat leaked, the big ship was rolling in heavy sea off the north side, and Driggs and his companion only just managed to swing themselves on board ship, where they both spent the night.

Driggs was a six-foot-tall outdoors man in prime physical shape, but these first days shook him in ways he had not expected. And besides the stress to what, in a letter of July 25, he called his "iron constitution," a mysterious conflict with Sheldon Jackson troubled him. The nature of what occurred between the two men remains obscure. They had met in San Francisco, and perhaps even gone looking for supplies together. But now in the Arctic, Driggs felt neglected. One point of contention was that Jackson had spent time with the Revenue Cutter men at Cape Prince of Wales while the mission there was being constructed, but he failed to come ashore to Point Hope. In truth, even at Cape Prince of Wales, in addition to a few men from the *Bear*, the carpenters had been volunteers from whaling ships that were in the area. Meanwhile, Haverside had traveled ahead, and now with Anaqulutuq to advise them, the two white men went ashore to discuss where the mission should be sited.

Now in his third day, Driggs's uncertainties continued. He had been convinced, on the one hand, by his new white acquaintances to locate near Jabbertown, where there was access to drinking water in nearby ponds. On the other hand, Jackson had ordered Haverside (on what authority is unclear) to build near the village. There was sense in both arguments. Driggs wrote:

> I wanted the house on the other side nearer the station but [Anaqulutuq] advised me quite strongly to come over here. The location is about the best one on this side: about one mile from the settlement, and has the advantage of a good supply of drinking water, a thing the village does not boast. It is true that there is a long narrow swamp running thro the place but it also runs through the graveyard and is not fit for a man to drink. Dr Jackson writes me that it is too far away and that he will recommend the removal of the house into the village. (Driggs, Letter of Sept. 25, 1890)

In the end it was Anaqulutuq who settled the matter. Leading Driggs and the captain to a spot between the village and the white men's station and scraping a hole in the gravel, he revealed a puddle that would provide water at least during the summer. With this decision made, Haverside's men began hauling lumber inland. A desperate and exhausting period followed. Driggs wrote:

> While Capt H. was erecting the Mission I was present all I could be. My things were all on the south side and had to be attended to as there was no one to look after them but myself. The station men were asleep at night. On the other hand the Natives were prowling around, it being perpetual day. It was very fatiguing and hard work to walk seven miles down to Tigara through a coarse gravel, then a mile across to the house,

> standing around all day then walking back at night to sleep in my little enclosure. (Driggs, Letter of Sept. 25, 1890)

It rained, snowed, and hailed. While the carpenters raised the mission, Driggs shuttled between house site and beach, lying awake at night in his flooded enclosure. After some days, Haverside invited Driggs, Anaqulutuq, and his wife to sleep on the *Oscar and Hattie,* where the rough weather trapped them for twenty-four hours. Anaqulutuq's wife was badly seasick, and Driggs, who could see how far the surf was rising, was consumed with yet more anxiety about the things he'd left exposed on the south side. Happily, Marlin hauled Driggs's boxes to safety. His initiative paid off. Marlin and his colleagues would need the contents of those boxes before winter set in.

On the following day, when Driggs and Haverside came ashore, the shell of the mission had been completed. A strange scene followed. Fearful that he had built Driggs's house on the wrong site, Haverside "requested me [Driggs] to walk around to the side where no one could over hear. He then asked me if I would be willing to assume all responsibility—in the selection of the site when the Mission was created." Driggs reassured him that he "did not see any thing wrong as it was undoubtedly the best site that could be chosen taking every thing into consideration" (Driggs, Letter of Sept. 25, 1890). Haverside then asked Driggs to write him a letter of recommendation to Josiah Knowles, the powerful director of the Pacific Steam Whaling Company. In return, Haverside promised to take Driggs out to the newly arrived *Bear* to visit Jackson.

In fog and high water, Haverside struggled a mile out from the *Oscar and Hattie*, Driggs strenuously bailing the leaky dinghy. Jackson was still in bed when they arrived—even though, as Driggs tartly noted, it was "after seven." A chilly scene took place between the exhausted doctor of medicine who was about to face his first winter in the Arctic and the comfortably insulated doctor of divinity who would, after his cruise on the *Bear*, be heading back to D.C.

"I presented my [teacher's] certificate but had no time to converse," wrote Driggs to headquarters, after which point he insisted that he should, for safety's sake, do the return trip in one of the *Bear*'s dinghies. "So," Driggs concluded, "my interview ended in nothing" (Driggs, Letter of Sept. 25, 1890). It is difficult to know exactly what Driggs wanted from Jackson. Permission to locate near white men? An acknowledgment from Alaska's newly appointed Agent of Education of Driggs's teacher's certificate? Whatever its cause, the ill feeling between the two men troubled Driggs's first year, and although the relationship slightly improved when Jackson visited Point Hope the following season, Driggs continued to feel deeply critical of Jackson's neglect—most particularly over the next years on the issue of Point Hope's promised reindeer herd, which failed to arrive until 1908 (see Chapter 15).

Back on land, Driggs continued to sleep rough. His house was ready, but his supplies still lay on the south beach and he dared not abandon them. The stress intensified. The doctor's nights remained cold and sleepless:

> Exhausted by my long and severe walks...the worryment and exposure that I was enduring was telling on my health for even the cast iron constitution that I possess could not in the long run stand the strain. My back has not yet recovered the strain it

> went through. I'd seen my things were being gradually ruined. Yet I was powerless to prevent it. (Driggs, Letter of Sept. 25, 1890)

Driggs's luggage held large and varied supplies. He had 1,550 feet of lumber and compendious equipment for medical, domestic, school, and missionary life: clothes, books, educational supplies, Bibles, drugs and medical equipment, a year's supply of food, oil lamps, lamp oil and chimneys, outdoor tools, photographic gear, an iron stove, and the fifteen tons of coal he had purchased in British Columbia. Without these essentials, both his first year's work and his own survival would be in the balance.

It rained and blew, and Driggs was still sleeping in his stockade until he was finally "forced to beg a place to sleep in the [whaling] station." Here Driggs was put up "with a rough board bunk where I lay nights with one thickness of a sail cloth under me…and one thickness over, the rain dropping from the roof onto my bed. Even that was preferable to trying to sleep [among his boxes] under the sail. The nights that I could sleep were few and far between" (Driggs, Letter of Sept. 25, 1890).

Still, Driggs could not move until his boxes could be transported—either by land across the peninsula or by water around the point. The seven miles overland might have been possible by dogsled had there been snow. But the snow did not come until September, and so Anaqlutuq promised that he would bring some skinboats on the first day the water was negotiable.

"At last such a day did arrive," wrote Driggs. And an astonishing account of his next ordeal followed:

> I hurried to the tent of the "sound Native," and told him the chief had arrived. I loaded the chief's boat first and sent a white and colored man [McKenzie] with it. Suddenly my heart seemed to jump into my throat as I heard Anaqulutuq say, "Hurry up, the wind is beginning to blow." I started him off and then hurriedly loaded the second canoe and started to walk across land…so as to overtake the first and arrived at the mission just in time to see it upset in the surf…. The men were thrown into the water, a wave passed over the white man's head before he could regain his feet, my things were thrown out, my trunk went floating off but was afterwards regained, my little handbag that contained my papers and other things was entirely lost, also quite an amount of my supplies. I felt desperate. Just the accident that I had waited so many weeks to avoid had overtaken me. Then the second canoe came in. I rushed down and grabbed its side, determined that it should not meet the same fate as the first.
>
> I told the men to keep the stern out, but they did not understand. A big roller came, they all jumped and ran. I grabbed the side of the canoe, the long tow line had been carelessly left hanging over the side, and the water wrapped it tightly around my legs. I felt the canoe receding and knew that my only chance was to hold on. I could feel the cord tightening about my legs. I let go the boat. It would pull me off my feet, then the next wave would throw the heavy weight on top, and end my existence. All the others had ran away. I could not let go with one hand and draw my knife to cut the line.
>
> My only chance was to hold on. I must have exerted a strength of three or four hundred pounds. Then another wave came and I felt the tension relaxed, and jumping back, then men returned, and together we pulled the boat up out of the reach of the waves before it had time to recede. (Driggs, Letter of Sept. 25, 1890)

Had Driggs's next eighteen years been less remarkable, the doctor's trial by storm, his losses, and, not least, his struggle with the surf, ropes, and skinboat might read less portentously. The physician had arrived on this ancient but changing peninsula, in several of whose transformations he would participate. Driggs would change, too, and this second landing was a part of his initiation.[3]

Angry with Jackson, Anxious About Trade

Throughout this turmoil, Driggs continued to be angry with Jackson. His resentment was, in part, misplaced. Whatever Jackson's shortcomings, he did in fact visit Point Hope. Ten days after Driggs's arrival, Jackson wrote in his journal: "On July 21, the storm having abated, I went ashore to inspect the school building, which was in process of erection... Captain Healy sent his carpenter and a numbers of sailors ashore to assist in the work... This is the second of our new schools in the Arctic" (Jackson 1890–1900).

Jackson then walked the mile separating the mission house from Point Hope village:

> While at Point Hope I visited the Native village Tig-e-rach (Finger); but few of the people being home. I also visited the cemetery. The dead, tied up in deer and walrus skin blankets, are laid on platforms above the reach of dogs and wild beasts. The present population is about 300. [after the sea fight at Cape Seppings]—leading hunters killed and a famine set in which carried away half of the remaining inhabitants... (Jackson 1890–1900)[4]

Driggs moved into the mission house in late August and was seized by further anxiety:

> the whale boat that had part of my goods on board was upset in the surf and my losses were a great deal heavier than I could afford. I even lost the sail that Captain McGregor of the *Narwhal* loaned me. I am exceedingly sorry about the sail for Captain McG was very kind... (Driggs, Letter of Sept. 25, 1890)

Driggs now wrote from the safety of his house, but he faced a new emergency. It was autumn and he urgently wanted to mail his letters before the ice cut him off. The *Bear*, on its return journey, was his best chance of a carrier. Ideally, he would also get aboard for some final consultations. This was not to happen:

> I am writing this in great haste—A ship is in, and I am in hopes to get this letter so I can give you a full detail of all my transactions...
>
> I ran six miles and spent a night on the beach trying to get my letter off on the *Bear*, but she sailed at 3 a.m. much to my surprise. I tried to attract the attention of the officer but probably they did not see me. (Driggs, Letter of Sept. 25, 1890)

With these ordeals still hammering in his system, Driggs closed his correspondence for the season with a brave gesture of optimism. He was consoled by one thing at least: "My house is quite a comfortable one, and I am ready for my scholars any time they arrive back from their summer hunting season" (Driggs, Letter of Sept. 25, 1890).

There were still worries and discomforts. The chimney leaked and his rooms were drenched each time it rained. He had lost supplies and was also unsure how he was going to subsist. On the other hand, he had already been encouraged by both Native and Euro-American friendships. And while Driggs preferred not to live in the village and saw the Jabbertown men as his more natural companions, he never expressed fear. The one note of pathos lay in his erroneous assumption that he would be joined by a fellow worker: "Be sure and send me up a good man next year, tell him he need not be afraid of the Natives... He must bring up lots of underclothing. For outside clothing and shoes he will dress the same as the Natives do, in furs" (Driggs, Letter of Aug. 27, 1890).

Driggs's other anxieties were about property obligations and the vexed issue of trading. First, there was Captain McGregor's sail. While Anaqulutuq was making a final attempt to land, he had thrown the canvas overboard. Regret is followed by unaccountable shame at having borrowed the sail in the first place: "It is true I will pay for it but would much have prefered giving it back, it seems such a poor way to repay his kindness." The loss of the sail then elides with fear that he will fall into the habit of trading—an activity forbidden to Episcopalian missionaries (see Appendix C).

To comprehend this, we must return briefly to San Francisco, where Driggs had spent two weeks in the summer. Fresh off the train, with very few funds beyond a small advance from the missionary society, Driggs had to equip himself for both his school work and his personal subsistence. San Francisco, home to the Alaska Commercial Company, the Pacific Steam Whaling Company, and S. Foster and Co., three of the powers in the whalebone and fur trade, was booming. Driggs had to tour the chandlers, provisioners, and hardware suppliers alongside whaling men, prospectors, and other miscellaneous roughnecks for his year's equipment. Hard up but resolute, Driggs wandered between San Francisco's entrepreneurs and managed to make his necessary purchases:

> [T]he books and materials are all of the simplest kind and will last two years... I carry from S. Forster [sic] and Co about seven pounds of manila paper which will also be for the use of the school room. In winter there are two months and a half when there is no day and that will be the time when the scholars will attend best, so I ordered two lamps with extra chimneys and twenty gal.of oil for the school room, the stove and fifteen tons of coal are ordered, along with the flag pole...(Driggs, Letter of June 5, 1890)

Something of Driggs's innocence and the extent to which Point Hope represented unknown territory to the outside world is revealed in his discovery that: "I find there is no trading store at Point Hope and therefore had to increase my personal order from S. Forster [sic] and Co."

Driggs also believed that he might have to build the mission with his own hands:

> There are no tools at Point Hope for the fixing of the foundation and erection of the house, so Mr Knowles will send them up, also a thousand feet of cheap rough boards for me to make the benches and the small outhouse, all these boards I will have to plane by hand, it will give me a great amount of work and its fortunate that I will arrive there during the [summer daylight]...

> I expect to sail on the *Jeanie* tomorrow evening or Saturday morning and hope to reach (D.V) Point Hope by July 1st. From all accounts they are a very hard set I am going among; especially the whites. (Driggs, Letter of June 5, 1890)

On the surface Driggs appears serene. But his casual remark on the absence of a "trading store" at Point Hope would develop later into a moral crisis. Following the doctor's loss of the sail canvas come these distressed pages:

> There has [been] one thing worried me a great deal ever since leaving San Francisco... Mr Foster said that I would have to take articles for trade as there is no money used in this country. That what work and furs I would require, also my meat and fish supply, would have to be paid for in trade goods. I had made up my mind to do all my own work, and as for furs I would wear my regular clothing with a double suit of under wear, the idea of trading was distasteful, I therefore refused. (Driggs, Letter of Sept. 25, 1890)

This confirmed to the Board of Missions that he would follow unwritten Episcopalian guidelines on the issue of trading. But not all the missionary societies took this line on trade:

> Mr Stevenson who was going to Pt Barrow for the Presbyterians told me that Dr Jackson...advised him to take trade goods for use with the Natives. That he would find it an impossibility to get along without it. I also learned that Mr S was going to act as agent for Joseph's of San Francisco in the collection of curiosities. Knowles also spoke on the subject and I refused. He spoke again and said that it would be impossible for me to get along without it. I told him that I did not think it would meet with the approval of Dr Jackson or Capt. Healey. He then told me that they carry every year quite an extensive supply of trade goods for their individual use.
>
> I then informed Capt. Knowles that I could not afford to do so. Besides I did not care to assume responsibility in that line. He said not to let that bother me but to call the next day. I did so and he showed me a list which I looked over and saw nothing objectionable. He said that he would fill it out and send it to Point Hope and that any furs of any value that I might trade for I could send down to him and he would allow me one third. I thought by taking his offer it would pay for all [my fur clothing and payment for Native help]. At that time I was under the impression that I would have to erect the Mission myself... Mr Knowles probably profits. (Driggs, Letter of Sept. 25, 1890)

While Driggs had not fallen among rogues, these paragraphs suggest vividly the inexperience of a man thrown suddenly into a world in which commerce was a way of survival. Badgered to supply himself with a trading outfit, Driggs had been let into an alarming truth: every white man in arctic Alaska was a trader. This was the means of making one's way. Stevenson and Jackson did it. Indeed, as suggested by Driggs's canny remark about Knowles's "profit," many churches were involved inextricably with the traders about whose activities they had such moral scruples. Missionaries were being dispatched, after all, in an attempt to neutralize traders' influence on Native Americans. The Pacific Steam Whaling Company, of which Josiah Knowles was director, was never complicit in purveying liquor. Driggs was therefore shocked at what Knowles told him. Still, if trade could open the way to charitable work, then the profit motive might be justified:

> Again I thought that it might pave the way to a small hospital project I had in view on an exceedingly economical basis, all of their things as I had them arranged in my mind would not have cost over 75 dollars, I therefore told him I would accept his offer only that he must make the amount small.
>
> On the day I left he showed me the bills, he had put four times the amount on board that I had expected, it was noon and I had been informed that the tender would sail at 4 P.M. What was I to do, it was too late to countermand and throw up my agreement, Mr K had been very kind I did not wish to hurt his feelings, my trunks were on board, I was in a dilemma and did not know what to do, I did not have time to write the full particulars to the Board in NY, my intention had been to keep them always informed concerning all my transactions… (Driggs, Letter of Sept. 25, 1890)

Thus trapped with his trading outfit, Driggs suffered torment:

> After getting into the mission the subject presented itself in its true light. You can't imagine how much I think over it these days when I am alone, when if even a Native should pass by it would be a relief. But Tigara is deserted and its inhabitants will not begin to return until October. At least twenty times a day I find myself saying, If Mr Knowles had only left me alone. It is not that I have traded for I have not. And you can rest assured that I will not.
>
> What worries me is that I thought of doing so…I did wrong when I told Mr K that I would accept. The Almighty would have presented some other means for my getting along. I feel very much as David must have when he wrote the 51st Psalm, only our offences were different.[5] (Driggs, Letter of Sept. 25, 1890)

Here, for the first time, with an evocation of the Almighty, Driggs writes with missionary piety. But given that the doctor was, in the end, deposed for participating in commercial whaling, and in the light of later dealings that he tried to keep secret, it is painful to witness this early struggle.

The Almighty did appear to intervene. The storm that had delayed Sheldon Jackson's arrival on the *Bear* prevented the *Thomas Pope* from delivering supplies to its Jabbertown customers. Refusing to put out its boats, the tending vessel simply took off—in Driggs's view, an act of cowardice and moral dereliction. Days later, when the *Pope* was wrecked, with heavy loss of life, a way was opened for Driggs to offload his trade goods and thereby ease his conscience.

Following the wreck, the Jabbertown men proposed traveling to Kotzebue to pick up supplies at the Hotham Inlet. Driggs intervened:

> Happily…the aid that I was able to give them made the trip unnecessary, for I had told them they could have any thing they wished out of the extra supplies Mr Knowles had sent me. Only they must pay Mr Knowles themselves and that my name must not appear in the transaction in any way. (Driggs, Letter of Sept. 25, 1890)

That Driggs could transform his trade goods into a gift was a moral relief; and "that [his] name must not appear in the transaction" removed him further.

Driggs's moral predicament can be understood in the context of Episcopal Church thinking of the period, and the background to this is outlined in Appendix C. Here it is sufficient to emphasize two main points. First, while there was no specific prohibition in the Church Canons of the time against missionary trading, there existed a milieu of disapproval for the improprieties that trade, in particular with Native people, could lead to (Episcopal Church 2006). In this, as we have seen, the Episcopalians took a different line from the other Protestant churches, including the otherwise more austere Presbyterians as represented by Sheldon Jackson. Second, looking ahead to 1908, it was, in the end, precisely Driggs's entry into the baleen trade that led to his deposition, as recommended by Peter Trimble Rowe, Alaska's new Episcopal Bishop. In the context of this, it is sadly ironic to read Rowe's tribute to Driggs's fortitude in resisting the temptation to engage in trade when, just before the turn of the century, he wrote: "I was also glad to hear men, as far north as Point Barrow, say that Dr Driggs was the only [true] missionary in the country, because he attended strictly to what he was sent to do, minded his own business and did not combine trading with missionary work" (Rowe 1899:88).

Thus for the time being, the doctor had saved his reputation. But it was his act of friendship towards the Jabbertown traders that autumn which would, over the years, gradually lead him to join their culture. The story of this is given in Chapter 26.

Notes

1 Driggs's punctuation often expresses hurry and emergency, and while I have standardized the punctuation and spelling in many of his later writings, this first letter is as the doctor wrote it. Driggs's hyphenated "s" crops up throughout his correspondence. What this expresses is hard to interpret. Most of Driggs's correspondence was addressed to Dr. William S. Langford, Secretary to the Episcopal Board of Missions, and to Rev. Joshua Kimber, Associate Secretary, Church House, New York City. Driggs had only slight personal acquaintance with these and other Church House officials.

2 Haverside had traveled in the Arctic since at least 1884, when he and Charlie Brower were employed by the Pacific Steam Whaling Company to operate the coal mine at Corwin Bluff north of Point Hope.

3 Anaqulutuq remained Driggs's friend for many years, but the remark in English above is his only recorded utterance. Driggs did, however, record a significant moment when they traveled together and the shaman attempted to change the wind's direction. See below Chapter 19.

4 Three subsequent journal entries suggest Jackson wanted artifacts for his Sitka museum, the World's Exposition in Chicago, the seminary museum at Princeton, and also his personal collection. In July 1891 he wrote in his journal: "we anchored abreast of the [Point Hope] school house at 9.15 pm. As late as was the hour, several umiaks full of Natives came on board. I purchase 3 white bear and 1 black bear skin and 1 white cub. Also some small curios." From the outset, Driggs, along with other missionaries, had been engaged in trade for Jackson. Given his fear of being revealed as a trader, this arrangement contributed to the tension between Driggs and Jackson, and by 1893 Driggs had withdrawn from the arrangement. Jackson continued in his 1891 journal: "Dr Driggs and I visited the Native village. As the mission was short of provisions, I sold the doctor 3 boxes of Pilot Bread at $5.70, and left with him 4 boxes of bread to be traded with the Natives for a collection for the Columbian Exposition at Chicago in 1893."

In 1895 Jackson wrote from Point Hope: "Bought a large number of masks at 2 to 3 cents each." These can be seen today at the Sheldon Jackson Museum in Sitka. The scale of Jackson's trading is indicated in this journal entry: "July 27 1891...anchored off Cape Blossom. By afternoon there were 12 umiaks fastened to the ship and about 300 Natives on board—An active trade was had in black and brown bear skins and white and red foxes—I secured 12 black and brown bear skins, 32 white fox, 15 red fox, 10 lynx, 20 mink, 1 land otter and 1 child's parka. The Natives received in exchange for their furs, flour, powder, caps, lead and muslin drill, which they greatly need."

5 The verses that Driggs had in mind were: "Have mercy on me, O God, according to thy loving kindness... Wash me throughly from my iniquity and cleanse me from my sin. For I acknowledge my transgressions: and my sin is ever before me... The sacrifices of God are a broken spirit, a broken and contrite heart..."

CHAPTER 17
First School on the Arctic Coast

Dr. Driggs spent September weatherproofing his house, making desks and school benches, and cutting driftwood. The house was in two parts: a bedroom-kitchen and a school room that would double, as Driggs would later complain with some professional justification, as a surgery. The entire double-walled space, which would be heated until the following year with a single wood-and-coal-burning stove, measured just over forty by twenty-two feet.

For much of his first month at the mission, Driggs had the company of Anaqulutuq and his wife Mammaŋniiña, and in an strangely belated article published in 1906, the doctor described how the "chief" had stayed with him, and that this "was the beginning of a friendship that has lasted ever since." Driggs continued:

> The loss of the coal supply had to be made good ...so Anaqulutuq and I would shoulder our axes and go to beach cutting and stacking driftwood... We could not converse together except by making signs and faces: but soon we became expert enough to understand each other; besides, I was beginning to pick up a few Native words. The cutting of driftwood was not an easy task; but what was far worse, was that I had to board my assistants and be their cook...[1]
>
> Both my Natives had good appetites, but especially Anaqulutuq, his capacity for food being simply enormous. At last one day I decided to try the experiment of filling him up. So taking my largest iron pot and cutting up five ducks and then mixing them with a generous supply of potatoes, I made a pot full of stew...and soon the pot was empty... (Driggs 1906:7–8)

If Driggs disliked cooking, at least he had meat and the company of powerful and therefore helpful and protective Point Hope figures. This anchored his position in the village, and represented, apart from uneasy episodes with later missionary colleagues Edson and Knapp, the only instance of domestic life that Driggs recorded for next two decades.

By the end of the month, Driggs was ready to open school:

> At last that day arrived, bringing with it a snow that lasted nine days. Three or four families had returned. My school room was fixed. What wood I had accumulated was placed behind the stove. That morning I sat in my little kitchen listening eagerly for the first arrival. But as time wore on, no one came. (Driggs 1891)[2]

On these moments hinged a turning point (a phrase used by Driggs much later in his career; see Chapter 25). To the west lay the village. Five miles east were the Jabbertown traders. His old life behind him, the doctor sat poised to introduce his place of education. But nothing would happen without further initiative:

> So putting on my furs, I was starting for the village, when I spied a boy walking on the beach. I called him in and set him to work on the alphabet.
>
> That afternoon he progressed as far as the first eight letters, after which I presented him with two or three pancakes left from breakfast and told him to come the following morning. He has been since one of my best pupils.

News of pancakes soon spread to other children:

> By seven the next morning [the boy] returned, bringing two other boys and two girls. That evening after school they all waited around expecting something to eat. I had nothing to give, but promised if they would return the next day I would see what could be done.

Driggs understood the pull of food. Though the issue of trading still troubled his conscience, Driggs recalled a conversation with Jackson:

> He had advised that each [missionary] include in our next year's orders some sea biscuits, and at noon give each child one. The advice recurred to me and I made up my mind to put it into execution and see if any benefits were to be derived.
>
> So mixing flour, molasses and water together with baking powder, I made a sort of cake, which proved to be quite popular and aided me greatly in building up my school. It has been my only means of discipline, for I made it a rule that every scholar had to be in the school room at least two hours before closing or their name would not be entered as an attendant that day, and therefore they would not receive any cake. The rule has worked well and has gradually enforced an attendance that somewhat approaches regularity.

Before long, Driggs had established a routine and clearly relished his work in the schoolroom. His earliest accounts offer vivid and delighted glimpses of children in action. The mission is a place where they come for diversion and Driggs seems happy to look on with detachment. While he plays host, it is they who are at home; and while Driggs will be their teacher, it is he who is learning:

> My first scholars usually arrive at any time from six to seven in the morning with the exception of the long night when the latest I ever knew them to come was twenty minutes after nine.[3] I unlock the door as soon as I hear the first footsteps on the frozen snow. They always greet me with *Alapah* ("It's cold!") although they do not seem to mind the cold. On their first arrival they will gather around the stove; then in ten minutes you can see them running out bare headed, cutting slices of snow and eating it. I tried to keep the school room furnished with water, but they drink it up faster than it can be melted. I have never seen such water drinkers; they seem to look upon a cup of water as a luxury. Perhaps it is their long-continued diet of frozen meat and

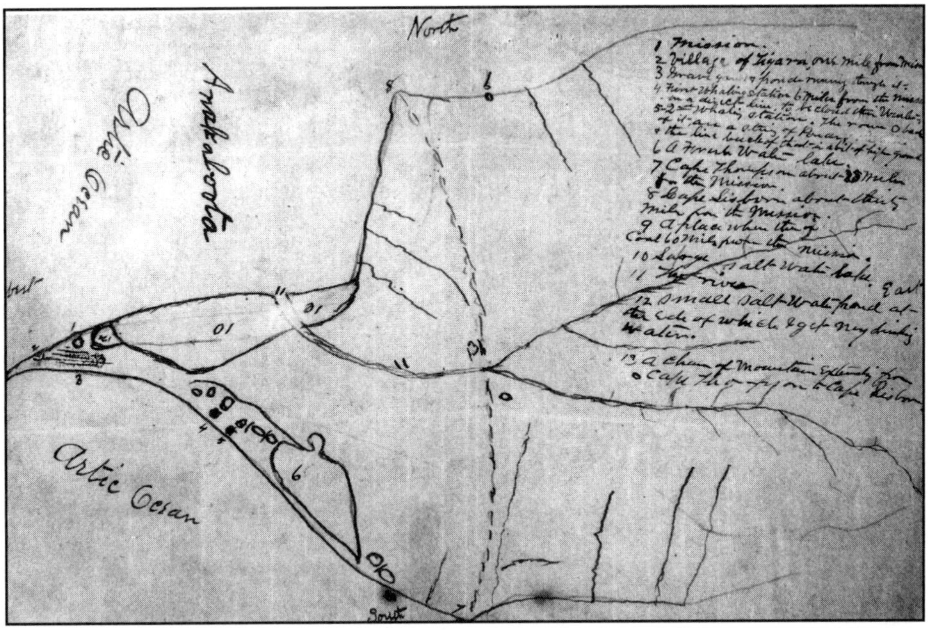

FIG. 29. *Map of Point Hope region drawn by Anaqulutuq with Driggs's annotations. The accuracy of the peninsula outline, also showing Capes Lisburne and Thompson and the Kuukpak River, is outstanding. Point Hope, Jabbertown, inlet, ponds, lagoons, and beach ridges are the major features. Archives of the Episcopal Church, Austin, Texas.*

> snow-eating that creates the thirst. The meat diet certainly conduces to severe nose-bleeding; every day you can see several scholars running around the school room with a plug of deer hairs sticking out of one or both nostrils.

Since the school was a mile northeast of the village, school attendance was partially dependent on the weather. On calm days the boys would hunt or play, along with girls, outside. In bad weather, they walked to the mission and spent their hours in warmth and comfort. A letter of April 1891 evokes those times clearly:

> The stirring up of the ambition of the children did not occur until February when blizzards set in, lasting into March. The storms are beyond description. The ice with its big mounds, extending as far as the eye could reach, was driven clear from the ocean on the north side. The Mission at times seemed taxed almost to its full capacity to withstand the fury of the gale; the snow flew by in perfect sheets; the Natives could not go on the ice, neither could children play out of doors, so they flocked to the Mission. I was quite uneasy concerning the little ones of seven or eight years of age, and hired men the first night or two to see that they reached home safely, but found the precaution unnecessary. Weather can not be too bad for an Eskimo.

Driggs had no teaching experience, and one advantage of this was that he did not feel obliged to follow educational convention. While Lopp and Thornton at Cape Prince of Wales tried to introduce school discipline, Driggs—presumably realizing that he had no choice—allowed the children almost complete freedom. They walked round, ran outside, lay on the floor, chatted; as long as they did some of what Driggs asked, he was happy to let things evolve. But as Driggs admitted, this flexible approach did not address fundamental issues. He still knew no Iñupiaq language. The children spoke no English, had no conception of what school meant or why Driggs wanted them there (a condition that Driggs may have partially shared!). The basis of his success lay in his amiable self-confidence, his affection for the children, and the realization that he could never engage everyone's attention simultaneously. In the light of this, Driggs knew he must improvise—one method being to invent games that involved some competition:

> During those big storms, while the attendance was larger than usual, I made extra efforts to interest the children… I had tried in every way to start a class in arithmetic, but they would not become interested… I had tried in various ways to instruct them in addition, but they would not learn. At last, one day, more in desperation than anything else, I remarked that not one should touch a slate until I was through. That excited their curiosity. Then taking the slates, I wrote on each one a few small examples, and said I was going to see which one brought me the correct answers first. Immediately the arithmetic class became the centre of attraction, each scholar being anxious to be the first. Even the visitor became excited and urged on their favorites. The rest of the children became anxious to study arithmetic. But it had its price. They must first go through the primer twice, then learn to write their names, besides count and write figures up to 100. From that time on I could notice the improvement. It is remarkable how readily these people learn... All that is necessary is for them to become interested.

The classroom visitor to whom Driggs refers was a common feature of his daily routine; and since anyone was welcome to watch what was happening, sometimes these spectators became temporary adult students. Driggs did not record how he divided his time during these early months. But what he does give us is a sense of the improvised and irregular rhythms of those long sessions. Besides arithmetic, there were exercises in reading, writing, drawing, and singing. The doctor described these in some detail:

> My singing class has been a great aid as it nearly obliterated the alphabet class…I wrote the numerals up to eight. The children laughed, but soon became interested, and tried to sing. I was struck by the very limited range of their voices. So calling each one up, tried them individually. As soon as they saw what I was going to do there was a general stampede. You could see little Eskimo heads sticking out from under the benches and tables; but not one tried to leave the room. As each one was called by name they would come out of their hiding place laughing, and make the effort; for they are very obedient. After practising their voices for one or two weeks, the first tune taught was the alphabet, arranged for singing, the first verse being the capitals written in regular order, the second the small letters written backward.

Here Driggs had arrived intuitively at a path into Point Hope culture. "Limited" as Driggs found children's voices, everyone had grown up singing, and the musical tradition of the village was always expanding. Anyone who introduced a new song—whether they were a Point Hope shaman, a child who had added to the repertoire of children's game songs, or as here, a missionary—was almost certain to make an impact:

> The tune proved itself very popular and was sung all over the village. I heard of one young man seventy-five miles away that had learned the alphabet by hearing it sung by some of the Tigara people. He had also procured some of my little slips of manilla paper on which simple words are written to drill the children before advancing them into the primer.
>
> The writing class was also interesting. I gave them two bottles of ink, but they always examined each and decided which they considered the best. That one is used, the other avoided. Each scholar has a pen bearer whose duty it is to dip the pen into ink. They also prefer choosing their own positions. Looking over in one corner of the room, I saw an object just rising above the table; it was a girl of fourteen. She was on her knees and elbows writing. In another place, a boy lay face downwards on the floor. He also was writing…
>
> The reading class has been the principal one and the progress made by many of the children has proved quite satisfactory. It has also aided me in learning the language, which is not easy to acquire. I can speak it to-day nearly as well as any of the [Jabbertown] whites. Yet I am not able to carry on a conversation of any extent. Not any of the Natives speak English, so, each word learned has to be through inquiry.

Aware, somewhat anxiously, that he had yet to raise the subject of religion, Driggs also declared that "[i]n the next winter's work, I have included Sunday instruction." Otherwise, Driggs appears to have been serene, and summing up his nine months' work, he wrote:

> Although the first part of my Missionary career seemed to be one of trials and perplexities, and at times discouraging, yet I have succeeded in establishing a school that I am beginning to look upon with considerable pride, and can safely say that never has there been any undertaking where I have found myself so thoroughly interested. There is scarcely a night but that I fall to sleep thinking of my children and what I can do in the future to benefit their condition…I can not look back to a single day since the first of last October when I have had the least feeling of loneliness...
>
> There is something fascinating in the work, and no matter how tired you may be there is always some object ahead…and which nerves you to keep pressing onward…

These thoughts prompted the next season's plans, and Driggs well understood that school attendance was dependent on both harmonious relations with his pupils and circumstances in the village. He had to be tolerant; but given that his character combined both mildness and determination, this came naturally. Driggs also understood both the irrelevance of punishment and the value of his daily provision of cake. Besides this, the doctor showed an immediate awareness of the population crisis. The previous spring's whale hunt had been a failure. The caribou had all but vanished. To the south there was famine and Jabbertown was filled with families in search of work and food. Point Hope's iglus were, as yet, far from empty. But if things did not improve the doctor might find himself in a deserted village:

> The population has been small this winter. As nearly as I can estimate, about one-half are away: some at Point Barrow, quite a number of others were wrecked at Icy Cape while on their way to the Mackenzie with Captain Bayne. Also there are a number of families at Point Lay, as well being scattered around at other places. The cause of their absence was the scarcity of food last year. But as those that were here this past winter have fared very nicely (the seals being abundant most of the time, also they have killed many bears and quite an amount of deer) therefore it is probable that most of the absent ones will be present next winter.

In contrast to this optimistic note, Driggs provides an initial, informal village census:

> In taking the census this month I counted everybody from the oldest down to the youngest and found that there were only 161.[4] Out of that number I have had 68 pupils. Among those 68 were two adults who ultimately deserted the school to hunt. Four other young men also deserted. All the rest I have retained as scholars. Three I have lost by death. One of these was carried off on the ice and has not been heard from since, and probably froze to death and was eaten by the bears. Another one lost her life from frozen feet. I was asked to amputate but the blizzards rendered it impossible for me to venture out in the dark. The third scholar died from sudden haemorrhage.

Apart from passages of overt crudeness, perhaps included to shock the Board of Missions (see also Chapter 18), Driggs has only sympathetic and positive things to say about Point Hope. He may find people "dirty." But they are clearly not dangerous or "savage." Most important, the doctor and his new acquaintances discovered that they positively liked each other:

> As a people I like them very much in all but their uncleanly habits. In my years of experience in the tenement houses of New York City I have never witnessed such sights. They are naturally well disposed and by education will ultimately become a splendid people. There are some perhaps who are disposed to evil, but they are very few…I have not the least personal fear and go among them freely, even into their houses and have always been well treated. It may be that they are learning to look upon me as their friend. The children seem to be very much attached to me, and I have certainly become so to them.

By the end of nine months, Driggs had visionary plans, which he would never fulfill, for Point Hope's development:

> One of my dreams for the future is my mechanical class. There is no natural material to work with, but cast up on the beach, is a large amount of driftwood that could be sawed into boards. The winds here are very high and blow almost constantly… Why can not I utilize the wind power to saw up the boards? Certainly no wind mill that I have ever seen could resist the blizzards. The wheel would have to be so constructed that it could withstand the heavy gales in summer and also be housed in winter…

Another initiative that never came to fruition was to remove everyone from their iglus and establish them in frame houses:

> With the first boards I will construct a workhouse, after which I will build houses for the Natives... The cost of the houses will simply be, two hinges, two or three panes of glass and a few nails. It would prove an excellent way to instruct the mechanical class, besides giving the Natives warm, comfortable and more healthy homes than their present mode of underground living, which is undoubtedly doing a great deal towards exterminating the race.

This is a chilling observation, which derives from the fact that in unventilated iglus, the epidemic diseases that afflicted Point Hope were impossible to stem. But the doctor was wrong in assuming that airtight and probably colder wooden houses would improve matters. Driggs then blandly adduced these disturbing statistics:

> Their death rate is out of all proportion to their births at the ratio of three to one since the first of January. Previous to that date it was yet larger. It seems a pity—for they are really the only fit ones to populate this section of our country. If Congress could be induced to send a medical commission up here to investigate the subject, or else a medical man appointed to each mission it would do a great amount of good. I do not doubt but that in the end it would achieve grand results in the way of saving the race from extermination.

While the initially exciting phenomenon of the school must have drawn children from traditional pursuits and reduced their contribution to community life, the busy and informal atmosphere of Driggs's school regime nonetheless led to friendships in wider Point Hope society. Driggs's relationship with his pupils' families, partly cemented by his medical interventions, became a further ingredient in his schooling enterprise. Point Hope people brought their children up to be tough, but while traditional learning regimes were arduous, children were seldom punished, and to abuse a child outside one's family would have represented a dangerous affront. In harmony with this milieu, it was in Driggs's character to treat everyone with decency. To shout at, hit, or detain "disruptive" children was not in his repertoire. The doctor therefore seems to have made no enemies in the village, and his social initiation appears to have been rapid and harmonious.[5]

Confident in the worth of what he had already created, Driggs could write with a commitment based partly on experience and in part on a somewhat exaggerated optimism:

> By my third year I expect to have everything in splendid working order. What books are required will be here; all improvements that are necessary will have been made. I will thoroughly have mastered the language, and it's my intention, God willing, to make it the Golden year of the Mission. The fourth year I wish to start the Mechanical class.

And the doctor could claim of his first nine months:

> Our Mission is the pride of the Point, being the first real house that most of the Natives have ever seen, they call it Iglopuk (big house). Any visiting Natives are always brought to the school. If I happen to go out of the room for a moment, the children gather around, then read, count and sing to the visitors. They take considerable pride in showing themselves off.

But school was only one aspect of Driggs's activity. He also worked afterhours as a physician. Some patients came into the schoolroom and waited all day for classes to finish. But many of his calls came from outlying areas. These Driggs visited, sometimes laboring fifteen miles on foot, only to find that the patient had moved or required painful and exhausting surgery; and while, over the years, his surgical skills saved the lives of many, the majority of the sick lay beyond his help. The doctor recorded one such visit to Jabbertown:

> At the first station (Koenig's) the colored man McKenzie died early in December. He had been laid up with rheumatism. My visits were of necessity few and far between. The last time I saw him he cried, and seemed very glad to see me. I tried to comfort him and promised to call soon and extract some teeth, but in a few days a heart complication put an end to all his sufferings. He was buried in the snow when his own dogs found and ate part of his head.

Despite wearying journeys and few days of rest, Driggs relished the exercise of his medical expertise. Patients in need and the landscape the doctor traveled through to find them are dramatically evoked in several of his letters:

> [T]he tribe east of here at Noatak are dying off very rapidly from some form of epidemic, probably catarrhal pneumonia. My own medical experience here has been quite extensive. At times I have five or six patients in a day. If the patients are unable to walk, they are brought in a deer skin bag on a sled; if they are too sick to come out I go to see them after school. When they live long distances I have go on Sunday.
>
> The long walks, anywhere from twelve to sixteen miles, wading through snow and then along the beach is anything but agreeable. It takes me about a day to recover from the fatigue. The thermometer always registers considerably below zero; each eyelash has its bead of ice... eyebrows and moustache also have their icicles.

As we shall see later, Driggs was also an enthusiastic hunter and this provided an additional incentive to travel. He shot wild fowl and the occasional caribou and longed for big game opportunities:

> My rifle is always my companion on these trips during the short days and the long night. I consider it best to be armed on account of the liability of meeting bears. Although I have seen but two living ones far out on the ice, I know they have been near the Mission as I have seen their tracks...

Of course arctic life would always be stressful. But now that he had settled, Driggs enjoyed coping with challenges that were seldom straightforward:

> The house also gives me satisfaction in all but the formation of ice which covers the walls, ceilings and everything else: even the kitchen and school room, which are kept comfortable through the day, have their limited share. Then when the thaw set in I was flooded for over a week and was forced to bore holes through the floor to allow the water to escape. However I have plans... One is ventilation...large attendance of people at the Mission and the steam arising from them, besides their breath, along

> with steam formed by the fires, all condenses on the walls and ceilings in the shape of ice. This lasts from four to six months…

Fuel gathering presented an overwhelming problem:

> October has been the most trying month... My wood supply gave out the first week of school, and not wishing to use the coal, I tried to get Anaqlutuq to bring more, but his only reply was Wait, the snow was too soft. By and by his wife should bring tremendous loads. There was only one thing left to do, and that was, each night after school to take a rope and walk half a mile up the beach, hitch the rope to a log and drag it through the soft snow to the Mission, then cut and split it for the following day's supply, making a double trip Saturdays...

Driggs also had support from Jabbertown:

> Both Mr Marlin and Hackmann have my full gratitude for when they heard of the trouble I was in. They came for a day and hauled wood. Mr Hackmann stayed with me for two days and sawed enough to last until the snow was suitable. Those trips… from the Station, then three trips across to the South beach and returning home in the evening necessitates a tramp of about twenty miles.

Another solution lay in barter with Point Hope people:

> One of the worst mornings a woman came with a sled load of wood and asked if I would give her a little flour for it. I was glad to do so, my wood having given out... I therefore gave her six heaping [measuring] dippers. She seemed much pleased at my generosity, and the news spreading through the village caused a perfect influx of dog sleds to the Mission. After they had brought enough to last through the summer, I had them stop although they were very unwilling, food being scarce. I should have allowed them to continue if it were not that it was making too serious an inroad into the flour I had laid by for my children.

Water was also scarce. Before freeze-up in November, Driggs could use the spring near the mission house that Anaqlutuq had shown him. But learning the difficulties of water acquisition and differences in local water quality, Driggs could now write discriminatingly:

> I have depended on snow through the winter... Twice ice has been brought, but as it was…inferior to the snow… The nearest good water lake is six and half miles from the Mission, but requires a walk of eight miles in winter. There has to be considerable discretion used in selecting the snow for melting. That which comes with the north winds is formed by the vapours arising from the ocean, the water being brackish. The snow which comes with the big blizzards from the S.E. yields the best water.

Driggs found neither the cold nor the darkness depressing. The winter, on the contrary, was filled with drama and spectacle:

> The last time I had seen the sun was on Dec 7th, but its total disappearance probably did not occur until the tenth. It was about the middle of the long night that boy came running late to school, saying that there was a bear on the ice in front. I immediately took my rifle and went out, as it is not desirable to have any of those troublesome creatures prowling around. No bear was to be seen, but looking far in the south, just above the horizon was the midwinter sun. The sight was very realistic and for a time I was under the impression that the sun had returned. The true return did not occur until Jan. 3rd, which gave us a night of twenty-four days…
>
> I lit the first lamps on Nov 12th and had to use them constantly up to Feb 9th. The day is made up of a twilight that travels from east to west and becomes more and more faint as the night advances... Nature tries to make up the deficiency with brilliant Auroras, also the Moon does its share by not setting for several days at its full… In November I witnessed the rising of three suns.[6]
>
> Taking the winter as a whole it has not been nearly as severe as I anticipated, and I have passed through it very nicely. Our coldest weather was when the thermometer varied from 27 to 30 or 31 below. The average was probably about 15 below. A few miles from here on the main land the thermometer registers very much lower.

In the fullness of these descriptions, it is easy to forget that they represent a mere seven months of work. By late April, when the whale hunt had started and the children were involved with their family skinboat crews, Driggs's self-confidence came to a climax: "It has often been my wish that this Mission was a larger one for I believe that the same principles worked out here would make any Mission success." And in five other letters that season, he made ambitious plans for opening a mission on the Kotzebue Sound—a project that materialized later, independently of the doctor. In the meantime, on July 4, Driggs started assessing school attendance: "My monthly average gradually improved up to the whaling season, which has played sad havoc with my school. All the adults and children go on the ice. However they will all return to me again." Presumably in compliance with a Board of Missions requirement, Driggs calculated aggregate figures for monthly school attendance, the maximum for the year having been 851 (March), the lowest 179 (June), with daily attendance varying between six and twenty-five students.

In passing, Driggs also recorded an important statistic about population decline and its impact on the whaling season. Not only were the Jabbertown stations recruiting from southern Iñupiaq regions, but given emigration from Point Hope to Barrow and Herschel Island, even a successful *umialik* such as Anaqulutuq was short of a skinboat crew: "Men were so scarce this year that the two [commercial] stations had to call on the tribes further East for men and women, even the chief [Anaqulutuq] was forced to look to the other tribe to fill out his boat crew of seven men."

As we shall see later in more detail, these observations throw light on changes in village demography and whale-hunting practice. Meanwhile, a more mundane but still gratifying task was to order school equipment for 1892, and Driggs's requisition for that year gives us a valuable impression of material realities in his classroom:

> 24 Primers and Readers, 12 geographies of the simpler kind, 12 US histories for beginners, 2 doz slates and a box of pencils, 24 bottles of ink small size, 2 doz pencil tablets,

> 10 lbs. Manilla paper, 6 doz lead pencils, 35 gal. oil for the lamps, 3 lamps and 3 doz extra chimneys with a package of wicks, 5 barrels cheap flour, 15 tons coal, 3ft galvanised stove pipe and 2 elbows.

In contrast to the anxieties he had expressed about trade, Driggs is most at ease when planning the logistics of his classroom venture:

> This I believe will give me a very good supply for my third year. The manilla paper I have found very useful in the school room, the galvanized stove pipe I hope will stand this terrible climate better than the other which has become destroyed the first winter... moisture and freezing [are] hard on slate as well as the lamp chimneys.
>
> The mission being built of the California redwood makes it quite dark during the long winter, and therefore the necessity for the extra lumber. Dr Jackson advised that at each mission, every day a cracker be given to all the children. [This] seems to have aided me very much this past winter, therefore I have ordered the flour and will do the baking myself. I would suggest continuance for the first four or five years, or until the children [can] appreciate the benefit of education...

And in one of his tantalizingly unfulfilled promises, Driggs continued: "There is much of interest here that I hope to send you an account of later on." But Driggs's letters were to become increasingly, sometimes defensively, thin. Accounts of people, places, and events become scanty and his generalizations vaguer. In these early letters, the doctor does drop fragments of local interest, but these are seldom followed up. So for example: "One thing that interested me recently was to find that the Eskimos are believers in a God whom they call Chunamikchoa [the moon spirit *tatqim iñua*?] and to whom the devout pray." Driggs offers no further information, perhaps believing that such material might convince the Board of Missions that he would, relatively soon, be able to replace an apparent Native monotheism with Christianity. In this he would be pursuing the replacement theory espoused by Jackson. Thus the doctor leaves us with a somewhat meager account of local belief systems about which, over the years, he must have learned a great deal and which, often inadvertently, he had begun to replace with what he had brought with him.

Notes

1 By the time of writing (1906), Driggs may have forgotten that he had represented Anaqulutuq as speaking some English in July 1890 (see previous chapter). It will be remembered that Anaqulutuq had been appointed "headman" by Commander Stockton and that one of the qualifications for this must have been his ability to communicate with Euro-Americans.

2 All subsequent quotations in this chapter are from Driggs's letter titled "Leaflet on Alaska, Point Hope, April 21, 1891." An edited version was published as "Dr Driggs' Work at Port Hope (sic), Alaska," *Spirit of Missions* LVII:99ff.

3 But Driggs also wrote: "[After] housework, my breakfast was eaten between seven and half past in the morning, the next meal between eight and nine in the evening, no intermediate lunches, for always everything left from breakfast is begged away."

4 Contrast Sheldon Jackson's exaggerated figure of three hundred Natives on board ship in the previous chapter.

5 This contrasts with the experience of the missionaries to Cape Prince of Wales and Point Barrow who had traveled to Alaska with Driggs in 1890. Driggs commented nine months into all three missions: "I was sorry to learn through the Natives that the Mission at Cape Prince of Wales had been attacked, also letters received from the two gentlemen, Mr Lopp & Mr Thornton, mentioned the fact. They remarked that at the time of writing they did not consider themselves under any probable danger. The Natives also report that there has been trouble at Point Barrow." Harrison Thornton was murdered at Cape Prince of Wales on August 19, 1893; the motives for the killing are uncertain, and presumably only tenuously related to attempts at classroom discipline.

6 These are "parhelia" or "sundogs," caused by the refraction of light through ice crystals in the atmosphere.

CHAPTER 18

The Hidden John Driggs

The Early Life of John Beach Driggs

Some years after Driggs's death, an oddly misleading account of his arrival at Point Hope circulated among Episcopalians.[1] According to this story, Driggs was met by armed Natives who refused to let him land and it was only the intervention of the U.S. Revenue Cutter that convinced the warriors that "Dr Driggs was a different sort of white man than the men whom they had met on the whaling ships, and that he had come to heal their bodies and bring happiness to their hunger-bitten lives" (Goodman 1932). This sentimentalized view of Point Hope people with its alarmist representation of "savage warriors" is as surprising as it is wrong. Why should anyone fabricate such a tale when Driggs's own account of his arrival was already so dramatic?

The legend is nonetheless interesting in that it belongs to a culture of misrepresentation to which, as we have seen, even Michael Healy in his relations with Ataŋauraq and Jackson, as recounted in *Alaska* (1880), partially subscribed. For Driggs, too, was an eccentric whose story was not entirely straightforward; and this aspect of his character is revealed only in writings whose content never reached his employers at the Board of Missions.

Driggs's papers consist of two groups of documents. The larger group contains letters to the Board about mission business. A second, very much smaller, collection contains letters to a friend in Delaware and a private journal for the years 1894 to 1901. These latter papers are all that remain of a considerable private correspondence.[2] From these two collections two different characters emerge. The first, as expressed in reports to the Board, is governed by the proprieties of his profession. In the second, a freewheeling bohemian, in flight from middle-class convention, emerges.

John Driggs's parents were descendants of English and German immigrants to America who had removed to the Bahamas in the mid-eighteenth century, and thence to Cuba, from where they exported sugar and molasses to New York. John Jonathan Beach Driggs was born in December 1852 on a small family estate in northeastern Cuba. Despite a substantial American presence at the time, Cuba remained under Spanish control in the mid-nineteenth century. Around 1854, while Driggs's father Samuel stayed on in Cuba to dispose of the estate, the Driggs family, including the two-year-old John, returned to the United States, where they settled on a farm in Delaware. Samuel, having briefly joined them, then set off on a disastrous series of financial misadventures. Eventually, when Samuel had lost all his money, the oldest daughter, Henrietta, was forced to seek work. Prefiguring Driggs's career, she traveled to the Colegio Maria Josephina Hooker, an Episcopalian orphanage in Mexico City, where she worked until she was displaced some time in the 1880s by a local insurrection.

John Driggs, meanwhile, took a degree at Yale and went on to study medicine at the University of the City of New York, where he qualified in 1880 and opened his practice among immigrant populations on New York's Lower East Side. Dedicated and serious as he was, Driggs's early medical practice was intermittent, and he appears to have spent long periods visiting Henrietta in Mexico, where he was tempted to settle.

But by 1886 Driggs had entered a new milieu, and had the Episcopal Board of Missions pressed the doctor for a complete career record, Driggs would have revealed that between 1886 and 1888, he had worked as physician and house manager to the American Opera Company, an itinerant group of performers who traveled the Midwest, New England, and southeastern Canada. But by 1888, the opera company had disbanded and in 1889 Driggs was practicing as a physician in Delaware, ten miles from the family home near Smyrna. Three letters of the late 1880s allude to this period, but nothing of this operatic experience, and no reference to music and theater emerges from his letters to his church employers.

Driggs's private letters give us a glimpse of his involvement with the theater, and his first surviving letter is from Minneapolis in December 1887 to Fred (Dwight?), a medical friend in Delaware:

> My duties keep me quite busy as I have to count up the house every night as well as attend all the sick, one of our stage hands has been unfortunate enough to break his fibula while we were in Omaha, we take him along with us from city to city... What my future plans will be I can not say yet, altho I must say that money matters look very blue.

A few weeks later, on January 6, 1888, he was in Montreal awaiting news of "cold cash" from New York to keep the show on the road. "I wish," he continued, "there was theatre in Smyrna." And in the slightly stilted language with which we become familiar in his later correspondence: "I'm thoroughly schooled in that line of business." Responding to Fred's suggestion that he join his medical practice in Delaware, Driggs said that this could only be for the summer, so as to be "off again in the fall"—perhaps for Mexico. Certainly Driggs was looking for more in life than a physician's career offered. He seems also to have disliked big cities. He gets lost in Montreal: "it's worse than Boston: quite cold, but not as bad as Minneapolis," where "the side walks take the cake": they are mere "sheets of ice and there are no horse cars are running." A wonderfully ironic prefiguration of a career that two years later would take him to the Arctic!

Dissatisfied with medicine and anxious about money, Driggs now turned his mind to religion, and in March 1889, he wrote to the bishop of Delaware to say that for four or five years he had been thinking of missionary work, and that he now sought a foreign mission, perhaps in South America, "or else up to Alaska" (Driggs, Letter of Mar. 31, 1889). Supported by a recommendation from his local priest, Driggs was conditionally accepted for a newly created posting at Point Hope. But ten days later, having broken this news to his family, Driggs abruptly, and in a somewhat ingenuous manner, then braked and retreated: "There is one difficulty that had arisen to my going by the first of May and that is my Mother. She tells me that she is growing old and would most likely not live to see me again and therefore begs me not to go away"(Driggs, Letter of Apr. 10, 1889).

In two labyrinthine paragraphs, Driggs next reveals, perhaps for the first time, his professional loose ends and psychological complexities:

> [O]f course I am of age and can do as I please but then I will remember the tearful scene the last time I bade her goodby and the joyful reception I received on my return after an absence of nine years...then the 5th commandment presents itself forcibly... on the other hand I wish to go and the accounts of Alaska in the 'Spirit of Missions'... makes me feel that it would be a great field in which to begin and the offers you made were perfectly satisfactory, but I am going to ask you to kindly reserve my appointment for a few months, I feel by doing so that I will be able to gain my mothers full sanction which would be a great consolation to me in after years should I return to find her no more...
>
> You may think that I ought to have known of the objections before writing, all I can say in return is that I live about ten miles from my parents and have been so in the habit of not asking the advice of others, but doing as I thought best for quite a number of years past, that on the occasion I did not mention my intention of going away until the latter part of last week... (Driggs, Letter of Apr. 10, 1889)

Despite this rambling confession—which could scarcely have impressed its readers with the doctor's reliability—and the fact that Driggs was in competition with twenty other applicants, the Board of Missions accepted the delay. A sturdy physician who was also prepared both to teach school and to preach perhaps represented more than the Board could have asked for. Driggs's appointment was confirmed and his departure arranged for May 1890.

To qualify as a teacher in a government school, Driggs had to take exams. He did this with good grace, and in May 1890 received his certificate, having achieved reasonable, although not perfect, grades in Orthography, Reading, Writing, Mental Arithmetic, Written Arithmetic, Geography, English Grammar, and U.S. History—though he received 100 percent for physiology and hygiene.

But why did a potentially successful East Coast physician, with more apparent interest in theater management than in Native American children or Protestant evangelism, choose to travel so far beyond the western frontier to start a school for a people who were thought to be savages? Judged by the yardstick of nineteenth-century propriety, Driggs's motives were perhaps as contradictory as they were diverse. On the one hand, his altruism is borne out both by work among New York immigrants and by his subsequent eighteen years of dedication to Point Hope. Nor can his missionary enthusiasm be entirely discounted—though, like his Iñupiaq converts, Driggs's main interests lay in the outdoor life, and his religious orthodoxy, as we shall see, was not immune from the influence of Native belief systems.

But there were at least two other forces, economic and psychological, behind his interest in a remote placement. We know that Driggs practiced medicine only intermittently in the 1880s, and when invited to join Fred in summer 1888, he made it a condition that he would be released by the autumn. If restlessness accounted for Driggs's distaste for a mainstream career, there were also career limitations that lay beyond his control. By the time he qualified, there were already too many doctors in America, and most of them were in the northeast. As one medical writer put it: "in every community, there are three or four doctors where only one is needed… Many worthy ones languish, and those who flourish must do so by great skill, great tact and great industry" (Cathell 1883). But Driggs was also clearly bored by East Coast life. According to Cathell, in a vastly successful publication that was the standard work on medical etiquette, the doctor needed more than

> skill, tact and industry… A physician is judged by the company he keeps… Never have companionship with Irregulars…avoid this and every other soiling contact… Take care to be neat…wear a clean shirt and a clean collar; for if you dress well, people will employ you more readily, accord you more confidence…present genteel dress and polished manners, for…reputation is a very large part of a doctor's capital. (Cathell 1883)

In stark contradiction to such precepts comes a letter from Driggs in July 1891 when he is already in the Arctic. The letter is to Fred, and it opens with one of the Driggs's most spectacular glitches: "The letter…mentioning the loss of your sister was a great surprise, and you can't [sic!] rest assured that you have my utmost sympathy." Condolences thus expressed, Driggs unbuttons further:

> I live entirely alone but do not feel the least lonesome. You would like the life here, it suits me very well as it is *entirely free from the restraints of civilization* [emphasis added].
>
> On Christmas day I took my first long walk on the Ocean and then returned home and cooked my Christmas dinner by roasting a hunk of venison with baked sweet potatoes. I also built my first mince pie, and as I worked over the latter I thought How that if old Fred were with me, how much he would enjoy it, and I pictured to myself the old boy passing up his little plate for a second one of my generous slices and I took pleasure in the thought. (Driggs, Letter of July 1891)

Here, for the first time, we hear Driggs speak openly about his more personal mission to the north. Both footloose and at home in the Arctic, the doctor's life is "entirely free." The man who would bring Sunday and churchgoing to the traditional rhythms of Iñupiaq time spends Christmas on the sea ice, and follows this up with a tremendous dinner. There is no nostalgia for Christian society. It remained, paradoxically, his mission to bring "civilization" to his hosts. But the constraints of his own culture tie him no longer.

"School teaching is fun," he writes in the same happy letter, "the children are very obedient and not mischievous. I teach them reading writing and arithmetic. Also singing." Driggs's work brings him serious and philanthropic pleasure. And perhaps it was that mix of fun and dedication that made the doctor's early contact with the Iñupiat so uncomplicated and serene. Insofar as his own well-being was concerned, Driggs appears to have chosen his new life wisely. And it would not be until about 1907 that this hidden Dr. Driggs would be unmasked by his Episcopalian masters.

Driggs the Man

In addition to Goodman's account of his arrival, a further untruth about Driggs was broadcast during his second year at Point Hope. The allegation came from one Gilbert Borden, a disreputable ex-whaler who was superintendent of the Point Barrow Arctic Refuge station. Borden had arrived in Barrow carrying a trade outfit that probably included liquor. He soon quarreled with Brower, who was now whaling in Barrow, and he was also upset to discover that Healy had appointed the Presbyterian missionary Stevenson to be his assistant at the refuge station. Worse, Stevenson was going to utilize the station as a schoolhouse for Eskimo children. It needed only one more rascal to ignite this situation, and obligingly, in autumn 1890, Peter Bayne intentionally ran the schooner *Silver Wave* aground some distance from Point Barrow. When he arrived in the village, Bayne, with Borden's help, concocted a scheme to defraud the *Silver Wave*'s owner. Eventually, after Healy intervened, the schooner was restored to its proprietor James McKenna, while Borden drifted into obscurity (Bockstoce 1986:309–15).

In the interim, however, Borden had written to the secretary of the U.S. Treasury alleging that a "powerful conspiracy" involving Stevenson, Healy, and the Pacific Steam Whaling Company had been ranged against him. In the course of this attack, Borden issued a barrage of accusations. One of these was that "on reliable authority" (Bayne!), Driggs had ordered twelve barrels of molasses from San Francisco for the purpose of distilling alcohol in Point Hope—a perversion of the fact that Driggs had bought molasses to make schoolroom cookies.

The accusation came to nothing. But the event, as it reflected on Driggs's later career, is instructive. Just as Borden clashed with Stevenson, so, reading between the lines, it appears that Driggs must have fallen out with Bayne in Jabbertown. Driggs had written in August 1890 that he felt threatened by two Jabbertown whalers. Given Bayne's allegation, one of these enemies must surely have been Bayne—who had perhaps got the worst of a dispute with Driggs. There is no record of this, although in 1896 when Bayne returned to Point Hope, Driggs recorded a revealing dispute:

> Elungyeu—a Kivelynia native—came to me with a written contract, which Capt Bayne had given him, desiring to know if it was good. As it was not signed I told him that it was "not good until Capt B- signed it." The next day I requested Capt B- to explain his action in giving a native a worthless written promise. From his written reply it being evident that he intended to commit a fraud upon the native I forbade him the house and all intercourse with him. (Driggs, Point Hope Parish Records, January 1896)

This isolated note records the first of only two occasions when Driggs confronted a Jabbertown trader (see also Chapter 26). But the implications within the doctor's narrative suggest both Bayne's doubtful reputation among Point Hope's Iñupiaq community and a previous history of animosity between Driggs and the "Captain," perhaps, as here, stemming from completely antithetical moral positions.

While Driggs was untouched by Borden, he too would eventually fall prey to the illusion that his private business would remain, as he later wrote the word, "secrit." If the beginning of his Point Hope career was marked by the achievement of a personal autonomy, the end of his career came when that same freedom was expressed in ways that the church saw as incompatible with his position. Driggs never distilled liquor. But his own documents show that he did import whiskey. Native testimony also records that Driggs's drinking, like his tobacco chewing, was no secret, though both were viewed by the Board as improper in a missionary. More than this was Driggs's fatal assumption in the early years of the next century that he could run both a mission and a commercial whaling operation without being discovered.

The second of Driggs's letters to Fred might have shocked a reader in missionary society for its boyish preoccupations. Writing in March 1892, Driggs opens with an account of school life before once more expanding: "Here and the surrounding country is almost a hunter's paradise for both large and small game. The latter I can kill close to the house during the proper season"—another proclamation of his untrammeled existence. Next comes the first of Driggs's mildly crude observations: "One old lady came as a visitor to the school recently and brought her wooden thundermug along. I caught her sitting on it in the school room and made her take it out" (Driggs, Letter of Mar. 13, 1892). This is followed by a description of an old man urinating in the ceremonial house:

> During the long night I attended their mid winter festival—the first evening was their religious exercise and feast of whale meat eaten raw. The next night was a dance, also another one a few nights later. Although the assembly house was one mass of ice and bitter cold inside, yet the leading ladies and the gentlemen of the orchestra were naked above their waists. The musical instruments consisted of a sort of drum made by stretching the covering of the young whale's liver over a narrow wooden ring, made of driftwood. A man was taken short, so one of the ladies kindly held a keg with the head knocked out and into which the old gent found great relief. I am going to reserve a full account of these events, for when we are again together...

Driggs's letter provides one of the few eyewitness accounts of Point Hope *qalgi* life, and it is frustrating, given that the houses would soon fall into disuse, that the doctor confined himself to so few observations.

Driggs then gives a close-up of a woman suffering from venereal disease: "I have had to examine a number of the women for venereal disease contracted on the sailing vessels, and have noticed that they are bald headed in that portion of their bodies." This is Driggs, sited somewhat precariously on the border between physician and student wag, while in his next remarks about Koenig's wife Pausana, he is the cheerfully bluff doctor: "Koenig's squaw presented him with a son a year ago last September, and by the looks of things she is about to give him another presentation. He is an ex-Salvation Army man." And the final sentence in this letter contains a heartening description of Koenig in his role as husband: "He thinks," wrote Driggs, "a good deal of his Eskimo wife and intends always sticking to her."

Driggs's last surviving letter to Fred is dated March 12, 1893, and provides an overview of his social, domestic, and professional life. The letter opens in the spirit of humorous stoicism:

> There is a big blizzard on the rampage and I guess that somebody will have to come around and dig me out tomorrow. The snow has been trying its best to cover over the house this winter and so far it has done quite well. On the south side the drift lies about three feet deep on the roof. (Driggs, Letter of Mar. 12, 1893)

This sounds like Driggs's Sunday or a day when the weather prevented the children from coming. Secluded by the snow, time passes, and the doctor just sits and contemplates the weather. Then, in the same letter, he records his first case history:

> There has been an epidemic among the Natives, probably caused by eating a whale that was washed on the beach last fall... The first symptom was an impaired vision which persisted through the sickness and late on into the recovery. Jaundice was marked in most of the cases along with an obstinate constipation. No fever perceptible in those cases I examined. In two of the cases the kidneys were affected so that the urine was scanty. There was also loss of appetite, along with debility, and when the patient was on the road to recovery the throat became affected, the patient complaining of painful deglutition. On examination, I could not detect any thing more than a pharyngitis. Two or three of the adults died, but the rest of the cases went on to a tardy recovery.

Here, finally, having expressed youthful high spirits and happiness, is the physician speaking, and his detailed account of botulism or salmonella poisoning reveals the serious medical practitioner. In contrast to Driggs's usual vagueness, the vigor and precision of this passage leaves us hungry for what else Driggs might have written. But he was too busily engaged. And he may have believed rightly that his employers would have been indifferent to historical and ethnographic thick description.

But what did interest both Driggs and his Delaware correspondent were the arts of photography and big-game hunting. In July 1892, Driggs informed Fred that "the photographic machine" he had sent had arrived, but that he needed more glass plates and paper. Fred is enjoined to send "a good amount...by express, as I hope to make a good many pictures." Now in March 1893 he writes: "I have not done much of photography yet, for the weather has been very bad. All last summer it rained, and during the fall and winter it has been snowing. A good day is a rare treat." A number of Driggs's photographic portraits survive in *Spirit of Missions*.

In his last letter to Fred, Driggs outlines a trip to the Kotzebue Sound (not taken), which will give him an opportunity to hunt: "I will travel in one of the Native boats, camping out for nearly two months. There is plenty of game all along the route. I hope to get a chance at something big with the new repeating rifle I am expecting when my supplies arrive" (Driggs, Letter of Mar. 12, 1893). And wolves and bears provided a chance for some innocent boasting:

> The wolves have been trying to put in an appearance this winter, and have approached rather near the house, but I have not seen them. I have not the least fear of a wolf, so long as there are only one or two together. But a fellow had better look out if a whole pack should happen along, although there is not the least danger for there are too many hunters out all the time. The bears have been cheeky enough to come towards the Point, with the consequence that twelve have been killed so far this season, and plenty of others either scared off or wounded.

Hunting whales was of course something that a missionary could not do, as this would involve either going Native or involving himself in commerce. But Driggs took a close interest in both the Native and the commercial whale hunt. Alluding to whaling preparations in spring 1893, Driggs signals support for both efforts:

> There is a small schooner wintering on the lake. The captain [Bayne] expects to whale this spring, so I guess the bowheads will have a very interesting time of it next May. Altogether, Natives and whites, there are to be twenty-one or twenty-two crews out this spring. I hope they all make a good catch for they need it bad enough. It is either make or break this year with them all. (Driggs, Letter of Mar. 12, 1893)

Another aspect of Driggs's role in the community emerges towards the end of the letter. The doctor opens his door one evening to "a buck and his squaw on their way back to the village" who are seeking shelter. Having missed Joe Tuckfield, whose schooner at the inlet had become an important place of refuge for travelers, Driggs's house was their next chance of shelter. "They were," wrote Driggs, "lucky finding the Mission in this blizzard, for there is not much of a hole left to crawl through to gain admission to the house." Like his account of Christmas dinner, this letter concludes with a touch of humor: "It is now half past six in the afternoon and my appetite is beginning to take an interest in the seal liver which is in the oven, also the corn and potatoes which are thawing out on the stove. If you were only nearer I would invite you to sharpen your teeth with me" (Driggs, Letter of Mar. 12, 1893).

As in Driggs's remark about bowheads in the earlier paragraph, there is something mildly sadistic, and peculiarly feral, about the doctor's invitation to a tooth sharpening in this passage. Driggs reveled in his freedom and loved big, wild country. The wolves, bears, whalers, "bucks," and "squaws" that fill his pages had an appreciatively assimilated companion. Missionary he was. But the doctor was a little wild himself, and it no doubt took some cunning to conceal this wilderness character from the Board of Missions until early the next century. While Driggs had escaped the "restraints of civilization," he nonetheless continued to pursue official duties with scrupulous commitment: a virtuoso balance which, for the time being, suited all parties.

Driggs unfortunately could not maintain this balance between missionary propriety and private interests forever, and this will become clearer in later chapters. The "hidden" Driggs, who was in flight from "the restraints of civilization," would develop more complex involvements than a simple attachment to the free life of the hunter. For one thing, the autonomous life about which he expresses such happiness would take him closer to shamanism than his employers could have understood or tolerated. And the same attachment to unsupervised freedom would lead to an association with commercial whalers that would lead to his dismissal. The story of this endgame is told in Chapter 26.

Notes

1 The story probably came from Rev. Frederic W. Goodman, who worked in Point Hope between 1911 and 1925. Born in Worcester, England, Goodman was appointed Archdeacon of Arctic Alaska in 1925. He served for three periods at Point Hope during the furloughs of A. R. Hoare (1911–12) and W. A. Thomas (1922–23), and from 1925 to 1943 as resident priest. His contributions to Point Hope life included an interlinear translation of parts of the Prayer Book (1922), a hospital building (1926), and refurbishment of St. Thomas's Church (1928) (Goodman 1932:3).

2 I am grateful to John Driggs's collateral descendents in Delaware who made these documents available. The rest of Driggs's papers were, as one of his great nephews put it, "chewed up by silverfish."

CHAPTER 19

Driggs: Consolidation in the Village, 1892–93

As he opened his correspondence in spring 1892, Driggs remained confident, but some of his early euphoria had worn off, and his first letter of the season expresses anxiety about having to close school early when the whaling season started.[1] Given some free time, Driggs took up his pen to describe his classroom practice. Two things stand out. First, Driggs was now using the Iñupiaq language to teach English. Second, he had a flare for improvisation: "I repeated short sentences in the Native language, and had them translate it into English on the slates. The advanced class was one [in] which I repeated short sentences in English, and had them repeat after me, translating into their own tongue" (Driggs, Letter of June 4, 1892).

The letter proceeds to show Driggs at his most candid. Now that he was established, he felt secure enough to discuss his difficulties in some detail. He had too little time to deal with large numbers of students. Native sickness, hunting, and home responsibilities cut school attendance; his equipment was inadequate. To cite one small example, he ran out of slates. "I was forced," he wrote, "to give up the writing class as fragments of my six slates became exhausted." There were more serious difficulties. As Driggs progressed in his knowledge of Iñupiaq, he became aware that his students had no opportunity to practice English.

Teaching the spoken language was not, however, Driggs's main focus: the aim of his mission was to teach literacy. To become Christians, with direct experience of Holy Writ, the children must be taught to read the Bible and the Prayer Book. Driggs wrote:

> My reading class is the principal one and there are twelve scholars who have advanced into the second readers, all of whom are doing well… One of my boys has advanced so well that he is now correcting me in my words and in that way I am discarding all the incorrect ones learned at first. His name is Attungana and he has not been absent from school since the first of last October up to the present time, with the exception of about three weeks during the winter when he was confined at home with the rheumatism. (Driggs, Letter of June 4, 1892)

Elijah Edson, who would teach the same classes in 1895, offered a much less optimistic picture, and convincingly showed how much Driggs was obscuring the reality of the situation. But Driggs certainly knew his limitations: "They are trying hard to learn and it often makes me feel sorry that I can not devote my whole attention to the first fifteen or twenty. I am sure their advancement would be quite rapid…" (Driggs, Letter of June 4, 1892).

It was a difficult balance. Driggs wanted the children to attend, but he also understood that they had responsibilities at home. Even little boys were expected to bring in squirrels, fish, and birds while the girls were kept even busier with sewing, food preparation, child care, and the management of dogs.

But the schoolhouse had a function that went beyond the children's Euro-American education. As Driggs had described to Fred, the school had become a meeting place. Not only did the mission serve as a clubhouse for children, it was a gathering place for the sick, elderly, and hungry. The school had become an important new social center: "You can form some idea of what I mean by the lack of time, when my school room is crowded every day with scholars, patients, and others that I classify as visitors, but who as a rule have some favor to ask" (Driggs, Letter of June 4, 1892).

While the mission had become a kind of *qalgi*, the visitors were never there to dance.[2] They came because they needed help. In his New York practice, the doctor had become used to cramped, disease-infested dwellings. Now Driggs was receiving the sick into his own overcrowded quarters:

> There has been considerable sickness and my medical services have been in almost constant demand. Some of the cases being of a very important nature as far as the future of the tribe is concerned. There are two difficulties which I have to contend with in the treatment of disease. One is the lack of a room to treat the cases in. To see them in the kitchen, the one small room which I have to live in so many months out of the year, is often highly objectionable from the nature of the case [a probable reference to venereal disease]. To treat some of them in the crowded school room is entirely out of the question. Another difficulty is to get them to take their medicines as directed. (Driggs, Letter of June 4, 1892)

This important and revealing close-up of Driggs's professional immersion and the needs of a community that daily confronted him evokes very clearly a society that trusted what the doctor had to offer, and expresses a dependency that was new to Point Hope. And adding to the manner in which the mission had become a new kind of community house is Driggs's next statement: "Their leading physician [shaman] is at present under my care." There were about ten shamans working in the village at the time, and it is likely that Driggs refers here to his shaman friend Anaqulutuq. Whether Driggs himself was regarded as some variety of shaman is difficult to know. Point Hope people called him *Taaktaiḷaq*, a term from *taakti* ("doctor") and meaning "one without a doctor"—suggesting perhaps that Driggs was the only Point Hope resident who didn't have a physician. Like John Simpson on HMS *Plover* (1852–54)[3] and Drs. Rosse and Call on the *Corwin* and *Bear* in the 1880s and 1890s, Driggs, who was himself almost always in good health, was certainly in demand, and as a mark of the new era, even shamans were applying to him.

Whatever the doctor's views on shamanism, his letters of 1892 make clear that he did not see traditional religion as a barrier to his work. Nor did Driggs complain of opposition from tribal practitioners. His problems lay in time and logistics. The work was too much for a single man, and in 1892 he was forced to reduce his regime from six to five days a week. Saturdays would now be devoted to fuel gathering. Besides, he wrote, "the children seem to appreciate their Saturdays just the same as the boys and girls in New York. They are beginning to feel the restraint of coming to school." And "I also found it impossible to organize a regular Sunday school, although I have spoken as freely as my knowledge of the language will permit, & sincerely hope that some of it will prove to be like the seed sown in fertile soil" (Driggs, Letter of June 4, 1892).

However much freedom Driggs allowed in the classroom, school was a distraction from children's main business. Most of them, as Edson claimed later, attended simply to get a biscuit. This was probably an exaggerated view, but Edson's remark does underline the severity of the subsistence crisis and thus the lure of even a few bites of carbohydrate.

Driggs's remark about Sunday school is his first to the Board about introducing Christianity, and he couches it almost as obliquely as he might have done to his students. Driggs was confident in his secular role as teacher-physician. But as a preacher he was a novice. His original application to the Board of Missions had, after all, been for a job as an auxiliary to an established mission. Now for the first time Driggs expresses the assumption that his role as educator was preliminary to when a "regular ordained minister should arrive" (Driggs, Letter of June 4, 1892).

Also for the first time, and in the context of the contact crisis, the doctor raises the issue of moral education:

> Had [Sunday school] instruction been given a few years back, it would have gone a long ways towards the saving of the people, and disease that was introduced last summer would in all probability have been prevented. These people want a thorough religious and moral training. (Driggs, Letter of June 4, 1892)

This is as near as Driggs gets to referring to venereal diseases, against which religious education would, he thought, have provided some defense. Presumably the "considerable sickness" to which he had alluded but could not name was venereal. And here, in contrast to the light-hearted manner in which he wrote to Fred, his serious tone suggests that Point Hope's very survival was at issue. In this connection, Driggs guardedly refers to sexual behavior, and in a passage that it is difficult to interpret he appears to conflate traditional mores and contact with white men. First he sketches this somewhat confusing account of traditional custom:

> This month all the single girls of over fourteen years of age, have procured husbands with the exception of two, and they have been married in the past but finding that they did not care enough for each other a separation was declared. Marriage and divorce are both esteemed very lightly. The exchanging of wives and polygamy is common. In one case I believe I was influential in stopping a man from taking a second wife. (Driggs, Letter of June 4, 1892)

There is nothing in the ethnographic literature that describes group betrothals, so the first sentence of this paragraph remains a mystery. Years later, Driggs officiated at a collective marriage ceremony, but this is not what he means here, and the terms "marriage," "husband," and "divorce" are presumably euphemistic. Driggs's point was that the structure of local custom (which moralizing visitors viewed as overpermissive) was, with the advent of the whalers, unraveling into a promiscuity in which women were bribed or enticed into sexual encounters. This is what Edson describes later as prostitution (Edson, Letter of July 3, 1895). Driggs notes one instance of this, in which a twenty-year-old woman "had the reputation of having had nearly thirty husbands"(Driggs, Letter of June 4, 1892). In a complex scenario that is impossible to interpret, she was ostracized in the village and committed suicide.

What Driggs, in his understated way, appears to describe here is a kind of social chaos. And while we could not know it from his letters, these years of the mid-1890s were the worst in Point Hope's history (see especially Chapter 22). Perhaps to protect himself from the imputation of professional failure, Driggs suppresses information about the epidemics that arrived in 1893. But he was communicative on the subject of subsistence, and realistically describes the depletion of animal resources:

> The whaling season is a complete failure and the people are feeling very much discouraged, no whale meat to lay by for next winter, no whale bone to trade the ships for ammunition. They depend on their rifles principally for their food supply. The bears were also almost a failure. Many were seen, but only two were killed by the Tigara people, and three by the Kivalina people. The amount of fox skins held is quite small. Undoubtedly the flour I shall exchange with them for wood will be a great help and wish I could do more. (Driggs, Letter of June 4, 1892)

Despite hunting successes at Kivalina, famine had encroached on this southerly area, and the following year (1893) would see an influx of refugees from Kivalina to Point Hope, a pattern that would be followed in turn by Point Hope migrations (see Chapter 22). In the face of this crisis, and with the population shifting unpredictably, it seemed urgent to the doctor that he move his students ahead faster:

> I shall hurry forward the older scholars who will desert school within the next three or four years, but I hope to advance them so they can read fairly well, understand simple arithmetic and geography, know how to write and have an idea of our country as laid down in the simplest history. With the young ones I shall be far more thorough. All girls who have arrived at the age of fourteen can not be relied on as good pupils. The boys will remain somewhat longer. My whole work is preparatory to the good work to be done in the future. (Driggs, Letter of June 4, 1892)

All this lay in the context of the weary and incessant labor of keeping the mission supplied with fuel. Driggs increased his annual order for lamp oil, chimneys, files, saws, and axe handles, poignantly describing their brief utility:

> These things are all necessities as a large amount of wood is required each season. The Natives frequently break the ax handles, for the cold renders the wood brittle. The saw must be frequently sharpened for the...ice soon dulls them. Lamp chimneys are all the time breaking, the children are continually running in and out, the cold draughts from the opening door breaks the chimneys. (Driggs, Letter of July 16, 1892)

But there were also small material advances to report. A hand cart for carrying goods to the mission had arrived. Also a new stove for the schoolroom, about which, charmingly:

"The children are struck with admiration...and remark *Okeupuk alapah pechuk* (Next winter it will not be cold)" (Driggs, Letter of July 16, 1892). But the labor was relentless and Driggs badly needed help:

> There is no spare time at this Mission. All last summer and fall I had to work hard every day, with the exception of Sundays. I have not been idle one day since moving into the Mission. I sadly need someone to be with me, as there is enough to keep two [or three] very busy all winter... (Driggs, Letter of July 16, 1892)

If the Board of Missions saw Driggs as the stabilizing element in this community under assault by outsider influence, the commercial whalers were sometimes themselves equally vulnerable. Driggs succinctly evokes this susceptibility: "there have been six ship wrecks so far, and one ship not heard from since leaving San Francisco. Also one schooner drifted seven hundred and twenty miles, fast in the ice" (Driggs, Letter of July 16, 1892).

In the face of insecurities of this kind, Driggs maintained an overt calm. "There has been much for me to be thankful for," he wrote, "and that our Heavenly Father will continue to bestow his blessings on this his Mission is the earnest prayer of John B. Driggs" (Driggs, Letter of June 4, 1892). Echoes of Dickens's Mr. Micawber aside, the doctor's sincerity cannot be doubted. He had yet to effect any Christian conversions. But through dedicated application, he had consolidated his secular role in the village and endeared himself to the community to which he had committed himself.

Notes

1 Driggs's attendance statistics for May show an aggregate of thirty-eight pupils, of whom twenty-seven are boys. This influx of males presumably happened when the sea ice closed and hungry young men converged on the mission. Attendance in June dropped to eleven pupils. Driggs wrote on July 16, 1892: "My total average for the nine months [1891–92] was 27 [and 31/38ths!] scholars a day."

2 The term *qalgi* referred largely, but not exclusively, to Point Hope's ceremonial houses. The word also denoted any other kind of meeting place. A hunting camp, for example, where people gathered to talk, could also be referred to as *qalgi*.

3 In 1852 Captain Rochfort Maguire wrote of a Point Barrow shaman called Erk-sin-ra who treated the same patient as John Simpson: "Dr Simpson's patient...is afflicted with the pain in his chest, and Erk-sin-ra is engaged with him beating it away with a drum... We knew that he had been attending Dr. Simpson's patient lately.... We were much amused at his informing us that the woman who had died, had not been treated by him, or the result would have been different—Another of Dr Simpson's patients who has received a deep cut on his shin was down and dressed on the lower deck. As this was a case beyond Erk-sin-ra's skill he was content to look on approvingly and on one of the officers remarking that Doctor S. wore moccasins similar to his, implying that there was something professional in them, his modesty was put to the blush and he appeared much flattered" (Bockstoce 1988:292, 317–18).

CHAPTER 20
The Missionary Edson—Fear and Trembling

Reverend Edson Comes to Take Over

The Reverend Elijah H. Edson was forty-one years old when, from Pittsford, New York, he wrote to the Board of Missions that he felt called to missionary service. A native of New York State and ordained in 1883, Edson was an experienced priest and, although married with children, relished the prospect of an arctic posting. Like Driggs, Edson was a healthy six footer, an outdoorsman, and, as he wrote in his letter of inquiry, was "possessed of certain qualifications not often found in the clergy of the Church" (Edson, Letter of Apr. 6, 1894). Presumably this referred both to his impressive physique and to his building expertise—both of which were important qualifications for Point Hope life. The appointment was welcome news to Driggs, who was due for a year's furlough, although Edson would work alongside him for twelve months before taking over.

Edson arrived under the impression that Driggs had made some progress in Christianizing Point Hope. "In regard to the bread for Holy Communion at Point Hope," he wrote optimistically,

> I have not been able to discover where it is made. But my wife last night taught me how to make bread, and I mixed today's baking... I shall take a quantity of yeast cakes with me, and expect the Dr and I will have good bread. Concerning my robes, I... have a good Cassock but my surplice is about worn out. Would it not be well that one should be kept at Point Hope, with the altar linen and Holy Vessels? (Edson, Letter of Apr. 6, 1894)

Unsurprisingly, he was horrified by the mismatch between his expectations and what he found on arrival.

On the steamer going north, Edson had met Leander Stevenson, the Point Barrow missionary, who gave him blood-chilling accounts of "the intercourse of Natives with the crews of the whalers, their awful corruption, physical and moral resulting"(Edson, Letter of Apr. 6, 1894). This intensified Edson's somewhat easily excited enthusiasm, and made him "long to be at work." And this he started immediately after his arrival in July 1894, when he conducted divine service for the first time in Point Hope. Driggs and Edson were initially delighted with each other's company. Edson, said the doctor, would be "a splendid companion," while Edson ("Thank God!") found Driggs well. Within days this view darkened. As Edson wrote later:

> I found Dr Driggs greatly worn by his four years' struggle, and badly needing the vacation the Board has granted him. At first he said it was only lassitude that he was suffering from, induced by the unusually warm summer, but the langor did not leave him with the return of colder weather. (Edson, Letter of July 3, 1895)

With memories of Point Hope's flood of 1893 (see Chapter 8) still vividly in mind, Driggs took Edson to view the eroded north shore, and the two men devised a plan to move the mission inland. Edson wrote:

> During the last four years autumn storms have driven heavy seas diagonally along this side of the point with terrific boring force, cutting off fully fifty feet of the shore in front of the house. The doctor believes it is only a question of a short time when, if it is not moved back, it will be washed off. He has selected a school new site about 300 yds inland upon a ridge slightly higher than the present location. (Edson, Letter of July 3, 1895)

This powerful description of the Point, which at the best of times has always been in a continual process of erosion and buildup, is a helpful reminder of the dramatic context of the village environment—and it is hard to avoid seeing it, at least during this decade of hardship, as emblematic of the human predicament of Point Hope's residents.

Edson and Driggs were of course right. Not only were Point Hope people at risk on account of the subsistence crisis, but the very foundation of their lives on the Point was in jeopardy. Thus despite Driggs's low energy and evidence of mutual dislike, the two men collaborated briefly, in the interest of Point Hope's future, with idealistic fervor. Following a meeting with Sheldon Jackson on the *Bear*, the two missionaries composed an analysis of the subsistence situation in which they called for an extension of Jackson's reindeer herding experiment. Encouraged by Edson, Driggs now rose to discussing longer-term initiatives, and here for the first time identifies the problems that followed the introduction of repeating rifles and explosive whale harpoons: "The 'bomb gun' in whaling and the 'repeating rifle' are fast exterminating the whales and other sea-animals upon which the Eskimo depend for food. Already the deaths among them from starvation have been alarmingly frequent" (Edson, Letter of Aug. 8, 1894).

Jackson had already suggested utilizing various missions in the distribution of reindeer to Alaskan Natives, and Driggs and Edson endorsed the somewhat paternalistically restrictive notion that "missions hold and manage the deer: from time to time distributing them to the most worthy of their people upon certain conditions" (Edson, Letter of Aug. 8, 1894).

Large tracts of grazing land required by the deer would mean moving the mission or building a new one, and there would be a spinoff for the missionary effort. Revealing their view that the mission was essentially marginal to Point Hope village affairs, the writers suggested that reindeer herding could bring the mission closer to the people. The introduction of deer would give the missionaries "prestige amongst the Natives, and they will soon pass from mere tolerated appendages to the villages to the very centers of their life" (Edson, Letter of Aug. 8, 1894). Enlarging on their vision, the two men also saw the reindeer project as a means of binding the new Point Hope church into a larger Episcopalian district that Driggs would explore on foot that winter.

The letter concludes with a warily optimistic self-evaluation. Following Thornton's murder at Cape Prince of Wales in 1893 and the collapse of Stevenson's school at Barrow, Point Hope now led the way in the northwest Arctic:

> the fact remains that this mission alone seems to have realised in considerable degree that which was hoped for it at its inception. The Natives here look to your teacher not only for mental instruction but for counsel in temporal affairs also. Within the last few days Natives from villages farther south have solicited your Missionaries to come to them and establish schools. (Edson, Letter of Aug. 8, 1894).

There was truth in this claim. But the success of the mission was of course an index of the deterioration of Native morale, self-determination, and local institutions. Driggs and Edson were paternalistically benign and their interventions had positive medical and material outcomes. In the longer term, this ancient and autonomous community was growing increasingly dependent on a culture that was being imposed from without. The mission by the mid-1890s had become a permanent institution and the Christianity of mainland America had come to stay, however the people might come and go and the north side of the Point erode.

Driggs and Edson Cross Swords

Driggs was anticipating his holiday with excitement, but his anticipated re-entry into mainland life was tinged with the hesitant, outsider feelings: "On my arrival in San Francisco next year, I would like a letter awaiting me, to contain a check for $150 to refit myself in civilization again. Every thing that I brought has been ruined" (Driggs, Letter of Aug. 4, 1894).

Edson had written of Driggs's languor. But the doctor did have the energy to make three journeys that winter, leaving Edson in charge. Edson wrote that Driggs's trips, "one of three weeks and two of forty days each seemed to benefit him for a time. But he is far from well. I trust the return voyage and visits amongst old friends will completely restore him" (Edson, Letter of July 3, 1895). "Far from well" was unlikely to refer to a physical disorder, and Edson's cautious diagnosis appears to allude to Driggs's mental health. Driggs never dwelled long on his own hardships, and so Edson's description of household drudgery, alone, is thus the more disturbing:

> The effects of the isolation of this field, upon the mind, and by reflex action, upon the body are terrible. To do one's housework, cook one's own meals, wash dishes, sweep, cut fire wood for living rooms and school, teach a school of 40 or more Natives without a common language, to be on a constant restricted allowance of drinking water, to obtain which a journey in summer of one and a half miles, and in winter of 16 miles, must be made. And worst of all, to wage a constant battle with the lice to prevent oneself from being overrun of them, for they swarm upon the Natives—and coupled to all this, for 10 months of the year the rigors of the Arctic winter, if this not enough to break any man down in five years I am mistaken.

> If the man who goes through all that is not be forgiven all shortcomings and failures of disposition, there is no pity in the world. Why do I write all this? To let you know the real conditions of this work, and, that when I beg you not to leave any man here alone after this year you may see the strength of the ground of my appeal. (Edson, Letter of July 3, 1895)

Driggs—who relished an opportunity to shock his readers with accounts of gnawed flesh and gangrene—had never mentioned lice, and Edson's comments on the subject foreshadow his later outcry against dirt and sexuality, which Driggs had also seldom discussed directly. Nor did Driggs complain about his household chores except in connection with their impact on his teaching regime. He seems, likewise, not to have been exercised about food and water shortages. All these difficulties were perhaps components of the life which his hard-won freedom necessitated and which he therefore philosophically accepted.

Edson's comments on Driggs himself are ambivalent. On the one hand there is compassion. But Edson also insinuates a critique. While the Eskimos lived in metaphysical darkness and the whites who had corrupted them were degenerates, the missionary was himself, wrote Edson, subject to "shortcomings and failures of disposition." Driggs had no one to complain to outside his diary—his only surviving journal, dated May 1, 1894, to September 1901. And there are in his journal no complaints about the sort of hardship which, in Edson's view, had so worn him down. Here are the entries that describe his first days with Edson:

> 15 July Steamer Jeannie arrived and brought Mr Edson my new companion, also my supplies.
>
> Sunday 22 First regular church service held in the Mission today by Rev. Mr Edson.
>
> 26 July Jim O'Hare, Mr Edson and I worked at the mission. Yesterday we changed the kitchen.
>
> 31 July the Bear arrives. Mr Edson and I dined with Capt Healy aboard the Bear.
>
> 1 August Dr. Jackson and a number of the officers of the Bear called. (Driggs, 1894–1901)

But Driggs certainly welcomed his new freedom to leave the mission. In mid-August he took off to Jabbertown, spending the next month fishing and camping with his trader companions. The party returned with a skinboat full of salted-down fish, and by September 8 Driggs was back at the mission. He wrote: "Dug a new well as Mr. Edson was using rain water. Communion service. Mr. Edson and I hold packed service, many can't get in."

The two men continued to collaborate, but the partnership remained uneasy. While Edson took over church matters, the doctor pulled rank in areas of social relations:

> 25 September: Anton Betts [Antonio Bettencourt] sent me a letter saying that the Natives were returning him their [retainers against the whaling season]. They said to him that I had told them any Native who worked for a white man would not go to heaven. I have never made such a remark to anyone and neither will I ever. I gave the letter to Mr. Edson to answer.

By October, Driggs was expressing irritation: "Mr. Edson seems to be growing absent minded. He does things then forgets all about it declaring positively that he never did so." And in November, his temper boiled over: "I told Mr. Edson this morning that I shall never consent to live with him after this year is up. His flatulancy is a perfect nuisance. Also his mannars [sic] are not those of a gentleman."

The best Driggs could do was to escape to the fresh air, and he visited Peter Bayne, who was living in the *Silver Wave* at the inlet. He ran into blizzards, froze two of his toes, and was holed up with Bayne for three days in another storm. But Driggs's trip refreshed him and on Christmas day he was again socializing with Edson in the company of Koenig and Hachmann. Christmas dinner was served at the mission, and that evening, in the spirit of a night out, the four men walked to the village and "went to the dance in Tigara." In other words, they squeezed their four-fold bulk into the *qalgi* and uncensoriously watched Point Hope's rituals. Given that information about this visit came from a member of Uŋasiksikaaq, one of the two ceremonial houses that remained in Point Hope, this was possibly the *qalgi* the four white men visited (Aġviqsiiña 1940:87).

Just before the New Year, Driggs started north on another trip, this time with Anaqulutuq and his wife. For Driggs this was a vacation. But for Anaqulutuq it involved serious hunting, although this was unsuccessful. His rifle killed efficiently, but without a harpoon's line and float, it was hard to retrieve seals. The new technology involving a float with retrieval hooks and/or a one-man *umiaq* with which to recover seals was not always available or possible to deploy—particularly on long journeys (Gregg 2000:147).

In his journal, Driggs described the weather as "disagreeable," and they were stormbound near Cape Lisburne. Driggs, remarkably, seems to have taken journal, pen, and ink with him, and he knew that January 27 was a Sunday. His entry for that day, flatly recording as it does, without comment, a shamanistic ritual, is perhaps unique in American missionary history: "Very stormy. I staid in iglu all day. M and A did the hoodoo act to change the wind. Mon 28 Very high north east wind."

Driggs and Edson had witnessed Native dances and perhaps shamanistic events on Christmas evening—Driggs no doubt had seen much more. Now, on a Sunday, without censure beyond the mildly pejorative phrase "hoodoo act," the doctor had sat through a privately conducted rite that was partly for his own benefit. Magical control over the wind was a routine Point Hope preoccupation and "wind doctoring" (changing the wind's direction) was not confined to initiated shamans. That it was "M and A" who performed the ritual implies that Mammaŋniiña was a woman with quasi-shamanistic power who routinely accompanied her husband in magical songs and other magico-religious activity. Driggs presumably just spectated on the séance, and whatever the outcome, the party made it round Cape Lisburne and returned to Point Hope in early February.

Driggs and the Two Shamans

If it's difficult precisely to interpret Driggs's relationship with his friends' "hoodoo" at Cape Lisburne, a second passage from his journal written some four years later confirms that he had indeed been convinced by shamanistic magic. It will be recalled that Driggs's association with the shaman Anaqulutuq had started—as had friendships with Jabbertown traders—from the moment of his arrival. And just as the doctor had shared quarters with Jabbertown men, so he had played host to Anaqulutuq and his wife in the early weeks of his mission residency (see Chapter 17). This, as Driggs noted about his relationship with the shaman, "was the beginning of a friendship that has lasted ever since." And later, as the doctor discussed his integration into village life, he wrote that "Their leading physician [Anaqulutuq] is at present under my care" (see Chapter 19). Whatever Driggs's comprehension of and early attitude to shamanism may have been, there can be little doubt that he had witnessed shamanistic acts many times. Still, it remains astonishing to read in a journal entry for May 2, 1899, the following affirmations: "Wind south. Changed to N.W. temp. +18 +34. [The shaman] Ayauniq changes the wind."

Here, following initial pragmatic observation of wind and temperature, the doctor moves directly to statement of belief in the efficacy of shamanistic power. If, as suggested above, Driggs's near participation in Anaqulutuq's "hoodoo" made his experience unique in missionary annals, this further comment, four years later, is yet more dramatic. For this simple journal statement reveals that Driggs the missionary was not merely sympathetic to shamanism but that he was able to collude with both its premises and its putative results—and in this respect his beliefs lay more or less parallel to those of Point Hope's early Christians, who similarly lived within two systems of faith. There is, unfortunately, no further evidence of Driggs's interest in the subject.[1]

Significantly, Driggs's own conjuration of the supernatural is followed by a reference to whaling: the connection between his initial "Wind south" and Ayauniq's intervention being that both Native and commercial hunters required a north wind to keep a water channel open in the sea ice for the whales to migrate through. The central passage in Driggs's journal entry reads here: "Ayauniq changes the wind. First two whales killed by Lieb and Nelson. Lieb's has 7 or 8 ft bone." That the doctor so casually inhabited both the shamanistic and commercial worlds suggests, at the very least, an unusual capacity for sociocultural coexistence. Unlike most resident white men, Point Hope people lived in a world in which shamanism and commerce were not mutually exclusive. Driggs, in this respect, might be classified as the complete Point Hope person who combined every possible regional activity. But this would be Driggs's last extant allusion to shamanism, and there is little more that can be adduced.

Reverting to Driggs's vacation at Cape Lisburne—five weeks at the coldest time of year and in the stormiest region of the territory—he returned to the company of Edson, whose fussy and somewhat highly strung ways continued to irritate him. Edson had, in the meantime, taken the opportunity to establish his presence with the white traders, and on Driggs's return, Edson visited Jabbertown for a few days. The meshing of relations between the mission and Jabbertown had now gone full circle.

In late February, Driggs was again on the trail for a third winter journey. This time he traveled with Hachmann and his Native wife's brother. This trip took him to the Noatak River on a "tour of inspection" with a view to setting up schools in that region. Just before reaching the Noatak, Driggs wrote with satisfaction that his reputation had traveled ahead of him, and that "Natives are anxious that I should start a mission school in this section." Somewhat wistfully, he added: "Yesterday was the first time I have seen trees in nearly

5 years… It seems good to be in the woods and out of the wind." A rare and poignant moment of contemplation.

The trip home was arduous. Driggs and Hachmann contracted snow blindness in the strong spring light. Then, in a crowded iglu near Cape Thompson, Driggs succumbed to a fever. He slogged home against a north wind, arriving at Jabbertown on April 1, where he convalesced at Koenig's—the roomiest and most comfortable house in the neighborhood. There was happy news awaiting Hachmann, although Driggs did not appear overexcited by it. "Mr Hachmann," he wrote succinctly, "has a son born during his absence."

Edson Alone

"Revd. dear Brother" began Edson to William Langford, Secretary to the Board of Missions, in June 1895. In preparation for taking charge of the mission, Edson briskly entered parish business and outlined his plans for moving the schoolhouse: "I have prepared the new foundation; cut and stored from timbers of a wrecked ship the necessary blocking to be placed under the house when raised from the ground whereon it now rests" (Edson, Letter of June 20, 1895).

Mindful of his scanty budget, Edson reassured colleagues at Church House that the work could be accomplished, "at very slight expense [of trade goods]: 1.50 lb flour, 3 lbs of gun powder and 5 lbs of shot." And again pointing a finger at the lackadaisical doctor, he continued:

> The house badly needs painting this year [Edson's underlining]. It has not been painted since its construction and now the boards are showing in many places where the original paint has worn off. Dr Driggs informs me that he has not ordered any paint.

Having itemized necessary materials, Edson turned his attention to the marital status of some leading Jabbertown families. Point Hope's first Christian marriage and baptism, each of them representing alliances between Tikiġaq and Jabbertown, was, for better or worse, a new stage in Point Hope's acculturation, events that Driggs could not have achieved. So on June 23, 1895, Koenig, with Driggs a lay witness, was married to Pausana, and their two sons, Frederick and Herman Carl Koenig, were baptized.

Driggs now went south, to the U.S. East Coast, and while he was away on his yearlong furlough, Edson ran the mission and socialized comfortably with the Jabbertown traders, in the course of which he had a significant confrontation with one Charles Sanborn, an employee of Peter Bayne. This story is important, for it highlights ways in which women and children were at times abused by Jabbertown traders: events to which Driggs had alluded, but perhaps not confronted and certainly not described. Edson narrated the incident as follows:

> On 10th of March 1895, Capt Bayne informed me that Sanborn had boasted of his intention to procure some whiskey and with it to get some girl from our school intoxicated and then criminally assault her, expressing his contempt for anything the Missionary could do either to hinder his deed or punish him. This was confirmed by Marx Lieb, a man in charge of whaling station No2 of the *Emily Schroeder* outfit.

> Under ordinary circumstances I would have considered Sanborn's words but the boastful vaporings of a bad-hearted man, but unfortunately the very crime he threatened has been twice committed here by white men, once by means of chloroform, resulting in fearful injury to the poor child. I also considered that we are thousands of miles from any court or Officer of law from whom we could invoke protection...
>
> Mindful of this fact, I called upon the said S and in the presence of Hesse and Lieb told him of what I had been informed, adding these words—"I care not for any slanders that bad men may utter of me, but if you do that dastardly thing before the Revenue Ship arrives here, I will shoot the offender. Moreover upon the arrival of the Vessel I shall lay this matter before its Captain, and where proven, demand from him such protection as his office allows him to administer..."
>
> The man has abstained from his threatened crime, and several of the better class here have expressed the opinion that my words will have a salutary effect upon the reckless element so largely represented here at present, and that visit this shore every summer. My object has been accomplished—the crime has been averted—but God forgive me—at what price! (Edson, Letter of June 20–23, 1895)

This, in several respects, is an archetypal story. As a case history, Sanborn's planned assault provides a concrete example of what occurred throughout American contact history, but whose detail is lost in hearsay and generalization. Of particular interest is the manner in which Edson's intervention both aborted the attack and positively engaged the Jabbertown community. How the Board of Missions responded to Edson's vigilantism and what might have been the consequences had he carried out his threat are unknown.

A year had passed since Edson had landed, and with Driggs on holiday, Edson's letters opened new areas of discussion. "Some things I shall relate are not pleasant," he warned. "I will write with utmost frankness." With a mixture of small-mindedness and compassion, Edson embarks on the description of Driggs's health, in the course of which he celebrates to his own hale condition:

> I am glad to be able to say that, excepting a persistent rheumatism, I am well, thanks to God! During the Doctor's absence [on his trip to the Noatak] I carried on the school alone for two months as best I could, conducted Divine Service, and for the space of one month preached to the Natives in their own language. (Edson, Letter of July 3, 1895)

Given the extreme difficulty of Iñupiaq syntax, this last assertion is hard to credit. But at least it confirms that Edson, like Driggs, took language acquisition seriously. How to transmit biblical teachings posed a separate problem. Without an Iñupiaq version of the Bible or Prayer Book, no proper conversion was in prospect.

Edson then writes a long description of the school, and this provides one of the most informative passages that we have on the subject. What Driggs had hinted at, the brutally honest Edson exposes in cold detail:

> The conditions here are so very different from those in civilization, that this school can not be judged by the standards or results obtained there. Here is no co-operation on the part of the parents. The children attend when they feel like it, and when they do not, they stay away. Many of the adults do not want the children to learn to read

> and write, for they are shrewd enough to see that the result must be the decay of their superstitious beliefs and ceremonies.

This is realistic and observant. The village had its own life, and its rhythms were quite separate from the mission and its ideology. If the school impinged on Point Hope life, this was at the convenience of local people when they were so disposed—and then only up to certain limits. Driggs must have shared this insight. But while Edson was not a man who could live with opposing views, each of which might simultaneously be valid, the more tolerant Driggs could achieve precisely this. He had settled in Point Hope to offer schooling. But because he liked the people for who they were, he understood their resistance. If some children did not attend, this seemed as natural to him as their attendance was gratifying. Driggs also well understood the next subject of Edson's discourse: "The 'hard crackers' given out each day are a moving force in securing a large attendance. That was demonstrated when our supply of crackers gave out before the end of the term, followed promptly by a considerable loss of scholars." And more dramatically, of some Christmas presents sent by the Episcopalian Women's Auxiliary:

> It had been the custom to reserve Christmas gifts until the end of the term, occasionally giving a few of the children a peep at some of them. But desiring to emphasize the lessons of the Nativity, I requested the Doctor to have them distributed on C-mas eve, preceded by an address suited to the festival. He complied, but warned me that the probable effect would be to diminish the attendance at school after C-mas, and the result bore out his prediction.

In a digression from school business, Edson then turns to the most fundamental issue facing Point Hope, which was the sheer survival of the village. In the formal language of a report, and using figures assembled by Driggs and Koenig, Edson outlines Point Hope's population situation, using childbirth, infant mortality, and emigration statistics:

> This tribe is dying out rapidly. There are not at present more than one half the people that were only eight years ago.[2] The mortality among the children is frightful and they are not a fruitful and prolific people. From 1891–1895 there were born to Native parents 30 children; of this number seventeen have died. There were also born at this time eleven illegitimate children having white fathers, six of whom have died. Fourteen children have died the past year. Infanticide is very common.

Edson made this assessment at the outset of the epidemic years, and as we have already seen, Point Hope's population was in crisis on account of both sickness and outward emigration:

> We expect the school will never again be as large as it was the past year, because of the dying out of the tribe, and further on account of the announced removal this summer of nine or ten families to Point Barrow and the country eastward of that place, near the mouth of the McKenzie River, in which region the whaling company has its principal station [on Herschel Island] and is trying to build up a Native settlement.

Bleak as the situation was, Edson's pessimism was not, in the end, to be justified. While the population was in jeopardy, the school Driggs had established was the first in a continuous line—including a government school at Jabbertown—which has remained unbroken into the twenty-first century. Continuing his critique of the school, Edson launches an account of classroom realities: a stark but vividly detailed account of children's abilities and attainments. There were, wrote Edson, five or six who could

> read fairly well in the *Second Reader* and one, possibly two in the *Third Reader*. Four can work fairly well in addition and subtraction, and this last term have made a start in multiplication. There is quite a large number of them who can copy any sentence written upon the blackboard, forming letters beautifully. But pronounce a simple sentence to them and they cannot write it at all, because they can not spell... Some of them recognize a good many printed words but do not know the meaning of many they can pronounce. These [as though still contradicting Driggs] are the facts concerning our most advanced pupils. The rest grade down in all degrees of ignorance to absolute lack of knowledge in the English language.

This is in blunt contrast to Driggs's habitually vague summations such as "My reading class is the principal one and there are twelve scholars who have advanced into the second readers, all of whom are doing well" (Driggs, Letter of June 4, 1892). To preempt unfavorable publicity, the *Spirit of Missions* editor wrote on Edson's manuscript "omit last sentences." But it was true that even the more advanced pupils only comprehended very simple English and no one could read or write.

Edson had some insight into the reasons for this. The first lay in the absence of school materials: the mission simply needed more books. But cultural and environmental issues were yet more significant, and Edson succinctly outlines these:

> the conditions of their homes are such that they <u>cannot</u> study there, for they have no light in their underground houses and because a Native house consists of one room, not one of which is more than 25 ft square, while most are only 10 ft square. In this one room from 8 to 20 people live and move and have their being, court their sweethearts, marry, and beget their children and rear them, in short carry on <u>all</u> the functions of life and duties of housekeeping. (Edson, Letter of July 3, 1895)

This analysis is densely realized, though it highlights Edson's personal recoil from the open sexuality of iglu life as much as it does the darkness and confinement that made homework impossible. And nowhere in mission papers is the discord between Native reality and normative Euro-American expectations so clearly illuminated. It was one thing to isolate children during daylight hours within the walls of the school. Here it might, optimistically, be assumed they could be refashioned, through a literacy-based curriculum, into American schoolchildren, and they would return to their families carrying with them the seeds of their learning. But such, wrote Edson, was the contrast between school expectations and the conditions of home life, that if children learned anything during their schoolday, it would be lost when they went home. School could not therefore transform them until domestic life was also changed. Driggs's earlier plan to rehouse the village would be a necessary step towards bringing the village in line with the school. Only then could progress at school be fed into the community. This, in the first decades of the twentieth century, both at Jabbertown's government school and in Point Hope, is an approximation of what happened.[3]

For the time being, children arriving at the mission experienced the classroom as a completely foreign space, and naturally they could at this stage only use that area in ways that approximated activity at home. To their credit, therefore, Driggs and Edson adapted their "instructions" to a situation that was always fluid. Whole-class activity was not going to work, and so both teachers devised individually focused methods. The detail in Edson's next description is as brilliant as the despair is excessive:

> A scholar takes his book and sits down at one side of a little rough board table, the teacher on the opposite side (for to sit beside him would without fail insure a liberal migration of body lice from pupil to teacher). Then with finger or pencil the teacher points out the word which the scholar speaks if he knows it, or if he does not, the teacher tells him it and shows him that it is made up of letters and tells the name of each, and so on through a sentence or two or perchance a few paragraphs. Then that scholar is done for the day. He may have waited four hours for his lesson and will wait perchance there more for his cracker. If he comes the next day possibly he will remember what he was told the day before, but probably will not. While waiting for his turn to recite, and afterwards for his bread, he whittles with his knife—laughs—talks—walks around the room or goes out doors whenever he pleases. (Edson, Letter of July 3, 1895)

Alarmist as he sometimes was, Edson wrote with clarity, and his remarks on the issue of time-keeping are similarly arresting. In even the smallest Euro-American communities, school hours were regulated by the clock and made public with bells. But

> these Natives have no way of measuring time. There is and can be no regular hour for beginning school. Were we nearer the village and the school furnished with a bell, we could call them, but we are a mile from the village and are without a bell. It may be asked why not run up the flag? That is impracticable for several reasons. It would not last two days in our winter storms. It could not be seen one half of the time because of storms; the flag man would have his fingers frozen bending it on to the halyards. We must wait until there are enough children present to warrant us beginning, and then others drop in when they can, or feel like it. (Edson, Letter of July 3, 1895)

The flexibility with which Driggs and Edson adjusted to Point Hope time was matched by the absence of any attempt to impose discipline. Edson continues, whether consciously or not, in Shakespeare cadence:

> Order, relation or subordination—that which lies at the foundation of all learning—there is none.[4] Insubordination there is also none because there is no attempt at order. At Cape Prince of Wales where it was attempted to establish order it resulted in the murder of the teacher Mr Thornton. And at Point Barrow where Mr Stevenson attempted to secure order the scholars simply would not go to school. Our praise is that we have no order—but then we have a school. When the noise become too great, he calls out loud enough for all to hear: *Ahbau nipu-tu-tin pe-chuk*: "Talk loud not."

There were of course more complex issues behind Thornton's murder. But even within the cheerful anarchy of Point Hope's classroom, the line sometimes had to be drawn:

> There were only three administrations of discipline during the term. The Doctor expelled one married woman because she would not abandon her habits of prostitution. I withheld one boy's cracker one day because he carelessly broke a window in the school room; and later on I sent a boy away from school for making obscene pictures on a slate and showing them to the girls.

Edson's special interest lay, naturally, in Christian education. Driggs had cautiously felt his way towards some religious teaching in his third year. But now Edson added prayers and hymns, into which the children threw themselves enthusiastically:

> There was nothing new in our methods the past year save opening the school each day with prayer, and closing it with prayer and a few songs. The children were also taught two or three Christmas hymns, which they rendered very well. Also the Missionary hymn "From Greenland's Icy Mountains." They learn tunes very readily. There are several voices among the boys that any choir master would be glad to have. The Mission needs a small "chapel organ."

Edson's view of what the children should learn conformed to what most North American missions promoted. The teaching of English was fundamental to the Protestant initiative. Once "savage" minds had grasped the language of the "civilized" majority, they would be open to hearing the Gospel and reading the scriptures. Edson suggested an extra dimension. Just as Driggs had called for maps of the Holy Land so as to offer the children a geographical and historical basis for religious teaching, so Edson sought to challenge both a limited worldview and what he saw as the anti-truths of shamanism with American geography and simple physics:

> The school ought to be furnished with the following: 1. Plain strong seats and desks. 2. A globe, and map of U.S. showing cities by name. 3. Some simple apparatus for showing some of the fundamental things in Physics. Also 25 First Readers and 15 Second Readers. For not until these young people learn that no man can make the winds blow or change the conditions of the weather, can they be freed from the thralldom of their medicine-men, or the Gospel get a lodgement in their minds and hearts.

The unanswerable question with which some missionaries must sometimes have struggled was how to discriminate between the claims of Native shamans and the miracles described in the Bible. As a self-described rationalist, Edson perhaps more than Driggs could not accept that an act of "hoodoo" could have any impact on the atmospheric pressure that generated wind. But a verse from the Gospel such as "What manner of man is this, that even the wind and the sea obey him?" (Mark 4.41) must have made them weigh carefully the difference between heathen attempts at supernatural power and God's intervention into the world of physics. But as we have seen earlier in this chapter, Driggs had sat passively at Cape Lisburne as Anaqulutuq sang his wind songs, and still more revealingly remarked in his journal that another shaman had moved the wind round from the south.

Edson made other practical contributions. In July 1895, responding to a directive from Washington and in line with a congressional program to give private holdings to missionary churches, Edson surveyed the land around the mission house. The object of this was to help missionaries create a *cordon sanitaire* of up to six hundred and forty acres around their buildings against the possibility of Native attack. But with the Sanborn affair still fresh in his mind, Edson more realistically saw this appropriation as a measure that would "enable them to warn off any adverse settlers who might be disposed to locate near the Mission" (Edson, Letter of July 27, 1895).

The survey helped Edson prepare the ground for moving the mission. Tools, lumber, and food for the workmen arrived in July, and Edson, with Driggs and a group of Jabbertown helpers, moved the building in the first week of August. Edson employed and took signed receipts from Joseph Tuckfield, C. Wall, James O'Hare (his mark), John Hackman, George Walker, Henry Hagbery, and Henry Koenig. Most of these received about four pounds of flour, fifteen pounds of shot, one box of primers, two pounds of coffee, and three and a quarter pounds of tobacco. Why the missionaries employed Jabbertown men in preference to Point Hope Natives is not clear, and it is slightly odd to witness the white men in receipt of the same goods that they themselves habitually traded with Eskimos. Some of Edson's payments no doubt went later towards goods exchange between Jabbertown and Point Hope.

Following Driggs's departure for home on the *Bear*, Edson embarked on some practical initiatives. To promote hygiene, he bought fifty pounds of "common soap" for distribution in the village. Then, before school opened, he constructed three new tables, each with space for books and slates, which "at once, without slightest friction, produced order and an excellent degree of attention in the study of lessons" (Edson 1895–96). Admirable as this contribution may have been, the subtext to Edson's claim once more suggests that it was he, not Driggs, who started to bring "order" to an otherwise chaotic situation.

Notes

1 It remains merely to be said that the shaman Ayauniq was a practitioner of this period and that he survived the epidemics, dying around 1910, a date that can be approximately determined from Asatchaq's childhood memories of Ayauniq: largely on account of an episode in which the shaman tried to pass off his own urine as home-brewed whiskey (Lowenstein 1973–89)!

2 Even Driggs was not resident in 1887. The detail may have come from the Jabbertown men who arrived that year.

3 Point Hope iglu dwellers continued to light and heat their homes with oil lamps through to the early 1960s. But paraffin lamps and, by the mid-1960s, locally generated electricity now made reading and writing possible. Nonetheless, most people were living in wooden cabins by the 1950s; the last iglu dweller, Nanny Uyatauna, moved out of her (electrically lit) iglu in August 1973. But while home study was thus made possible in the early mid-century, post–eighth grade students in the 1950s and 1960s were forced to leave the village for BIA schools in southeast Alaska and Oregon, until high school provision in the village was instituted in 1974.

4 "Observe degree, priority, and place/Insisture, course, proportion, season, form,/Office and custom, in all line of order" (*Troilus and Cressida* I.3)

CHAPTER 21

Sunny Teachings: Death and Resurrection

Driggs and Edson: Religious Instructions

Driggs had refrained, perhaps out of a sense of tact towards the host community and a degree of uncertainty about how to proceed, from pushing too hard at Point Hope's traditional beliefs—and perhaps also because he suspected that it would be pointless to make too early an attempt. In 1889, Commander Stockton had recorded the Point Hope creation story and written how the "long winters give them an opportunity to keep alive their traditions" (Stockton 1890:195). Similarly, Driggs, in a short book published in 1908, wrote of Point Hope's shamans that they "have undoubtedly rendered a service to their people in the past by acting as their historians in preserving their traditions" (Driggs 1908:53–54). These views did not preclude evangelical work. They do, however, express the thinking of open-minded Episcopalians. Partly for these reasons, it was not until 1892 that Driggs found anything to say about Christianity. And then his comments, at once regretful and optimistic, were that he "found it impossible to organize a regular Sunday school," although "I have spoken as freely as my knowledge of the language will permit, and sincerely hope that some of it will be like seed sown in fertile soil" (Driggs, Letter of June 4, 1892).

Driggs was not happy about his lack of progress. But rather than indulge in self-recrimination, he wrote with familiar and understated vagueness, "There has been much to contend with, which is undoubtedly the history of all new Missions," and he weighed these difficulties against what was more positive: his admiration for the Point Hope people. "I have been particularly favored in the people I labor among...although we look upon them as savages, yet they have traits to be admired"(Driggs, Letter of June 23, 1892).

But Driggs was never entirely passive, and he did admit in 1893 to some conversations on the subject of religion and subsequently to opening Sunday school for about forty students in the same year. Having "gained a slight insight into the beliefs of the people," Driggs now felt that both parties might be living in mutually recognizable religious worlds: for "some of them, perhaps all, are believers in a future existence." This shared metaphysical belief might offer ground for conversion and in this connection, Driggs told the following story:

> Last month the mother of one of my pupils died, and he told that previous to my arrival, an older brother and sister had died, and that ever since his mother had wept a great deal and wished to follow them to [their unknown land]. They believe that four

> or five days after [death], that some departed member of the tribe will visit the grave and conduct the spirit to its home, and so this woman was happy as she [died]. She believed that her children would soon arrive and welcome her to that eternal home where life's sufferings have passed. (Driggs, Letter of May 16, 1893]

Driggs uses this anecdote to illustrate an assumption that traditional Point Hope religion and Christianity had something in common, unconscious of the fact that his student's story was an expression of a millenarian cult (*uivvaqsaat*) that had arrived from the Kobuk and would reach both Point Hope and Barrow in the 1880s. Since *uivvaqsaat*, as we shall see, contained some Christian influence, Driggs was therefore right in conflating what he assumed to be traditionally Native with the Good News that he should have been conveying. All this will be explored in more detail in Chapters 23 and 24.

The doctor's point of entry for his Sunday classes consisted of a mixture of Bible stories, pictures, and maps of the Holy Land. It is hard to imagine how Point Hope children might have interpreted what they heard and saw. The maps cannot have meant anything, although nineteenth-century biblical illustration, with its robed men and women and bearded kings and prophets, must have provided pleasantly mystifying entertainment. Driggs wrote:

> I have found the pictures, especially the "Sunny Teachings," of great aid. I first show them the picture; then read from the Bible. Then turning to maps, I show them [where] these events took place, and tell them the story.
>
> These instructions are in the Native language, which is not an easy task. Yet the children understand me and seem to be deeply interested. It would be impossible to have a more attentive class. That these instructions may [lead] these people unto Him is my earnest wish. (Driggs, Letter of May 16, 1893)

It is wonderful to contemplate the gentle Driggs with his maps of Israel and Judea surrounded by a mesmerized crowd of Iñupiaq children as he inducted them, in polysynthetic morphemes, into Judeo-Christian religious geography. Clearly they liked it. Driggs wrote: "a number of scholars have already well advanced enough to commence reading the Bible with my aid, and a number of others who will soon do so."

The following year Driggs struck a more pessimistic note. The population of 1892 had gone down by one fifth, and Driggs wrote in the spring of 1894:

> The work is gradually making progress, but I find that the shifting population is a difficult problem to deal with. There are many that go away and others that come for a season… Each fall it is almost like beginning over again.
>
> In my religious instructions, I can not yet notice there has been any very great impression created. It seems hard for them to appreciate that there is a loving Father in heaven, whose children we are, and whom we should serve. I have begun teaching our church services, having them read the Psalter, and also responses to the Litany. Perhaps it is somewhat beyond their capabilities at present, but there are a few advancing so nicely that they will be able to read with more ease… (Driggs, Letter of June 8, 1894)

To confess no "very great impression" was brave. And certainly according to Edson, Driggs had made virtually no progress. Language difficulties, wrote Edson in 1895, represented the main problem:

> From the first we have found it very difficult to turn our thoughts into their words because of two things. 1. Our meagre knowledge of their language and 2. The paucity of the language in words that express religious and spiritual ideas. They had no idea of God, or of a Heavenly Father. (Edson, Letter of July 3, 1895)

And then Edson told a story that must have seriously shocked the Board of Missions:

> One of their old men went to a whaling station and asked the Captain [Bayne] this question: "What is this the new white man tells us about God and that we have a Heavenly Father? Is what he tells us true?" When answered that it is true, he next asked: "Then why have you not told us that long ago?" To which answer was made, "Oh! I knew a man would come bye and bye on purpose to do that."

Driggs, as missionary, had so far achieved nothing—this, at least, was Edson's message.

What Edson may not, of course, have seen, were the quieter aspects of what Driggs had accomplished. Edson, who was something of a performer, evidently enjoyed making a splash, and his arrival in 1895 caused a stir. We have observed Edson's preparations for Communion. On July 18, 1894, the rites of the church were celebrated in Point Hope for the first time. To witness this event, an astonishing 145 people crowded into the schoolroom. The subject of Edson's sermon was "God, and all men His children by nature" (Edson, Letter of July 3, 1895). Driggs interpreted—though Edson (so he claimed) was soon preaching in Iñupiaq.

Once the ceremony had taken place—Edson's ecclesiastical robes conforming usefully no doubt to the draperies of Bible illustration—Christianity had become a Point Hope institution. The social aspects of Driggs's schoolroom in its character as *qalgi* had acquired dramatic ritual status. The new religion was now ready to compete, at least on the level of theater, with shamanism.

Soon after Edson spoke publicly about a "Heavenly Father," the phrase appears to have taken on local significance. We have seen how important the notion had been to the man who took it to Bayne. In a letter of the same period, Edson also describes the experience of a young woman (a "witness and martyr") who had refused both to take part in spouse exchange and "who replied to the enticements of a white man—'*God will see me*'" (Edson 1895–96:63; his emphasis).

While we may question the authenticity of the reported statement, there are also grounds for its approximation to truth, but perhaps as an expression of religious syncretism rather than as evidence of conversion, for the concept of a loving and watchful heavenly father might very well have coincided with Point Hope's relationship with local powers.

The Iñupiaq spirit world was of course forever at hand. But whereas the Christian God was taught to be a benign manifestation, for Iñupiat people the spirits were generally an ambiguous and often dangerous presence. Taboos demanded by the souls of ancestors, animals, and spirits of many kinds impinged on every action, and probably the most significant of the local powers was the moon spirit (*tatqim iñua*), who provided the game animals and maintained a sometimes punishing vigilance over human behavior. Just as *tatqim iñua*, gazing down through the sky hole, might observe and avenge a taboo infraction, so the Heavenly Father appeared to be keeping an eye on sexual behavior, of which Edson, in particular, vociferously disapproved.

The convergence of a Heavenly Father who disliked non-Christian sexual activity and the Moon Spirit of local myth who had himself been a sex criminal cannot go unnoticed, and the coincidence of these beings may have given a strangely paradoxical momentum to missionization. Of course, the young woman's comments may have come from Edson's imagination or some simple misunderstanding. Or she may have rebeled, as women sometimes did, against a particular spouse-exchange arrangement and Edson supplied the motivation. A third possibility is that she did say "God will see me," although the construction of the sentence leads one to suspect that Edson had lifted or modified it from a pietistic tract whose language it resembles.

It would be misleading, however, to differentiate entirely between Edson's and Driggs's attitudes to Iñupiaq sexual life. Like Edson, Driggs disapproved of Iñupiaq concubinage with white men. But while part of the emphasis of Driggs's critique was medical, the sexual issue lay at the center of Edson's religious views:

> They are polygamists, and worse than that, trade wives with each other for short seasons, the women entering into the project as heartily as the men. Their marriages are entirely of convenience. It is not uncommon that a woman has had several living husbands as they take one another 'to have and to hold' as long as each party is satisfied. (Edson, Letter of July 3, 1895)

But on the complex subject of relations with white men, Edson grew more insistent: "From dissolute whalesmen and traders on this coast they have learned the worst vices and contracted the worst diseases of the whites."

For Edson, then, venereal disease represented an aspect of moral disorder. It was for this reason different and perhaps worse than flu and measles. But to his justified outrage at white men's predations, Edson seemed also to be expressing a personal antipathy:

> They are licentious. I could relate many things that have come to my notice and knowledge proving this, but they are too vile to express. I will only say that there are but few women young or old in Tigara Village who do not prostitute themselves to white men for bits of tobacco and other trifles, and married women are worse than those not married, their husbands in some cases consenting providing they share in the rewards. Nor will a wife object to her husband using other women *providing he lets her know his intentions*[Edson's emphasis], if he does not inform her she will pout for a few days and then it is forgotten or forgiven; and all this is done not only with the knowledge of their children but nearly always in their presence.

Compounding all this was the alcohol problem:

> There is now on this Point the very white man [Bayne] who taught them to make the apparatus for distilling an intoxicating liquor from flour and molasses. I had occasion to talk earnestly with this man about the evils of his life and learned it from his own

> mouth. Now there are but 3 or 4 houses in the village in which brewing is not carried on during the winter, and drunken sprees are very common.

Edson's response to the whole mixed community's conduct finally boils over in a shriek of revulsion, but there is no evidence that he stimulated in the Board of Missions any remedial action:

> If there is a worse spot on earth I have never heard of it. I know it will seem incredible to you, yet it is sober fact. *And our mission as at present equipped is powerless to correct these evils* [Edson's emphasis]. We need Christian women here but they would require hearts of granite to sustain them in their work. Both the Doctor and I have been obliged to rebuke women for their wantonness, and speak to them as a man should never be obliged to speak; but if we did not do it no voice would be heard in protest of this evil. Fathers have prostituted their daughters on the ships, and their wives too, and did not lose standing in their community because of it. One of the principal men in the village and who has been highly spoken of by some U.S. Officers who have visited these shores pimps for the whale ships.

In Edson's view, one root of all "these evils" lay in the absence of a perceptible ethic in Iñupiaq religion:

> As to their religious beliefs—they firmly believe in their immortality; but I found them without any conception of God and without a name for Deity. They have a tradition of the creation of their own country, but the creator was a man like themselves—only with a raven's beak projecting from his forehead. They believe in an Evil spirit whom they call "Tooni-na" (*tuunġaq*);[1] he is the author of all their woe, misfortunes and their sickness. They fear him exceedingly and endeavour to propitiate him with gifts and ceremonies, hence they are devil worshippers. They have none of the skill of Indians in medicines, but rely entirely upon the incantations of the medicine men, who, they believe, can change the adverse winds, bring whales, seals, walrus and fish and secure for them success in hunting.

To add a further layer of negativity, those Point Hope people who did take an interest in the new teaching did so, believed Edson, for the wrong reasons: they thought that the white man's God might help them in their hunting. As Edson wrote: "Congregations number from 15–50 according to weather. A large number expressed the desire to learn to pray and to observe the Lord's Day. But it very soon became evident that they desired to pray for greater success in hunting." "Purity of morals" lay close to the center of Edson's vision, and it is not hard to detect in his language the desire to disinfect—morally, spiritually, and physically—these contaminated people, "and induce them to abandon their filthy and immoral habits."

Edson, unlike Driggs, was repelled by the "filth" in which Tikiġaq had so far lived cheerfully, which to Edson was both physical and moral, the one no doubt being an expression of the other. And if Christ's blood represented the ultimate purifier, a lot, on the gross level of hygiene, might be accomplished with a good supply of soap and towels: "They disgust me every day in things too horrible to relate… But the work is hard—oh so hard! and perplexing. A gleam of light and therefore of hope is their daily call for soap and towels. Send us soap and towels in plenty."

In feverish existential arousal, Edson wrote, as he expressed it, in "a condition of fear and trembling." He was, he exclaimed, "alone—yet not alone": God was with him. And no doubt he was so excited to be in this commanding solitude of responsibility that it was a relief not to have the imperturbable Driggs there simply getting on with the job.

Edson: "A Devil Man"

Edson had wanted to hold communion every Sunday, but for practical reasons, he settled for once a month. Then early in 1895, his services stopped on account of a grotesque neo-shamanistic event, an account of which had traveled from the Noatak and which Edson described as follows:

> Shortly after Xmas a Native from the country east of this region came to this village bringing word that a Native in his country had died and on the third day after his death come to life again. That during the interval between his death and revival he had been in the place where the Eskimo go after death and there had received a message to the Tigara Natives which this man had come to deliver to them: "That it was bad for them to attend our services and that they must not do it any more." (Edson, Letter of July 3, 1895)

A number of elements converged on this episode. First, while the "third day" detail clearly comes from Christianity, the period of suspension between death and resurrection coincides more or less with a similar four- or five-day interlude between two stages of the death process in Iñupiaq eschatology (see Chapter 13). In that sense, the story represents a culturally compound religious phantasia. Second, like the story Driggs had told about the grieving mother, this was also probably a reflex of the *uivvaqsaat* revivalist movement, which preceded the arrival of missionaries in the Kobuk and Noatak area. (Robert and Carrie Samms established the Friends Church in Kotzebue in 1897. And even then, their early teaching is unlikely to have contained much Christian eschatology.) Third, somewhat in the genre of inter-regional shamanistic rivalry, it represented a long-distance assault on Edson, perhaps, in his character as a supposed counter-shaman. Edson, in having made a colorful spectacle of the new religion, had after all rendered himself a conspicuous new target. And given how emphatically he also sermonized on the subjects of alcohol and sexuality—which for some people represented profitable economic activities—it seems likely that the author of the resurrection charade must have had some interest in the continuation of these activities. That possibility was not lost on Edson:

> Shortly after arriving at Tigara my name was added to [the story], to the effect that I was a "Tooni-nu-man," a devil man and "ah-choor-ak," bad. In our instructions we had told them it was bad for them to make *ton-ga* (whiskey) or to drink it. Also that it was "bad" for them to allow their women to go on-board ships for immoral purposes, or for the women to prostitute themselves to the white men on shore. Thus our teaching had hit at their fondest sins. Some of them in private admitted that what we had said was true and good. Others, it seems, have resented the teaching. Evidently

> they have held me only responsible for the teaching and considered that the Doctor as only the interpreter.

Edson eventually worked out that the messenger from the south must in fact have been suborned by a Point Hope man who both distilled liquor and procured women for the whaling ships. Edson thus correctly interpreted the story as an attack on himself and his teaching. But it was Edson, not Driggs, who was being threatened ("It was from this Native's house that the story went forth to our Natives with the addition of my name to it"), and he well understood that recent history was stacked against his interventions. The most striking aspect of the event perhaps lay in the fact that a parodic Christianity had been contrived in order, as Edson remarked, to maintain "their fondest sins." But in another way, the incident perhaps expressed a Native outlook on the contact situation as a whole. If white men had introduced the ambiguous pleasures of liquor consumption and profitable but dangerous inter-ethnic sexual commerce, the new white man's religion, emanating as it also did from Euro-American culture, should also logically belong to same cultural package.

The second mishap took place later when Edson was again ascribed shamanistic power. It was spring 1895, and Point Hope's whale hunt should have started. But in the absence of a north wind to open the shore leads, hunting was delayed. Edson wrote:

> During the first part of the whaling season the winds were unfavorable for many days… A white man whom the Doctor has had at times to do chores around the Mission (James O'Hare a ship wrecked sailor who has declined to back to civilization and has made his home in Tigara, having taken a Native wife) came to the mission House and told me in the Doctor's presence that some of the Natives were talking about me, alleging that I was the cause of the bad weather. And also that it was because I had come here that so many children had died the past summer.
>
> He also told me what I had not known before: that my name had been associated with the [resurrection] story from the East. I then asked the Doctor why he had not informed me of that. He had only told me of the alleged death and resurrection. He replied that he had "thought it best not to do so." Then in confusion O'Hare said: "I supposed you knew that or I would not have said what I have today."
>
> However the winds became more favourable, whales came, and now the Natives have a supply of meat. I must confess that the knowledge of these matters made me feel uneasy respecting my safety, but that has largely passed away…

Once again Edson finds himself at the center of an intercultural tangle, this time connecting him to the belief that shamans could maliciously stop game animals from entering the region (*niġġutaiḷi*). Suuyuk, whose murder sparked the Ataŋauraq feud (see Chapter 9), was one such practitioner and there were at least two other episodes during the contact period. In one story, a Kobuk shaman who was working at Jabbertown in spring 1900 was invited to open the sea ice, to find that his efforts were being, somewhat paradoxically, obstructed by a rival Point Hope shaman (Lowenstein 1973–89: told by Umigluk David Frankson, 1977). In a separate story, elaborate intershamanic Point Hope hostilities ended with the death of the malicious shaman before the sea ice would open (Lowenstein

1973–89: told by Asatchaq, 1976). Driggs himself is not recorded as having been accused of *niġġutaiḷi*. The more flamboyant Edson would more easily have created a spiritual milieu that might be interpreted as shamanistic interference.

Of added interest here is the conversation of the three white Americans. While Driggs and O'Hare express foreknowledge of the accusations against Edson, Edson had been unconscious of his role. The episode provides an intriguing vignette of two experienced Euro-American Point Hope residents observing a moment in history where traditional Point Hope beliefs and the new religion, in mutual misunderstanding, dramatically converge. Clashes such as these had apparently never happened during Driggs's earlier, more cautious ministry, and after Edson returned south, there is no record of a recurrence.

Notes

1 While this was the term for several kinds of local or disaffiliated spirits, it was sometimes used with reference to the more powerful spirit of the moon.

CHAPTER 22
Disease in Point Hope and the Great Sickness

> Taking it all through I have had quite an extensive medical practice, more than I can well attend to a great deal of the time.
>
> —John Driggs, 1892

> There is a large Native colony at Point Hope, but here, as elsewhere, they are rapidly decreasing; sicknesses and epidemics, from which we are more or less immune, are the death of many a Native...
>
> —Ejnar Mikkelsen, 1909

John Driggs, Physician

Driggs had seen death in plenty at Point Hope by the time he wrote to his old medical friend in March 1892, but the worst epidemics had yet to come. When the Danish explorer Ejnar Mikkelsen visited briefly in 1908, Point Hope's population had dropped to half its 1890 level of about one hundred and sixty (Jackson, *Travel Diaries*, 1890–91). By the time of Mikkelsen's visit, the worst was over. But the years 1893 to 1905 were perhaps the most distressed in Point Hope's nineteenth-century history.

Driggs's assignment from the Board of Missions had been to set up a school and to Christianize the village. And while the Board assumed that he would, in the manner of a reservation agent, attempt to stabilize Native life—and that this would lead Natives to good health—the doctor had been employed as a missionary teacher and not as a physician. Driggs's career spanned a momentous period of medical history, which coincided with major advances in scientific medicine. Between 1879 (just before Driggs entered medical school) and 1900 (shortly before he was deposed) came the research by European scientists that provided the foundations of microbiology, which would take surgery and disease treatment into the modern era. Much of this work was relatively slow in reaching America, but even by 1890 Driggs must have been aware of Pasteur's discoveries (his development of a rabies vaccine in 1888 would have delighted any doctor surrounded by dogs and foxes), Koch's identification of the tuberculosis bacillus in 1882, and most significantly the experiments of Joseph Lister in the use of antiseptics in the 1870s.[1] All these advances, particularly Lister's application of carbolic-acid sprays in surgery and wound treatment, would soon revolutionize medical practice.

Driggs and his patients were, however, beneficiaries of one crucial development in American public health. While Civil War doctors had no knowledge of germ theory, the filthy conditions responsible for high death rates after Civil War battles shocked American physicians into more hygienic care regimes. As in Florence Nightingale's Crimea, hand- and instrument-washing, clean dressings, and the ventilation of sick rooms gradually became standard procedure. But Driggs did carry a range of up-to-date surgical, dental, and diagnostic equipment. Among his tools were the laryngoscope, the ear speculum (Eskimo children being liable to ear infections), and the stethoscope, which had been perfected by American technicians in the early 1850s. Given that pulmonary diseases loomed large in the epidemic cycle, the stethoscope was one of Driggs's most powerful investigative tools.

What Driggs carried in his pharmacopeoia is speculative, since a nineteenth-century doctor's choice of remedies was discretionary and dependent on what was available. Medicine had progressed from the "heroic" period of bleeding and cupping, but the remedies used by scientifically trained doctors were often indistinguishable from those of rural apothecaries and herbalists, and were no more effective. A glance through some seven hundred and fifty entries in King's *American Dispensatory* of 1898 gives a fair idea of the botanical preparations Driggs had to choose from:

- Aconite: for "asthenic or adynamic states"
- Garlic: as a stimulant, diuretic, expectorant, and "rubefacient"
- Anemone: for eye diseases, secondary syphilis, whooping cough
- Belladonna: for impairment of circulation; large doses paralyze, small doses stimulate the nervous system
- Cannabis indica: for nervous depression, irritation of the urinary tract, wakefulness, menstrual headache, stitching pains in the heart, hallucinations
- Camphor: in large doses a narcotic and irritant; in small ones a sedative
- Cydonium (quince seed): for insomnia, fevers, morbid watchfulness, muttering delirium, quick, irritable pulse
- Digitalis (foxglove): for febrile diseases, inflammations, insanity, neuralgia, whooping cough, palpitation of the heart, epilepsy

Other remedies would have included cocaine, opium, ipecac (an emetic root from Brazil, for dysentery), atropine (an anti-spasmodic), liver pills, medicinal wines, and preparations of mercury, strychnine, and arsenic. Raw ingredients were dissolved in alcohol, iodine, and spirits of ammonia. Ether and chloroform were in standard use for anesthetics by the late nineteenth century, and Driggs is recorded as applying ether before surgery.

As described in Chapter 16, Driggs lost most his drugs when he landed in 1890 and had to wait a year for replacements. But whatever he salvaged would have made little difference. The tuberculosis, pneumonia, bronchitis, measles, scarlet fever, meningitis, and whooping cough that afflicted Point Hope were all beyond cure. Bedrest and fluids were the only remedy for influenza, though this often remained fatal. Syphilis and gonorrhea were as intractable in Europe and the American mainland as they were in north Alaska.

Driggs thus worked during a period in which scientific rationalism was attempted before medical science was advanced enough to support it. And while Driggs might no more have cured patients in New York than he could in the Arctic, perhaps it was the pre-scientific nature of Point Hope's belief systems that enabled him to be relatively successful in these difficult circumstances. With demonstrable surgical skill and a bold assumption of authority, Driggs generated faith in his therapeutic skills. This, as with John Simpson's midcentury practice at Point Barrow, represented a version of what people expected of shamans. Both medically and religiously, Driggs had introduced new forms of magic, which, like shamanism, sometimes worked, but mostly didn't. Injuries requiring surgical intervention, often helped by anesthetics, Driggs could often deal with successfully, and some of his alcohol- or opium-based tonics also offered temporary sedation. But these may have represented the extent of his medicinal therapies. "The pharmacopoeia was a bag of blanks," the medical historian Roy Porter has written:

> As of the late nineteenth century the few medicines that were effective included mercury for syphilis and ringworm, digitalis to strengthen the heart, amyl nitrate to dilate the arteries in angina, and little else... The old time American doctor of the late nineteenth century was said to get by with calomel, opium, buchu (a diuretic), ipecac (an emetic and evacuant), and Dover's powder (used for colds and chills). (Porter 1997:674–75)[2]

There is no record of Driggs using any of these remedies.

Driggs and the Shamans

It was customary for Euro-Americans of the nineteenth century to look down on what they believed were the pretensions of shamanism. Driggs's attitude to shamanism was, as we have seen, ambiguous (see Chapters 18 and 20). But more important is the picture, sketchy as it is, of the manner in which Point Hope adapted Driggs's medical interventions to their own tradition. Both hostility and acquiescence, as we saw with Dr. Simpson's work in Barrow, may have greeted Driggs's endeavors. But given the Iñupiaq laity's ambivalence towards its shamans and often intense intershamanistic rivalries, this was in the nature of things. An astonishing incident that took place on the *Bear* just ten days after Driggs arrived in Point Hope further illustrates the impact that surgery could have on Native people. Driggs, after his difficult landing, was still camped on Point Hope's south shore. And as the *Bear* stood in at the Point

> the boats were lowered and some of the men went ashore to help build [Driggs's] schoolhouse... One of the Natives here had had his hand shattered with powder during the winter, so that the bone of one of his fingers protruded. Our doctor [Call] told him it must be cut off. This he did not relish and said it would hurt. [Healy] told him that the doctor would give him something to put him to sleep so that he would not feel the pain. He was still incredulous and hesitated, but finally said, "You good man, I do what you say." The doctor then chloroformed him and fixed his hand all right. The

> other Indians [sic] had never seen anything like it before and thought him dead. There was quite a stir among them as they came and looked at him. But when he awoke their surprise and utter astonishment was beyond description. The patient was delighted and followed the doctor and captain calling them "good men." (Mrs. Michael Healy, July 1890 article in *Boston Transcript,* in Jackson 1890–1900)

Of most interest here is the combination of successful surgery with a display of quasi-shamanistic theater, in which Dr. Call appears to kill and revive his patient. If doctors could perform such miracles, then Driggs's medical status would also be assured. There is a photograph of the new "village headman" Anaqulutuq, who befriended Driggs, on the deck of the *Bear.* If he, a shaman, had witnessed that anesthetic, Driggs would have had a powerful advocate.

Two anecdotes from later in Driggs's career illustrate ways in which the imported American medicines could sometimes be culturally assimilated. Both instances come from Sheldon Jackson, who recorded in his journal how some Point Hope people had adopted medicines into their amulet system—in one case without having ingested anything:

> Doctor Driggs gave one of the sick Natives some powders to take. Meeting four months afterwards, the patient was profuse in his thanks, saying that the medicine had completely cured him…and ended by pulling the package of powder out of his pocket to show that he had not lost them.
>
> At another time, meeting a funeral procession, it was stopped by the widow, who wanted to tell the missionary how much his medicine had relieved her late husband. And as a token of their appreciation, the corpse had the bottle in his hand, taking it to the grave with him. (Jackson 1890–1900)

Given that these remedies belonged to what Roy Porter has described as a "bag of blanks," they presumably worked as talismans as they would have done within the traditional system. In this way, Driggs had an entry, no doubt unwittingly, into local shamanistic custom. And later in his career when Driggs described interactions with shamans, it is clear that they used the doctor as they did each other, pragmatically and competitively, as the need arose. In cases such as the following, Driggs's medical and religious initiatives became awkwardly tangled:

> A young man called and asked if one of shamans attended church. I remarked that I thought he did occasionally. He said that he did not and then went home and accused the shaman of making his child ill by not attending our service.
>
> The following day the shaman called and said if I would cure the child he would appreciate it very much and would be a better church attendant in the future.
>
> He stuck to his word as long as the child was sick but was not quite as regular after its recovery. (Driggs, Letter of June 28, 1900)

Driggs's missionary status seems here to have complicated therapeutic relations. First a young father, who, like most of the village, embraced both shamanism and the church, claims that a shaman's non-attendance of church made his child sick. The offending shaman is then shown also to inhabit the two worlds, and agrees to attend church if

Driggs will cure the child that he has harmed. Finally, the shaman comes to church until this specific action has, through his contact with the new institution, made the child better. After this he can stop attending. Church, then, is represented as a therapeutic point of entry that has joined the traditional world of magic and also provides a context for Driggs's medical power. The shaman's attendance thus becomes both an act of healing and a "fee" to Driggs, who has performed as both doctor and priest.

Driggs's next story shows how such events could snowball. Having demonstrated his skill with one tribal doctor, it then becomes possible for a second shaman to apply for the same treatment:

> Living in the next iglu was another shaman. He also had a sick daughter and had tried all his arts on her but in vain. He was feeling very much discouraged for he was afraid she was about to die and required me to attend her. I found the girl very ill and delirious with typhoid fever. But after a protracted illness she recovered. It was the first case of typhoid I had seen in this section. (Driggs, Letter of June 28, 1900)

Thus Driggs's practice was absorbed, as though by chain reaction, into the community of shamans, and Driggs's missionary status helped provide authentification for his position as a healer. But given that Driggs could offer no explanation for visitations of sickness, traditional theories of causality remained current until about 1920. There are, for example, several stories from that period about a shaman called Isuaġalaaq, whose actions were claimed to have killed Kiḷigvak in about 1923 (Lowenstein 1973–89). Much later, Masiiña claimed to have gone on a spirit journey to Russia where, in March 1953, he killed Stalin (Asatchaq, in Lowenstein 1992:196). Theories of spiritual causality may, indeed, have been reinforced by Driggs's identity as both missionary and doctor. His surgical skills and his ownership of medicines, which his patients used as amulets, also conspired to suggest that he was a new kind of shaman.

Disease in Point Hope, 1892–1904

While Euro-Americans, albeit with some acquired immunities, were relatively defenseless against the major diseases, Alaska Natives were, in truth, no less susceptible than had been the Arawaks, Caribs, Aztecs, and Mayas who had been laid to waste by the Spanish in the early sixteenth century. The diseases that erupted throughout the Caribbean in 1492 had reached their northwestern limit at Point Hope, and Driggs himself had, coincidentally, traveled an almost identical route: from Cuba where he was born, through North America, and up to the far northwest of the continent.

Driggs was presumably aware of the limits to what he could do, and until 1893, when the situation began dramatically to deteriorate, he reported very little about Point Hope medical issues. Two years into his work, and already confronted by serious disease in the village, Driggs was still being quite secretive on the subject. "There has," he wrote vaguely, "been considerable sickness and my medical services have been in almost constant demand" (Driggs, Letter of June 4, 1892). Driggs was nonetheless capable of a more disquieting note, suggesting first that "the future of the tribe" was in peril and also that he had no suitable consultation space. "Another difficulty," he explained, was "to get them to take their medicines as directed." Still, Driggs, who profited from the reputations of Revenue Cutter physicians Rosse, White, and Call, had established himself securely by 1892. Not only was the laity consulting him in large numbers, but, as mentioned already, Anaqulutuq, "the leading Native doctor," was under his care.

Driggs's next mention of disease comes in 1893, when he reported to the Board of Missions a food-poisoning epidemic. A dead whale had washed up on the beach, and this "was stored away for winter use":

> Then during the long night they commenced eating the meat and with that commenced an epidemic of sickness. I warned the people against it, but they are like children, they wanted the blubber... In those cases where the meat was very freely partaken of, a fatal termination resulted within a week, in other cases, the primary severe symptoms were passed through...recovery being slow. (Driggs, Letter of May 16, 1893)

This disturbing incident has its historical interest in that food poisoning did not, of course, represent a contact malady, though postcontact scarcities may have rendered the community less cautious about eating contaminated food. But here again, Driggs provides a disappointingly incomplete picture of the event and fails to mention the number of people who fell sick or died—though his use of the word "epidemic" suggests that it was a major incident.

By 1893, when the whale-meat poisoning happened, Driggs had become an authoritative public figure who walked regularly to the village and went his rounds as John Simpson had done at Point Barrow. In a letter that spring, he describes forcing his large person into the smallest iglus:

> I first have to climb to the top, then lower myself through a hole in the ground, after which, dropping on to my hands and knees I feel my way along the underground passage until I arrive at a depression in the earth, then turning my head upwards I climb through a hole in the floor, which admits me to the dimly lighted room, in which the family live densely crowded. Daylight is admitted through a hole in the roof covered with a portion of a seal... (Driggs, Letter of May 16, 1893)

With better community health in view, Driggs was already thinking ahead to when he could rehouse the entire village, but his dream of building a lumber mill for processing driftwood never came to fruition. Still, he was right to suggest in the same letter that "crowded underground homes are not the proper places for the bringing up of young children," although the "better homes" that he visualized would have remained airless incubation grounds of sickness.

So far in his letters, Driggs had identified only a single infectious disease: one of his "scholars" had died from tuberculosis in 1892. But summer 1893 brought an onslaught that Driggs now openly reported: "there is a considerable amount of pneumonia and bronchitis among the Natives, caused by the most wretched summer weather which we are having. Nothing but rain and fog, a glimpse of the sun is a rare treat" (Driggs, Letter of July 26, 1893).

Dr. Rosse of the *Corwin* had also blamed bad weather in the Aleutians for pneumonia. Given that Natives had always lived healthily in these conditions, this was not the whole story. Now, however, cold, rain, and hunger exacerbated imported infections and even the imperturbable Driggs was shaken during that horrifying summer. In September he wrote: "There has been a severe epidemic which has proved quite fatal among the adult population. The people became thoroughly alarmed and went away taking their sick with them, in several cases leaving their dead either partially or entirely unburied" (Driggs, Letter of Sept. 6, 1893).

As we have seen, taboo demanded that the dying were taken outside, for the phenomenon of death was itself metaphysically contagious. The village was becoming, like the deserted iglus up and down the coast, a place of avoidance; and as in other Native communities, there was no time for the performance of death rites. Driggs now penned a passage of rare circumstantial observation, and in two swift paragraphs sets Point Hope within the landscape that had so disfigured the Americas since the late fifteenth century:

> Near one tent I found the father of the family laid out on the ground in the rain to die and no protection placed over him. At another tent a family had been deserted by his two sick sons, the mother being already dead and the father soon following her.
>
> Noticing on the beach a piece of a tent I went up to investigate and found a small boy under it, very ill. Then passing my hand further under the cloth to find out what the object was that he was nestling to, I found it was the body of his dead mother, while a short distance off lay his dying father near the unburied remains of his grandfather, and so while making my trips through the village visiting the afflicted I witnessed many a distressing sight. (Driggs, Letter of Sept. 6, 1893)

Sheldon Jackson, who came off the *Bear* in August 1894, elaborated on Driggs's story:

> There were five dead in that group. Three quarters of the adult population were sick and one out of every six died. There were not sufficient people in the village to bury the dead—and the bodies were left to be eaten by dogs and their bones are still seen all over the village. Dr. Driggs estimates that the death rate to the birth rate is at 4.5 to 1. (Jackson 1890–1900)

Besides outlining the heartbreaking circumstances of sickness and death, both reports also implicitly suggest the social and spiritual impact that this epidemic would have on a small and vulnerable community. It would be a long time before the village could re-establish itself, when family life had been shattered, its rituals marginalized, its habitations neglected, and its dead overrun and consumed by the dogs that were otherwise partners in subsistence.

Driggs anticipated the worst for 1894, and before the ships arrived to bring more "summer sickness," he wrote:

> My medical aid has been in considerable demand; in fact should I attend to that line of my work properly, I should have but little time for anything else. I have been forced to draw the line at making outside calls except in very rare cases, my duties at the mission keeping me constantly employed. (Driggs, Letter of June 8, 1894)

Driggs worked alone and had established himself as a leading mage on this peninsula of by this time impotent magicians. But the doctor also welcomed support from nonmedical quarters. Thus of the 1893 epidemic he wrote: "Last summer Captain Healy...upon learning of the sickness I had to deal with, sent a gift of medicines. It was timely aid for the demands made by the Natives had well-nigh exhausted those drugs suitable for the cases I was dealing with" (Driggs, Letter of June 8, 1894).

The year 1894 nonetheless brought some remission. Jackson noted no summer sickness, and when Edson arrived in July, he reported the village to be in "general good health." But in 1895, while the disease situation had stabilized, subsistence remained in crisis. "Already," wrote Edson, "the deaths among them from starvation have been alarmingly frequent." And more starkly:

> This tribe is dying out rapidly. There are not at present more than one half the people that were only eight years ago.[3] The mortality among the children is frightful and they are not a fruitful and prolific people. From 1891–1895 there were born to Native parents 30 children; of this number seventeen have died. There were also born eleven illegitimate children having white fathers, six of whom have died. Fourteen children have died the past year. Infanticide is very common. (Edson 1895–96:66)

Edson's numbers correspond to Jackson's earlier guesses. In 1890 Jackson had arrived at the figure of "about 300" for Point Hope's population, while in 1891 he wrote that the "village has a population of 161 which is about half of those who belong here" (Jackson, *Travel Diaries*, 1890–91).

Swinging between hope and the fatigue reported by Edson, Driggs appears temporarily to have lost initiative. Ordering medical supplies in 1895, Edson had found "Dr. Driggs medicine order very slight." But Edson's school attendance records for 1895 suggest that the population had stabilized somewhat. In 1893–94 fifty-four pupils had been registered with an average daily attendance of thirty-seven. This rose in 1894–95 to eighty-six enrolled, and a daily average of forty-four. In the context of these uncertain factors, Driggs, on the one hand, declared with prophetic optimism that "there will always be a village here." Edson, on the other, cried out dramatically that "they are happy in the death they are dying" (Edson, Letter of July 3, 1895)—which alluded as much, perhaps, to his notions about Point Hope's spiritual condition as to its medical well-being.

Driggs returned from his year's vacation to a flu epidemic in the village: "A wide-spread epidemic of influenza made its appearance at the end of last summer (1896). I was kept busy attending the sick. [This] retarded my work" (Driggs 1897).

But, as the doctor observed the following year, although Point Hope people might be ill or dying, their spirit was alive: "I often take pleasure in watching these people, for they are a happy set, free from the cares that beset their more civilized white brethren, yet they are showing a desire to adopt the ways of the white man" (Driggs 1898). Rather as Driggs had confessed that he had, himself, escaped "the restraints of civilization" (see Chapter 18), Point Hope people seemed to him poised gracefully between the same kind of freedom and a growing inclination towards modernization. There was almost certainly truth in the doctor's observation. And it was to his credit that he, as one of the agents of change, represented a benign presence.

The Great Sickness, 1900–1905

The whaling industry was in decline, but the number of white men on land had increased. Gold panners were moving into the Kobuk and Noatak valleys (a fruitless escapade), and, far more significantly, the discovery of gold on the beaches of Nome in 1898 lured thousands of prospectors. With both white men and displaced Iñupiat increasingly on the move, distance from those areas offered no immunity:

> During the winter [of 1898] the chicken-pox appeared and visited the majority of the tribe; then at whaling season [1899], the mumps came as an epidemic and scarcely a Native escaped...very nearly every case developed complications. It was the first time I had seen either of these two diseases [here], and I think they were introduced from Kotzebue Sound, where prospectors [have been] wintering. (Driggs, Letter of June 13, 1899)

Otherwise, wrote Driggs, "there has not been much sickness outside of ordinary colds. But few deaths have occurred, so, for the second time, I am glad to report that the birth rate has exceeded that of deaths" (Driggs 1899). This encouraging note had, however, faded by 1900, as Driggs was able to link what he observed at Point Hope to the broader movement of Native peoples. Up until this moment, Driggs had not openly suggested a causal relationship between the whalers and Native epidemics. But now, wrote the doctor, "although several hundred miles from here," it was clear to him that new diseases—measles, whooping cough, chicken pox, and scarlet fever—were coming from both old and new sources. By 1900, Nome had mushroomed into a frontier town of some twenty thousand people, and as General George Randall, who commanded the U.S. Army in the area, reported, health conditions at Nome, including measles, typhoid fever, and smallpox, were in crisis (Mitchell 1997:131). This was the beginning of what Natives retrospectively described as Alaska's "Great Sickness."

With chicken pox and mumps at Point Hope in 1898–99, the arrival of new infections in the village anticipated and reflected two fresh dynamics: larger and more mobile populations of white men on land, and the increased variety of infections that they brought with them. The one mitigating factor during this period was the presence in some communities of missionaries and doctors who could in some degree moderate the results of an onslaught.

A paper by Robert Wolfe has described the spread and effects of the Great Sickness, and while much of Wolfe's account focuses on the Yupik and Athabascan regions, three of the most relevant examples that follow are taken from his study. Point Hope's chicken pox and mumps of 1898–99 brought no fatalities, and according to Driggs's parish record, none of the eighteen deaths during those two years resulted from mass infections.[4] Point Hope's two successive epidemics, however, reflected similar but much worse events on St. Lawrence Island in 1900. There, measles arrived from Siberia in June, infecting the entire population, though with just two fatalities. But three weeks later, flu brought by whaling ships supervened, and again, the entire community succumbed, "many with complications such as pneumonia, bronchitis and entercolitis." By August, forty-four people had died, while another twenty died of the tuberculosis that followed (Wolfe 1982:105). Unlike the catastrophe that visitors from the *Corwin* surveyed on St. Lawrence Island in 1888, this time alcohol seems to have played no part.

Wolfe continues with a description of Dr. Joseph Romig's experience at an Athabascan village on the Kuskokwim:

> We used to go slopping along in the mud…through the stricken village of tents and cabins with our rumbling dead cart hauling out the unfortunate victims, while all around in the dark could be heard the fatal cough and groans of the sufferers… The filth of the village was increased unmentionably… The most pitiless wind and rain continued throughout all this time. [The villagers] were entirely depressed by their awful condition and could not make a move to help themselves.
>
> The story is just about the same for every village. There were not enough well ones to care for the sick…The misery of the people seemed to be complete. They were cold, they were hungry and thirsty and sick, with no one to wait on them. The dead often remained for days in the same tent as the living, and in many cases they were never removed… Children cried for food and there was no one to give it to them… (Wolfe 1982:111–12)

Wolfe's account of social breakdown, social and economic collapse, physical helplessness, and community disorder is repeated in the following account of a village on the Yukon. There, Dr. White, the physician traveling on the *Nunivak*, reported:

> About twenty miles above Pitkas Point we stopped at a village of five houses occupied by twenty-nine people, most of whom were sick with either influenza or measles… The place was dark, damp and dismal, the fireplace was cold, and an odour of rotting fish permeated everything… A quarter of a mile above here was another village of three houses and some ten people. All of these were sick with measles and influenza, and many had died. The surviving members of this community were half starved and helpless, and after burying the dead and leaving food and medicine for the living we were compelled to proceed on our way up the river. (Wolfe 1982:111)

Wolfe's paper also summarizes an episode at Barrow that Charles Brower describes at length in *Fifty Years Below Zero* (Brower 1960:225–29). In summer 1900, fifty boatloads of Nunamiut, so far not contacted by Euro-Americans, arrived at Barrow to trade. Brower was struck by the strength and athleticism of these inlanders. They ran foot races against Barrow and, not having been affected by tobacco, trounced them. Then a whaling ship arrived. Seeing a number of coastal Eskimos and white men go down with flu, Brower assumed that the vigorous inlanders would not be infected. But as they started to feel ill, the Nunamiut took to their boats. Within a few days all two hundred had perished.

One story brightens this appalling scenario. At Holy Cross on the Yukon, there were missionary doctors who quarantined their mixed Indian and Eskimo schoolchildren and treated measles and influenza with squill (an expectorant derived from a dried bulb), laudanum, and spirit of ether. Only one out of twenty died, and none of the children under the care of a priest at the mission fishing camp died. "Organised health care, food and shelter," as Wolfe concludes, could significantly reduce fatalities.

It would appear from one of his letters of 1895 that Driggs may sometimes have ameliorated the effects of disease at Point Hope—presumably quarantining, diet, or careful nursing, though the doctor provides no detail. But certainly his patients had confidence in his powers: "the Natives do not like the idea of my being away and occasionally ask what they shall do if taken sick during my absence, as I am the only one that can successfully treat their sick" (Driggs, Letter of Sept. 13, 1895).

But Driggs did not, at least during the summer, operate entirely alone. With the Revenue Cutter's support, the doctor worked quietly to provide that convergence of "health care, food and shelter," which was possibly all that could be done to mitigate these multiple compounded crises. There were drawbacks, however, to the simple introjection of charity, and in June 1901 Driggs perceptively drew attention to the danger of creating a dependent subpopulation:

> [T]he relief goods which the *Bear* landed last summer [were] dispensed among the widows and orphans and the old and infirm. I made the effort to give the aid to the needy in such a manner that it would relieve their distress, while at the same time I would not be doing the people an injury by encouraging them to become paupers and beggars. (Driggs, Letter of June 4, 1901)

This last comment represents one of Driggs's most incisive analyses of his relationship with the village, combining as it does his habitual altruism with an impressive comprehension of the social dynamics of aid. The same letter outlines the range of the relief initiative:

> This work has not been confined to this tribe only, but included many members of other tribes [Kobuk, Noatak, Kivalina] who had come here for relief... I found these goods of considerable aid in treating the sick, as they placed me in a position where I could supply the proper diet to those who required it.

This multidimensional approach provides insight into the impact that Driggs, the church, and the Revenue Cutter service were able to make over a wide region. But as at St. Lawrence Island and the Athabascan interior, the Great Sickness represented many repeated sicknesses. For example in 1902:

> We have had a rather long [episode] of the grippe [flu] this summer, or more properly speaking, two attacks. After the whale ships arrived, the first epidemic broke out. Then later on after the middle of August, a number of Natives came down from Point Barrow and brought with them a fresh supply of the disease. (Driggs, Letter of Sept. 2, 1902)

In harmony with Wolfe's "health care, food and shelter" rule, Driggs and others responded to individual crises with both generosity and imagination:

> One evening in March, [1903, a couple] asked for shelter. Their infant scarce two days old had been born in a snow house. The wind was cutting the snow house to pieces and there was no room in the grandmother's iglu, besides which there was sickness there and a child had recently died. So they were forced to come to the mission. I let them

> have the small room and gave them a stove, that had been made from two powder tins and [a pipe] from a wrecked whaler. (Driggs, Letter of July 3, 1903)

Improved communications exacerbated the spread of the Great Sickness but also gave everyone a relatively clear picture of its extent. In June 1901 Driggs reported: "The past season has proved a very busy one, there being an unusual amount of sickness. A little over one-sixth of the population have died. Heavy as the death rate has been, it does not begin to compare with other places where whole villages were nearly depopulated" (Driggs, Letter of June 4, 1901).

Point Hope had by no means escaped. Driggs's scribbled statistics record eleven deaths between July and September 1900, "all of the grippe (flu) brought by the first steamer." Twelve children died in December 1900 of whooping cough, and in April 1902, ten children and four adults died from the measles that, according to Driggs, the trader Stein had brought with him:

> Last fall, a whaling schooner [unnamed] that had wintered near the Mackenzie River landed some women and with them came the measles. The disease never having prevailed here before soon spread, not sparing any one, young or old and proving unusually severe. It has left its affects: one or two small places losing 50 percent of their population. With us, the death rate was about 12 percent, principally young children and the aged. The infant population being almost wiped out. (Driggs, Letter of July 3, 1903)

Driggs's work was sometimes complicated by both the distribution of Point Hope population and by his own movements. When measles erupted in early autumn, Driggs was fishing on the Kuukpak River and in the following paragraph the doctor provides one of his rare but detailed circumstantial observations:

> I was up the river at the time, messengers were sent to hasten me home. We were having very severe storms. The ice on the river and lagoon was not strong enough to bear me and the messengers had no sooner delivered their message, than they lay down sick with the disease. So I had four cases to care for in the small sodded hut I was in. Other messengers were sent but they could not reach me. One boat being capsized in the lagoon. It was ten days before I could return home. (Driggs, Letter of July 3, 1903)

Few of the doctor's letters provide such a vividly textured account of life in the region, and this brief passage not only offers a snapshot of sudden disease onset, but also provides a disconcerting picture of the hazards of travel in a dangerous and unpredictable environment. As if the appearance of measles were not enough, the sheer difficulty of travel over young ice and in autumn storms further conspires to build a sense of extra threat.

Driggs went on his second furlough in July 1904 and was replaced for the year by E. J. Knapp. Writing that year to the Board of Missions, Knapp described a continuing population hemorrhage:

> [F]ifty of the Natives, men women and children, have gone away on whalers as far as the mouth of the Mackenzie River, and many will not return for a year or so. Who can tell how many of them will ever return? For life on board these whalers means moral

> degradation to both men and women, especially the women, and in all likelihood physical decay and death. (*Spirit of Missions* 1904:727)

This exodus represented almost one third of Point Hope's population and indicates the seriousness of the subsistence crisis. But there was no remission from sickness even in the far Mackenzie Valley. There, according to the Catholic Father Giroux, a measles epidemic of 1903 had killed one fifth of the population—which would have included Point Hope migrants (Stefansson 1908:26).

In the midst of these disasters, Point Hope's survivors at least appeared buoyant. By winter 1904, the sickness in Point Hope seemed to have burned itself out, leading to a period of comparative good health. Echoing Driggs's remark of 1898, Knapp wrote:

> They are a careless, cheerful, happy people, their faces always rippling with smiles.... During the past year, while there has been some sickness among them and several deaths, the health of the people has, in the main, been very good. They have not suffered want of food or clothing, and their supply of wood and oil for fuel they were able to eke out… (Knapp, Letter of July 29, 1904)

The community had been drastically reduced and life remained a struggle: "The sick among the Eskimo people enlist one's keen sympathy," wrote Knapp. "Their homes are so devoid of all the accessories that bring relief and comfort. Every Sunday there was always dispensary work to be done, and very frequently during the week the people, big and little, would come to the mission for medicine." But, despite occasionally critical cases, "usually the complaint was of a trifling character" (Knapp, Letter of July 29, 1904).

Driggs, on his return in 1906, took a census of the village. "The population," he wrote, "was the smallest ever at 121" (Driggs, Letter of May 25, 1906). As we have seen, Point Hope's population in 1800 had stood at more than thirteen hundred (Burch 1981:14). And while early-nineteenth-century wars may have halved Point Hope's numbers during this period (Wells and Kelly 1890:10)—an overestimate according to Burch—the village population of 1906 would be its lowest ever.

But while the new century had begun so disastrously, Driggs was by 1908 able to record that the population had started to rise. "This winter," he wrote, "the mission work is simply being held together. Each Sunday there's a good congregation at worship, averaging 125 out of a population of about 150. The day school has thirty scholars" (Driggs, Letter of Jan. 26, 1908). Hard as life continued to be, Point Hope had survived. And these early years of increase provided the basis of a growth that has continued to the present century.[5] Ironically, it was the doctor who would, himself, come to grief in those very months when Point Hope had emerged from this critical era.

Notes

1 Driggs was fascinated by rabies and in 1902 wrote a detailed account of a recent epidemic that reduced the Point Hope dog population from over two hundred to thirty-two. He also recorded eight rabies-related human deaths in Point Hope and Kivalina and described the behavior of rabid dogs, foxes, and reindeer.

2 Porter also quotes the memoir of a nineteenth-century American physician, Arthur E. Hertzler, *The Horse and Buggy Doctor* (1938): "I can scarcely think of a disease that the doctors actually cured during those early years. All they could do was relieve suffering, set bones, sew up cuts and open boils on small boys."

3 Neither Edson nor Driggs was resident in 1887. The detail may have come from a Jabbertown source.

4 In 1898 there were six deaths from tuberculosis, two from meningitis (including the mixed-race Herman Koenig, aged six), one from syphilis, one from "spinal disease," one "from liquor," one postpartum hemorrhage, and one from old age. In 1899, Driggs recorded only five deaths: two from old age, one from pneumonia, one from eating "the remains after distilling," another illegible. Although Driggs's records are often quite confused, these details almost certainly provide a reliable indicator (Driggs, Edson, and Hoare 1893–1909).

5 The population recorded by the U.S. Census of 2000 was 757; according to North Slope Borough statistics it now stands nearer 800.

CHAPTER 23
Millenarian Alaska: Return of the Spirits—1

Sometime in the mid-nineteenth century, about thirty years before missionaries arrived in the region, a millenarian movement took hold in the Noatak and Kobuk valleys and spread through northern Alaska, probably reaching Point Hope and Barrow by the 1880s. Evidence of a salvationist idea known as *uivvaqsaat*—Iñupiaq in external character, but possibly inspired by ideas from outside—has survived both in *uivvaqsaat* narratives and also in the biography of Maniiḷaq, an Iñupiaq prophet who was born on the Kobuk River in about 1830 and who died near Barrow ca. 1890.[1]

On the Kobuk and Noatak, the original meaning of the plural noun *uivvaqsaat*[2] can be translated as "those who have come round a bend in the river." The word had a different connotation in pre-contact Point Hope, where it described the navigation by sea of headlands such as the Tikiġaq Point and Capes Thompson and Lisburne.[3] In traditional times, the term had no supernatural or religious associations. With the arrival of millenarian hopes in the mid-nineteenth century, *uivvaqsaat* or "rounding a bend" in the Kobuk region came to suggest the return of ancestral souls, the supernatural appearance of goods, and even the arrival of the first white people, who unaccountably appeared from around river bends carrying new things and ideas. In Point Hope and Barrow, *uivvaqsaaq* ("coming round") was restricted to the notion of a heaven where souls would be reunited with their loved ones, while the plural *uivvaqsaat* referred to returning souls who would carry the newly deceased towards reunion with the community's entire spiritual population in a good place or a "heaven."

The prophet Maniiḷaq does not appear in any surviving Point Hope histories. But we know that *uivvaqsaat* played a role in Point Hope's religious life, because when John Driggs published his book about the village in 1908, he wrote on the assumption that *uivvaqsaat* belonged to old Point Hope tradition. Later in this chapter, we shall explore Driggs's understanding of *uivvaqsaat* and relate *uivvaqsaat* to how Point Hope people came to Christianity. But to comprehend the three-step shift in faith—from the old religion to *uivvaqsaat* and finally to Christianity—it will be helpful to see the *uivvaqsaat* doctrine in the context of Maniiḷaq's teaching and his career as prophet and "anti-shaman."

The Visionary Maniiḷaq and the Kobuk River Shamans

Most of the elders who recorded Maniiḷaq stories in the 1970s knew about the Kobuk mystic through conversations with their own elders; although many of them were alive towards the end of Maniiḷaq's career, very few of them saw him.

Why Maniiḷaq traveled north from the Kobuk to Barrow has not been explained. Perhaps, like many of his contemporaries, Maniiḷaq and his family were in flight from the subsistence crisis of the 1880s. Another reason also suggests itself: Maniiḷaq, as we shall see, was a shaman of sorts. But because he vociferously campaigned against Kobuk shamans and proclaimed his own divergent neoshamanistic teachings, it is possible that he left home both to escape harassment and to disseminate his own ideas. Several narratives allude to the route of his journey. But since he traveled due north from the Kobuk to Barrow, he came nowhere near Point Hope and there seem to be no Point Hope stories about him.

Maniiḷaq is, however, well remembered in Barrow. In 1960, at a Presbyterian church celebration, Lloyd Ahvakana described the Barrow Iñupiat as having been made "ripe for the Holy Spirit" by Maniiḷaq,

> who spent most of his time in the Kobuk valley before wandering toward the North Slope... Maniiḷaq's prophecy [was] that "uivvaqsaaq" was where the Iñupiat would go to meet their loved ones after their days ended. Ahvakana said this prophecy was made long before Caucasians came to the arctic, and when the missionaries started talking about heaven the Iñupiat "would begin to get excited about seeing their loved ones and talked about how happy they will be with them again."
>
> "Our people were primed for Christ's message, which Mitchimii [Leander Stevenson] brought [in 1890]," Ahvakana said. (Bingham 1998)

Stories about Maniiḷaq's early life suggest a traditional shamanistic role for him. But as his career progressed, Maniiḷaq differentiated himself morally and professionally from shamans. In the words of one Kobuk elder, "He was said to aqpigit–, not to be one who ever wants to be rid of anyone" (Newlin et al. nd: Told by Charlie Qixuban Lee) and while Maniiḷaq's first visions took shamanistic initiation form, it was this same experience that later storytellers interpreted as being prophetically Christian. These early episodes in Maniiḷaq's career contain descriptions of shamanistic initiation whose exceptional spirituality make them unique in northern annals. They are quoted here both for their own sake and for the extraordinary picture they offer of this transitional period in Native Alaskan experience:

> When Maniiḷaq was young, he is always referred to in relation to his mother. I cannot say, "He lost his father." Whether he actually did or not, I do not know. As a young boy, he helped his mother build a winter sod house, gathering and covering the structure with moss, making it as warm and cozy as possible for the cold winter ahead. She also took him snaring with her...
>
> One autumn, he decided to set a snare by himself. He asked his mother for a stick especially used in snaring. It was called an *ayaupiuraq*. He said to her, "Mother, allow me to try it. The bow and arrows you have made are my weapons. If there is a ptarmigan nearby, I will catch it."
>
> One day as he was out checking the snares, he saw a spot to sit that looked very inviting. Since it looked like such a comfortable place, he sat down on it to rest. As he did so, he heard the strange call of a small bird. He raised his head and listened carefully. "*Taatagiik, Taatagiik*," was what he heard: "Father and child, father and child," which can also mean "Father and son, father and son." Maniiḷaq repeated this word

to himself, "*Taatagiik.*" After the call had stopped, he stood up, checked the rest of his snares and went home.

Soon it became a habit for Maniiḷaq to rest and listen to the birdcall. When he realized that darkness was approaching, he hurried home… As time went on, it became difficult for Maniiḷaq to sit still at home. His desire to check his snares and sit on the resting place became great. He could not forget what he had heard and how strangely calm and peaceful he had felt upon hearing those words.

Eventually, he became sure that he could always hear those words at that very same spot. The words expanded to "*Taatagiik, taatagiik. Isrummiqsuqti, isrummiqsuqti*": "Father and son, father and son, the source of intelligence, the source of thought." He listened to this and sometimes went home without checking all of his snares.

As time went on, the call Maniiḷaq had heard could not escape his mind. When checking his snares, he rushed and quickly procured ptarmigan or rabbit, then headed for the same resting spot. Again he heard, "Father and son, father and son, the source of intelligence, the source of thought." Maniiḷaq was now certain of those words. On his way home, he became so carried away that he repeated those same words to himself as he walked along.

As spring approached, the days became long and soon Maniiḷaq did not come home at all on some days… The small game became easy to obtain. Animals were moving about, to and fro, directly in front of him. He would think, "I will catch that one," and would immediately do so.

One day when Maniiḷaq went home, his mother asked him anxiously, "I don't understand you. What do you do when you are gone? What causes you to come home so late?" Maniiḷaq was suddenly inspired with this reply, "Mother, I have been listening to the wonderful sound of a small bird." His mother inquired, "Well, what is it about that sound that is so pleasant?" Maniiḷaq replied, "*Taatagiik. Isrummiqsuqtik, isrummiqsuqtik*: Where is this source of intelligence?" He replied that it seemed to come from somewhere above him. He heard the small bird above him.

It was then that Maniiḷaq revealed what he had been listening to. He said that the sound came from above and that he felt so much reverence that he could not dare to even look toward this source of intelligence. His mother stared at her son and said, "You must be turning into an *aŋatkuq* (shaman) or something!"

"No, mother. I cannot say that I am becoming anything. However, I can now understand what the source of intelligence is saying. I listen with much pleasure and it tells me that no harm shall come to me. Mother, I am bringing home a ray of light each time I listen to it. I bring this light within my mind and thoughts by remembering those two words. However, they are the only two that I hear," Maniiḷaq reassured his mother. He added, "Do not worry. It is only the beginning. I want to listen. I want to learn. I know that something is helping us and that the small bird calls from somewhere, the source of which I do not know."

Maniiḷaq's mother was relieved and wondered, "What is it my son hears? I have lived for a long time, but never before have I ever heard of the source of intelligence." She searched her thoughts. "What is it that is so powerfully awesome that even the shaman has never even heard of it?" After this, she no longer worried about her son. (Newlin et al. nd: Told by Beatrice Anausuk Mouse)

In the light of these incidents it is easy to understand the mother's exclamation: "You must be turning into an *aŋatkuq* (shaman) or something!" Aside from the unusually subtle features of Maniiḷaq's experience and the gracefulness of movement between visionary exaltation and the demands of subsistence, many traditional features of initiation are here: the lonely young man, relations with a spiritual being, aural hallucinations, and acquisition of supernatural knowledge. All these contribute to our sense that Maniiḷaq had become a shaman. But in addition to this, Maniiḷaq's story corresponds closely with an Iñupiaq genre of legend in which boys who would become shamans were shown to live in poor and isolated two-person families (often termed *aanagiik,* "poor boy and grandmother"; Asatchaq and Lowenstein 1992:10) from which they would emerge as shamans. Stories like this were particularly common in the Kobuk and Noatak areas. But to Kobuk people from the 1890s onward, the "father-and-son" of Maniiḷaq's vision would also have acquired a Christian meaning—although Maniiḷaq himself did not live to see a missionary. The shamanistically inclined boy, who had fallen into a relationship with spirit powers, had also, in the view of the next Christianized generation, become the prophet of God the father and son.

In another story, Maniiḷaq's birth is given supernatural origin, which further emphasizes his double shamanistic and pre-Christian prophetic character. Here, Maniiḷaq's grandfather called up to the land of "those of the sky, heavens," and they sent Maniiḷaq to come down to earth, thus lending Maniiḷaq's identity a divine character. And whether or not Maniiḷaq was considered by his peers to be a shaman, he used his post-initiation status to combat shamanistic practitioners in the vigorous style with which shamans traditionally competed. But unlike the usual shaman disputes, which centered largely on issues of power, Maniiḷaq's quarrels often took the form of debates about shamanism itself, the social effects of taboo, and other shamanistically controlled practice. First, in contrast to the shamans whose power, as Maniiḷaq expressed it, "came from the earth," Maniiḷaq claimed a source of intelligence from *above.* This power, at once spiritual and intellectual, was benign, protective, and in harmony with the young Maniiḷaq's character, which—in Christianized retrospect—expressed a new ethical value. One song attributed to Maniiḷaq is said to have come directly from his "source of intelligence." And, like the words of the biblical David, Maniiḷaq's song proclaims a new spiritual theme, which claims to reverse the shamanistic dispensation:

> I will lighten the darkness
> I will make the darkness into light[4]
> (Newlin et al. nd)

The symbol of light, which like his source of intelligence came from the sky, is central to Maniiḷaq's struggle. Anticipating the Christian teaching of the Friends who would arrive in the Kobuk valley in 1897, Maniiḷaq claimed for himself an inner light that was inaccessible to dark powers. It is this light, as suggested in several stories, that sets Maniiḷaq apart and protects him. The storyteller continued:

> One evening, many shamans gathered to his home and using their powers of the world, they attempted to find his inner being, but Maniiḷaq was surrounded by such a brilliant light that the shamans could not find his inner being. Realizing that their attempts

> were futile, they gave up. After this, Maniiḷaq stood up and paced around the ashes of the open fireplace, "*Hi hii*!" he exclaimed, "So! The source of intelligence is going to do that for me." Saying this, he repeated it again and began his predictions... (Newlin et al. nd: Told by Beatrice Anausuk Mouse)

On another occasion when Maniiḷaq had made some of his predictions, the shamans grew "strangely quiet, their mouths closed as though they either respected or feared him." Not unlike a Christian preacher, Maniiḷaq did not exclude unbelievers from companionship with his "source of intelligence": "My source of intelligence," he told them, "is also watching over you, keeping you safe from all danger." And in terms that come closer still to evangelical language: "He said that his grandfather above was so powerful that he deserved the utmost respect and obedience. His power was greater than that of everything else" (Newlin et al. nd: Told by Frieda Anniviaq Goodwin).

And, as the storyteller continued: "You are deceived by one who is actually nothing. You believe in a useless trouble maker who only tries to make people live by the taboos. Why don't you believe instead in my grandfather above?" He often said this to the shamans.

The light that infused Maniiḷaq's imagination came both from within and beyond:

> He had found his divine guidance in his mind many years before. It was "the one above...." although he did not actually say, "the one above." There existed in his mind a place somewhere which glowed with a brilliance too wonderful to describe and was devoid of suffering. (Newlin et al. nd: Told by Beatrice Anausuk Mouse)

Internal light was thus identified with light from above, and both sources of illumination were associated with Maniiḷaq's ethical teaching: "'The people kill one another because of their ignorance. There should be no killing nor thievery to cause the starvation of one's neighbors.' Saying this, he often advised against all wrongdoing as he went about his daily life." Confrontations with shamans continued:

> Everyone, especially the shamans, had grown curious about the Maniiḷaq's power. ... When they had gathered at Paa, Maniiḷaq appeared calm as he rested in a prone position and told the shamans to summon their spirits. He did not become uncomfortable as the shamans chanted incantations and performed their rituals, summoning their spirits. They sang all night, offering oil and food into a fire, intended for any spirit which might be calling, "I am starving! I am thirsty!" This was how they summoned spirits. In the meantime, Maniiḷaq remained calm and undisturbed.

Again, as narrated by Christianized informants, Maniiḷaq is presented as a self-confident proto-Christian:

> When they were done, they turned and sarcastically asked him, "What is the matter? Are you too frightened to speak? Are you afraid now? Why don't you talk about the one you are always referring to as your source of intelligence?" There were many who taunted him and ridiculed him.
>
> Finally, he stood up and began to pace in a circle. "*Hi, hii! Yaiy*!" he exclaimed, adding, "My dear source of intelligence, you have blessed me with another day." In

> saying this, he was voicing his gratitude. When we wish to show our thanks, we say, "I am thankful that you have helped me."
>
> "*Hi hii*: you all will come to know and understand my source of intelligence. However, it does not matter what I say now, you will not comprehend my meaning. When the necessities of life become easier to obtain and survival becomes easier, then you will understand."

If Maniiḷaq is portrayed here as a forerunner of the Christians, the implication of the last detail is that the "easier" life is one in which Christianity properly belongs, makes itself available, and perhaps even helps to bring about. Euro-American prosperity and Christianity would, suggests Maniiḷaq, be part of the same dispensation.

But paradoxically, the most Christianized versions of Maniiḷaq's contests also reveal him at his most shamanistic. The following story takes place at Sisualik, a sandspit north of Kotzebue on which Iñupiat from a wide area converged each summer to hunt, pick berries, shamanize, and exchange goods. Here Maniiḷaq brought his local rivalries to an inter-regional stage. In the following story, Maniiḷaq is portrayed first in the character of a saint who has transcended human imperfection, but also as a ferocious shaman who threatens to gobble up his enemies. The scene opens as Maniiḷaq, elevating himself in serene and somewhat provocative separation, constructs a raised driftwood platform. Here he sits, as though in a historically removed and morally improved sphere, and watches the doomed, old world pass:

> [T]his was when people in Kotzebue killed others through the powers of the shamans. They argued and stabbed each other, and as he sat up on his platform, he thought about them with pity. Whenever he wished to observe them, he climbed up on the platform and cooled himself off as he watched. (Newlin et al. nd: Told by Beatrice Anausuk Mouse)

The story continues with yet more post-Christian coloration before reverting to the overtly shamanistic:

> As he watched them with pity, he sat deep in thought. He felt that an evil spirit was among them and controlled them, resulting in an ignorant people who felt no peace, only arrogance. One day when the children seemed noisier and more restless than usual, he made a ladder of a thin piece of wood and climbed up to the platform. Then he pulled up the ladder. He was sitting up there when two men came passing by. They were Uqummivayauraq and Ayaunivruaq, two shamans.
>
> Maniiḷaq sat and ignored them.... However, he noticed that Uqummivayauraq stooped down, then looked up and snarled, "Yes," he said, "Look at him hiding. He says, 'My source of intelligence,' which is probably why he does not want to mingle with us." The two shamans continued with such words of abuse but Maniiḷaq just sat there as if he did not even hear them. (How tolerant he was, just like Jesus.) He simply kept his mouth closed and ignored them. The two added, "What do you suppose he is *thinking* about now? It is us who are walking that actually think. Look at him just sitting there."

Now Maniiḷaq comes into recognizably shamanistic voice:

> Finally Maniiḷaq responded, "I could swallow (destroy) you both if I wanted." "Oh, I'm sure we shall have something to choke you with," was the reply. The shamans were referring to their helping spirits and amulets with which they could block the passage of anyone who tried to "swallow" them. After that incident...all the people who had gathered in Kotzebue were soon talking about the words exchanged between the two parties. At that time the two shamans were not able to comprehend Maniiḷaq's true power. This whole incident took place at the time when many people died [of disease] at Kotzebue.

Significantly, this exchange thus took place around the time of an epidemic, and as we shall see, both Maniiḷaq's teaching and *uivvaqsaat* doctrine were both, in their different ways, responses to the disease crisis. The oscillation between the idealized Maniiḷaq ("like Jesus") and his combative shamanistic character evokes the milieu of uncertainty and change that the epidemics created. The storyteller's implicitly Christianized view suggests a movement in Maniiḷaq of traditional values towards something experimental but as yet unformulated. And if in some way, which is hard to interpret—for storytellers born into the years of change *experienced* what Maniiḷaq anticipated and envisioned—Maniiḷaq did represent proto-Christian values, perhaps the stories about him go some way towards expressing the later phenomenon of postmissionary coexistence of shamanistic practice and early, sometimes nominal, conversions. If an idealized figure such as Maniiḷaq became an autochthonous "Christian," or at the very least dissociated himself from animism and the spirit world before the arrival of missionaries while retaining some of the trappings of shamanism, then why should other more ordinary individuals, at least for the duration of one postmissionary generation, not also adhere to both systems? While Maniiḷaq may be remembered by this generation as part victim, part hero and anti-shamanist prophet, perhaps his status among both the comfortably and uneasily Christianized next generation also helped to legitimize the pain and culturally diverse experience of that generation. This latter was after all the cohort that had the most difficult choices to make, and one way of maneuvering a choice between inherited belief and the new ways introduced by white men was to engineer their coexistence.

The Prophecies of Maniiḷaq

Maniiḷaq's sayings fall into roughly three categories: straightforward historical prophecies, assaults on the taboo systems that were instituted or enforced by shamans, and mystical utterances that, as we have seen, proclaimed a new kind of religious vision.

Like his attacks on shamans, Maniiḷaq's critique of taboo came from both intellectual and moral positions. First, Maniiḷaq was able to demonstrate that taboos were illusory and violating them had no consequence. Second, he attacked taboo as inhumane and particularly oppressive to women. It was shamans who enforced taboos and often made new ones, so Maniiḷaq's dissent was an assault both on shamanism and on a system that shamans upheld. There are several stories about Maniiḷaq breaking ancient prohibitions with complete impunity, thus opening the way to an era that would allow people to live with greater freedom:

> When a poor girl reached puberty, she wore a parka with a deep-faced hood which hid her face so completely that men did not see her for a whole year. She was made to live away from people in a winter home built by her parents… Maniiḷaq said he felt compassion and pity for the poor girl. He told the people, "The custom of shunning the girl will no longer be practiced." (Newlin et al. nd: Told by Beatrice Anausuk Mouse)

In a longer story, Maniiḷaq demonstrates the unreality of taboo regulations by breaking them in public and surviving unharmed:

> [O]ne time he went across to Sisualik. The people there considered it dangerous to eat fresh beluga maktak mixed with something. To prove that they would not die if they did eat it, Maniiḷaq purposely mixed the food in front of them. They fearfully moved away from him but, once again, he showed them that there was actually nothing to fear. He did not become ill or die.
>
> As he said about a previous demonstration: "There is nothing to fear. I have demonstrated the freedom that is to come…" (Newlin et al. nd: Told by Charlie Aqpaliq Sheldon)

Maniiḷaq's prophecies about the future were relatively straightforward. But his predictions of technological change seem also to have been sanctified by divine authority: "Everything will change," Maniiḷaq said. "This information I receive from my source of intelligence." Looking back from the twentieth and twenty-first centuries, it is as though material progress was being interpreted as an expression of and an accompaniment to Christianity. Most of Maniiḷaq's prophecies, however, appear to describe what we can now identify as air travel, motor transport, and the telephone. Here is some of the testimony:

> He said that people would even travel through the air. Hearing this, the people found it hard to believe him. So great was their disbelief that they laughed at him and said that he was speaking nonsense. After all, in those days, he was attempting to describe things of which they had never heard.
>
> [H]e predicted that strange visitors would come from the east and travel down the Kobuk through the sky.
>
> He predicted that people would travel in their boats simply by sitting, without the use of sails.
>
> [H]e predicted that people would speak through the air with ease. If a person spoke from a far off place, people would be able to hear him. (Newlin et al. nd: Told by Nora Paaniikaaluk Norton)

Other prophecies carried symbolic rather than historical meaning. The following two pronouncements (perhaps!) suggest climate change and the creation of permanent villages on the Kobuk River:

> The two consecutive seasons, whether they be two summers or two winters, have not yet arrived. It would be a time of great hunger, a time of famine.
>
> Ambler village becoming a great city, it is evident that this is coming true already. The village has already been established. (Newlin et al. nd: Told by Fay Uyubaq Foster)

One prediction is in a different idiom. This is Maniiḷaq's prophecy that a whale would emerge through the earth near the future village of Ambler on the Kobuk River. A storyteller recalled: "The spot where the whale is to surface is very deep. As children we passed it many times. It is so deep, which makes it all the more probable for a whale to appear there" (Newlin et al. nd: Told by Fay Uyubaq Foster). This, unlike Maniiḷaq's other predictions, has its basis in mythology. It relates both to stories about whale bones that had been discovered on hill tops and to narratives about whales that rose from dry land. Land-whale stories were also told in Point Hope. In these, a whale either surfaced in a village iglu or was conjured through the earth by a shaman (Lowenstein 1993: part 1). This land-whale prediction stands apart from Maniiḷaq's reformist thinking about a taboo-free world. Because the story derives from folklore and has its origin in an imaginative world from which he was separating himself, the whale prophecy, like his shamanistic "swallowing" threat, lies recognizably within the magico-religious tradition—which suggests that there were perhaps other more traditional aspects of Maniiḷaq's thought that have not been recorded. In a similar vein, it may well have been Maniiḷaq who predicted in about 1899 that Point Hope would be destroyed in a repetition of the 1893 storm (discussed further in Chapter 25).

Maniiḷaq and *Uivvaqsaat*

Maniiḷaq's references to *uivvaqsaat* come in two parts. On some occasions, Maniiḷaq identifies *uivvaqsaat* as an unknown people who will "come round the bend in the river." "Believe it or not," he is recorded as saying, "you shall receive visitors who travel swiftly along the surface of the water in a new way." These new people, he suggested, would "come from the east… They would come from the ocean rounding points and bends" (Newlin et al. nd: Told by Beatrice Anausuk Mouse).

These latter remarks are as arresting as Maniiḷaq's more apparently specific predictions about technological innovations. Here, in his remarks about new people from the east, Maniiḷaq appears to express an unusual intuition of geographical units and of continental space beyond the home territory. And what Maniiḷaq also seems to suggest here is the sense of a place from which changes will arrive.

Contained in the prediction is another insight. This is Maniiḷaq's assertion that there was a geographical reality "beyond the Iñupiat's limited knowledge of the world" (Newlin et al. nd: Told by Beatrice Anausuk Mouse). In the light of testimony like this, one is prompted to see even some of Maniiḷaq's more startling prophecies as expressions of an historical sensibility rather than just spontaneous intuition. Sixth-sense hunches about future transformations of Iñupiaq lifestyle may certainly have come into it. But if these hunches do reflect what a Kobuk man was saying in the 1870s and 1880s, they presumably also had some historical origin.

How might Maniiḷaq have come to the apprehension that white people would travel down the Kobuk River in high-powered vehicles? The absence of hard dates for Maniiḷaq's biography renders explanation difficult. But there is a relatively secure chronology for the Euro-Americans. No white men reached the Kobuk valley until the summer of 1883, when, for about two weeks, Lieutenant George Stoney explored the Kobuk delta, returning the next year in a small river steamer. Traveling two hundred miles in this, Stoney further penetrated the upper Kobuk by canoe, returning in 1885 with another steamer. Similarly, Lieutenant J. C. Cantwell, a member of Captain C. E. Hooper's Coast Guard team on the *Corwin*, reached the upper Kobuk on a steam launch in the summers of 1884 and 1885. But by this point Maniiḷaq had probably traveled north. And if the first steam launches were there for everyone on the Kobuk to see, there would have been no reason for Maniiḷaq to prophesy on the basis of what had already happened. It remains possible that at some point Maniiḷaq generalized from these historical events and used them to predict developments that his descendants would, in due course, see in reality.

Dates remain uncertain, but these early Euro-American voyages up the Kobuk probably did take place after the genesis of Maniiḷaq's ideas. Stoney and Cantwell were, however, not the first white pioneers in northwest Alaska, and we may assume there were stories about Cook, Beechey, Kotzebue, and other Europeans which had circulated among Iñupiat since the late eighteenth century. With the arrival of commercial whalers in 1850, white men in ships became routine coastal visitors. Kobuk people would occasionally have seen white men at Hotham Inlet and other points along the inshore waters. Maniiḷaq's semi-prophetic achievement was perhaps to draw from these episodes the notion of a historical momentum that was in the offing. His understanding of the *uivvaqsaat* "beings"—which in the testimony of others were identified as returning souls—was that they were thus either the white people themselves or else symbolic amalgams of approaching historical and technological changes embodied in white men, trade goods, ships, new ideas, and a language that would describe such phenomena.

Three statements attributed to Maniiḷaq suggest these convergences. First, that the *uivvaqsaat* were sometimes, though by no means always, identified with whites: "He said that [*uivvaqsaat*] did not have the same skin color as us…he said that people with light skin and different colored hair would come" (Newlin et al. nd: Told by Nora Paaniikaaluk Norton). Second, that the *uivvaqsaat* were not spirits but people whose descendants would have a technology in advance of that already achieved by nineteenth-century Euro-Americans: "He spoke of *uivvaqsaat* who traveled through the sky on vehicles which were run by fire. Today, we see those whom he referred to as the *uivvaqsaat*" (Newlin et al. nd: Told by Robert Nasruk Cleveland). Third, that the impact of *uivvaqsaat* on northwest Alaskans would be permanent: "He said that their lifestyle, which they were presently experiencing would change dramatically once the *uivvaqsaat* came" (Newlin et al. nd: Told by Robert Nasruk Cleveland).

Beyond Maniiḷaq's unique and exclusive pronouncements on *uivvaqsaat* as white men, there is no record of *uivvaqsaat* being given a historical identity. For most people—in the Kobuk, Noatak, Barrow, and Point Hope regions—*uivvaqsaat* doctrine was entirely about souls; and this, almost certainly, was the movement's dominant motif. One storyteller from the Kobuk seems consciously to differentiate between Maniiḷaq's version and the majority view: "But this word [*uivvaqsaat*] was used even before Maniiḷaq's time. At that time, it referred to the dead people rising again. It was said that on that day all the dead would rise again and it was also said that the day would be very great" (Newlin et al. nd: Told by Charlie Aqpaliq Sheldon).

According to this definition, the arrival of *uivvaqsaat* represented a nonhistorical "day of reckoning" that involved ancestor souls, not future white people. Here, in contrast to Maniiḷaq's more pragmatic suggestion, the *uivvaqsaat* were disembodied revenants, the souls of ancestors. These were not white people with trade goods or a new religion, but local people's spirits, all of them together, the recent and ancestral dead, in a crowd without number.

The idea of souls that would travel "round bends in the river" (grafted as this might be onto traditional soul theory) was new. But there was nothing new in the veneration of ancestral souls that traditionally traveled to a place in the south where they would exist in contentment. Other souls could be reborn through namesakes and thus reintroduced within a chain of semi-identical "same name" descendants. A minority became *tuunġat*, shaman's helping spirits. But there was nothing in traditional belief about the return of souls en masse; nor were ancestor souls conceived as comforters and spirit guides. These latter two manifestations of belief arose only in the late nineteenth century. Speaking of this consolatory aspect of *uivvaqsaat* theory, one storyteller described how "a widow began to weep for her husband. The dead man's brother approached her and said: 'Stop crying. Someday your husband will return ahead of the *uivvaqsaat*. They say that the *uivvaqsaat* are coming even closer'" (Newlin et al. nd: Told by Charlie Aqpaliq Sheldon).

The historical context of this scene was not simply that of bereavement, but of epidemics and famine. The conditions in northwest Alaska, with its game losses and depleted populations, were ripe for a popular idea that promised future earthly renewal or spiritual release from a life that had become intolerable. Like millenarian and messianic movements from first-century Palestine through medieval Europe—and not least among Native Americans of the Great Basin and Plains up until 1890 and Wounded Knee—*uivvaqsaat* arrived with a message of consolation to despairing people.[5]

As a number of storytellers suggested, *uivvaqsaat* was, in essence, a source of hope. It outfaced what life denied and defied death's finality. More than this: while disease and starvation threatened whole families and villages, *uivvaqsaat* promised the return of these populations with cross-generational increase: "It was said that the dead would some day rise again and that the *uivvaqsaat* would appear from the east. All the people who had died would come back as *uivvaqsaat*... It would be the time when we would all see one another again" (Newlin et al. nd: Told by Charlie Aqpaliq Sheldon). Much the same story would be carried to Point Hope, where *uivvaqsaat* also took root in the 1880s. And it was these beliefs that John Driggs would identify with traditional Point Hope religion and which perhaps laid a foundation for the introduction of Christianity. This will be explored in the next chapter.

Notes

1 Discussion of the origins of *uivvaqsaat* lies outside the scope of this book, but many images and symbols both in Maniiḷaq's teaching and in the wider cult correspond to those found in earlier spirit movements. Readers interested in exploring possible parallels to, or indirect derivation of, the Alaskan experience, particularly in relation to the Paiute and Klamath dance

movements that led to the better-known Ghost Dance in the Dakotas, will find a short list of references at the end of the general bibliography.

2 From *uivaqsaaq*–, "to round a point of land when travelling by boat" (MacLean 1980). *Uivvaqsaat* is also a place-name on the Noatak River (Burch, personal communication 2002).

3 Cape Lisburne (*Uivvaq*, "rounding place") was also termed "the far rounding place," while Cape Thompson was "the near rounding place." Point Hope lies between these navigational approaches.

4 "The Lord will lighten my darkness," 2 Samuel 29; "My God will turn my darkness into light," Psalm 18:28.

5 See note 1.

CHAPTER 24

Millenarian Alaska: Return of the Spirits—2

Uivvaqsaat in Barrow

Maniiḷaq never visited Point Hope, but he did reach Barrow, though how long he lived in that region is not clear, and by the time Leander Stevenson established his mission in 1890, many Barrow people had already accepted the *uivvaqsaat* message. As one Barrow elder who was born during the late-nineteenth-century epidemics said: "And so, it is said, those relatives, when they lose one of their own, those who have died will round the bend and return. They didn't think of them as dead. They would come alive again." These souls coming home were anticipated not in some distant future, but any day now: "Those *uivvaqsaat* are getting nearer, they said. 'Wake up, all of you!' [People] would wake each other because they were waiting for their relatives to come… 'The *uivvaqsaat* are getting nearer!'"(*Puiguitkaat* 1981:78).

Barrow elders born in the late nineteenth century also believed that *uivvaqsaat*, not unlike white men with their material gifts, would bring "everything for a comfortable life" (Spencer 1959:296–98), a notion that pulled together what Kobuk people, including Maniiḷaq, sometimes represented as separate ideas: that *uivvaqsaat* were the souls of ancestors who would carry the recently deceased to a good place, and that *uivvaqsaat* were white men who would arrive with trade goods.

Given local variation, the overall doctrine therefore contained both these notions, and local circumstance contributed to the range of emphasis. Like the company of souls that was imagined to sweep round bends, the character of *uivvaqsaat* fluctuated, split, and recombined into a number of different shapes. These variations across a territory, many of whose people were traveling to escape disease and hunger, coexisted and fed one another.

An example of direct transmission of the doctrine is highlighted by a story reported by a Barrow woman who was born around 1880. When she was a child, her family traveled inland and met some Kobuk people who told them that "Strangers are coming. We can follow them and we won't have to bring anything. They will have everything for us" (Spencer 1959:297).

The story of this encounter with travelers (who might have included Maniiḷaq) was related in English, so we cannot know the Iñupiaq word used here for "stranger." But Robert Spencer, who recorded the story, suggests that the word "strangers" (*anginyerit*) was a synonym for *uivvaqsaat* spirits. According to Spencer, these spirits would "bring back everything needed for a comfortable life" (Spencer 1959:296). One implication here is that the better conditions before contact with Euro-Americans would be restored by the very ancestors who had once enjoyed them. Here was one version of the salvationist idea. Another version might imply something more ambiguous. As in Maniiḷaq's vision of white men, so "strangers," in the Barrow view, might have two characteristics: They might bring things needed for a better existence. But they would also resemble the white men who would both introduce Christianity and disrupt Native tradition.

Uivvaqsaat in Point Hope

Uivvaqsaat ideas in Point Hope seem to have run along the same paradoxical lines. On the one hand, the *uivvaqsaat* would bring salvation. On the other, they could be ambiguous and intrusive "strangers" who both carried useful goods and wreaked social havoc.

A narrative from Point Hope illustrates this not entirely pleasant aspect of the millenarian idiom. Opening with a statement from myth, the tale is as follows:

> Old stories said the world was flat. The sky moved around on it with a big grinding noise at the edges. The earth was still but the sky moved round. There were "people" who lived above and below. There are two stories about the people who lived above. A woman complained about meat that disappeared from her cache every night. So two men went to keep watch and they saw two figures coming down from above and going into the cache. The men shut the cache and caught them.
>
> On another occasion two men were digging out the *qalgi* (ceremonial house). One of them complained that meat was missing from his cache. This was near the houses on the Point. When everyone was asleep, the watchmen saw two "men" circling down from above. They landed at the cache. One of the visitors (*iglaaq,* "visitor," or "stranger") said he felt something was wrong. This was not the way it had previously been [at the woman's cache]. The other "man" said everything was all right. They had seal skin bags over their shoulders. One went into the cache; the finally went in too. The Point Hope watchman ran down and pushed a heavy stone over the entrance and sat down on top.
>
> Then he shouted for people to come, and someone brought a light. They looked down into the cache and found the visitors up against the ceiling. The people went in and brought them down, held them tight and took them to the *qalgi.*
>
> The next day they decided to give a dance for them. The two Men from Above started to dance. They jumped very lightly. It was easy for them to do this; they were unlike ordinary people. The visiting dancers said they weren't used to so many people crowding the skylight. So all the people outside stood up from where they had been lying around the skylight. (Aġviqsiiña 1940:72–73)

Then, concluded the storyteller, they started a real dance. It was so fine that everyone forgot everything but the dance itself. Suddenly the two men jumped. They shot through the skylight. People tried to to grab them, but they could only just touch them. The two figures circled away overhead. These Men from Above had spoken in a language that Point Hope people "could understand."

Unlike most Point Hope narratives, this could not immediately be identified either as a myth (*unipkaaq*) or as a recent ancestor chronicle (*uqaluktuaq*). Its realistic focus on skin bags, meat caches, place-names, and village architecture on the one hand stamps it with mid-nineteenth-century character. On the other hand, its supernaturalism suggests the world of myth.

The classification of the story seems to have interested its narrator, Frank Aġviqsiiña, a Point Hope man born ca. 1875, for this was the only narrative among the many he told whose character he defined. The story, he said, is "*uqaluktuaq*, or true story. It is not very old. Some of the people living in iglus that are now in ruins saw [these events]." Given its combination of recognizable Point Hope material reference and of supernatural elements that lie outside traditional theory, the story is likely to be a local one about *uivvaqsaat.*

One detail from the story that has not been mentioned so far resonates with Spencer's material from Barrow. Spencer opens his narrative with the heading "Strangers in the Sky" (though he does not explain the exact meaning of this nor mention if the words represent a translation). Later, Spencer does quote a nineteenth-century woman who "heard voices overhead, speaking a strange language…and it was agreed she had heard the *uivvaqsaat* passing overhead" (Spencer 1959:297). Point Hope and Barrow were culturally close, and although the *uivvaqsaat* movement originated in the Kobuk and Noatak areas, "the people who lived above" in the Point Hope story correspond closely to Barrow's evidence of the "strangers" coming from above.

Another aspect of the Point Hope story must be mentioned here: while the storyteller described his narrative as recent, it had its relationship with traditional Point Hope (and pan-Eskimo) symbolism. For the sky visitors who shoot "through the skylight and circle away overhead" repeat a pattern from the origin myth in which the moon and sun spirits spiral into the sky through the *qalgi* window (see Chapter 13).

Aspects of Point Hope's other origin myth also appear to lie within this *uivvaqsaat* story. Point Hope and Barrow versions of the creation story are about the Raven Man (Tuluŋigraq), who brought about the formation of land and the seasons. But unlike the moon spirit, which continued to be active in human affairs, the Raven Man was a figure from "back then" (*taimmani*), only coming symbolically into play during the whale hunt and ceremonial house rituals when a successful *umialik* donned a cape of raven feathers in a symbolic conflation of his own capture of a whale and the Raven Man's creational harpoon strike. The Raven Creator remained otherwise in the more remote sphere of storytelling (Lowenstein 1993:3–50).

The Raven nonetheless came to life again in Point Hope's *uivvaqsaat* doctrine. According to Spencer, Barrow people believed that "the way the dead [*uivvaqsaat*] behave was ordained by Raven" (Spencer 1959:296). In Point Hope, the link between *uivvaqsaat* and the creator Raven was also emphasized: "[Raven] finally went off towards the south, intending to travel all over the earth and the sky 'fixing people up'. Eventually, he will return with the *uivvaqsaaq* (people who go around—unidentified people)" (Rainey 1947:270).

In this nineteenth-century addition to traditional narrative, the Raven is identified with several other semi-shamanistic trickster figures of mythological time who wandered the landscape "fixing things up" by killing the spirits that posed a danger to society. (But while Raven's acts of creative heroism would be in the long-term interest of human society, like most such figures of Native American myth, Tuluŋigraq is also a deceiver and a sexual predator: a spiritual demiurge driven by proto-shamanistic creative-destructive appetite a long way from the ideal man represented by the Christian Son of God.) The fragment quoted above, and the complex of traditional tales of which it seems to be a later derivative, sketches a pattern that corresponds to the circling and spiraling movement of the moon story. Like the tricksters of other stories, Raven wanders the vague primordial territory "going round" unsystematically as, in fact, the moon itself appears to do in the sky without apparently repeating its course from year to year. The difference in its *uivvaqsaat* inflection is that the Raven's millenarian journey has, in its circling motion, become more symmetrical and, more importantly, has acquired a new purpose.

This is vividly expressed in the following passage, which shows Raven to be the master of last things as he had been of the creation. Having circled the known world, Tuluŋigraq returns to the creational beginning that he himself had set in motion. At Point Hope, this was the land he had speared and raised from the primordial waters in the origin story. On this land, at some future climax of Point Hope's history, he would, according to *uivvaqsaat* doctrine, bring things to an end in the company of every soul who had lived in that society: "Eventually he will come back here, and he will have the *uivvaqsaat* with him. They call him Uivvaqsaaq, (The One who Goes Around). But nobody knows who these people [*uivvaqsaat*] will be" (Aġviqsiiña 1940:67).

Here the solitary, amoral trickster Raven has been partially Christianized but also reconstituted as *Uivvaqsaaq* (singular), "The One Who Goes Round," the leader of the souls who come round (plural, *uivvaqsaat*). Here, *uivvaqsaat* doctrine has acquired a visionary ethical component dramatically reformulated in the imagery of the Judeo-Christian apocalypse: "Raven (Tuluŋigraq) will have a great flame in front of him which will burn all the bad people. They will fight for ever while they burn. The good people will be taken to live with him. This story is told wherever there were Eskimos [in Alaska]" (Aġviqsiiña 1940:67).

Frank Aġviqsiiña, the Point Hope man who provided this detail, seems to have recognized the connection between *uivvaqsaat* and Christianity, suggesting that his parents' generation understood that *uivvaqsaat* had prepared the way for Driggs's mission: "[In 1940] they still believe this. Only they substituted Christ for the Raven" (Aġviqsiiña 1940:68).

As if this were not dramatic enough, the storyteller made another assertion: "Dr Driggs told them maybe Christ was in Point Hope once. But so long ago that they misremember the story." Whether or not Driggs meant this metaphorically or as an actuality (perhaps echoing what he knew of the Mormon belief that the resurrected Christ had appeared to Native American people), the storyteller suggests the doctor was understood to be in sympathy with *uivvaqsaat* doctrines.[1]

Another Point Hope cosmological belief that was absorbed into *uivvaqsaat* came from a theory about eclipses and the earthquakes that were thought to follow them. During one of the infrequent earthquakes on the peninsula, the beach ridges were said to "pile up." The same would happen with the arrival of the returning souls who were defined for Rainey as "Those who will return with fire when the earth piles up together" (Dives 1940:19).

This convulsion would, according to the storyteller, bring about a moral reordering: a "Last Judgement" (Dives 1940:19). And thoughts of resurrection, as expressed here in strongly biblical language, would bring a reversal of earthly suffering: "[The *uivvaqsaat*] would come back some time with fire, and the earth would 'pile up together'. Those who are bad...will be burned when the *uivvaqsaaq* come. Those who are crippled will be stronger than those who are now healthy. Those who are now strong will be weaker" (Dives 1940:19).

A comparable convergence of traditional thought and *uivvaqsaat* doctrine is expressed in an account of burial customs (Rock 1940:8). Here the pleasant but vaguely defined soul land of tradition where there "is always good weather" coexists with a more newly conceived place which resembles the new *uivvaqsaat* heaven: "a man [had been] murdered, and on the fourth day people heard him shouting from the sky that he was happy. So they believed he was in a good place" (Rock 1940:8).

Driggs arrived in Point Hope while *uivvaqsaat* fervor was still growing. Partly through Episcopal Church involvement in the Dakotas, and also because he had traveled west just after the Wounded Knee massacre, Driggs must have been aware of Ghost Dance activity. The events of 1895, which connected Edson with a "resurrection" in the Kobuk region, might also have alerted Driggs to the presence of some sort of revivalist movement in Alaska. But judging from his published account of Point Hope religion, Driggs seems not to have been aware that it represented a new, albeit composite, Native doctrine (Driggs 1908:59–67).

As in 1894, when he had tried suggesting a connection between Native spirituality and Christian ideas, Driggs again proposed a connection between the Iñupiaq and European religions. First, Driggs's transcription of *uivvaqsaat* as "Webukside" is, perhaps for the sake of Episcopalian readers, made to suggest "Eastertide" and "Yuletide." Similarly transformed is Driggs's account of the Raven creator, who is changed from shaman trickster to the "Great and Good Spirit" who will judge humankind from an ethical standpoint. Reporting on an eschatology whose themes he must have recognized from the biblical apocalypse, Driggs wrote:

> There is a firm belief among the Polarites [*sic*] that a time is to arrive when the world will come to an end, it being known as "Webukside" or the Judgement Day. Tuluŋigraq (Raven), the great and good spirit who was once on earth as a mortal, will be present to judge the quick and the dead. All are to be examined. The wicked, who through the sinful lives they led while on earth have not merited eternal happiness, are to rejected and consumed in the great fire which will finally destroy the world. Those whose good lives have earned for them eternal joy are to be saved; they are to pass with [Tuluŋigraq] into their future home, where they will live forever, free from all cares, or sorrows, or suffering of any kind. (Driggs 1908:59)

Point Hope's Raven creator was of course a traditional trickster figure who had no such role in the fate of souls. But just as Sam Rock, one of Driggs's Native church associates, had conflated the old and new heavens, so Driggs reported a similar amalgam:

> When a man dies, it is believed that after the third night some member of the tribe, who has made the journey before, visits the grave to conduct the new one to his home where he is to remain while awaiting the coming of "Webukside" [*uivvaqsaat*]. One the fourth day after a death, some member of the family strikes four blows with a hammer

> at the recent home of the deceased, which is a sign of farewell and means that spirit is not to return to that iglu again... (Driggs 1908:59)

Like Sam Rock, Driggs here combines traditional theory with *uivvaqsaat*. In the final sentence, Driggs alludes to the traditional belief that souls could linger in the village for four and five days before their journey east (see Chapter 13). At the end of this period, the surviving family would knock on the timbers of their iglu to encourage the soul's departure. But Driggs's opening sentence contains the new idea of a Point Hope *uivvaqsaat* spirit guide who would conduct the deceased to a new abode. This place would, however, be another intermediate dwelling, before all *uivvaqsaat* souls would be ready to "come round," the three days of Christianity here eliding with traditional Point Hope's four- and five-day "waiting period" after death.

Finally amalgamating the moon myth and the story of ghostly visitors with a revivalist version of the Raven story, Driggs concluded his *uivvaqsaat* survey:

> Like the Jews who have so long been awaiting the coming of their Messiah, so the Iñupiat have been waiting for the return of Tulunigraq for ages past.... When Tulunigraq had fulfilled his time on earth, he did not die as an ordinary mortal, but ascended into the sky, the people standing below watching... (Driggs 1908:60)

The storyteller Asatchaq likewise suggests the relationship between the Tuluŋigraq myth and *uivvaqsaat*. In one version of the origin story, Asatchaq gave the Raven harpooner a special song (*avataqsiun*) to accompany the float that was attached to the harpoon with which the Raven had struck, and thus created, the whale-like animal that would become Tikiġaq. As his weapon touched the whale, the Raven sang words that directly connected him to the *uivvaqsaat* cult:

> *Uivvaluk! Uivvaluk*!
> (The harpoon float goes round.)
> Tuluŋigraq will come round again! (Lowenstein 1973–89)

Just as whales come every year and the Raven whaler Tuluŋigraq was evoked during the hunt, so in the cult of *uivvaqsaat*, Tuluŋigraq in his partially Christianized role of savior would return like Christ at the apocalypse.

Asatchaq learned his stories from his mother, Niġuvana, and his maternal uncle Samaruna. Asatchaq's cousin Aviq (Dinah Frankson, born 1909), who listened to the same elders, gave the following account of *uivvaqsaat* theory. Here *uivvaqsaat* is imagined to be both a place and a windlike, spiraling spiritual motion. Aviq says nothing of souls, and yet her *uivvaqsaat*, while both a place and a windy movement, also consists of souls—for this traveling "land" *is* a "land of good people." Consistent with Point Hope testimony, the *uivvaqsaat* here literally "go round." They will metaphorically "come round a bend in the coast," but they will also appear revolving on their own axis:

> Niġuvana and Samaruna were old and knew many things... They said someday in the future the *Uivvaqsaat* would come. This *Uivvaqsaat* would be the land of good people

> and it would be rotating as it arrived. When it reached Point Hope, it would not touch the ground. It would continue in motion. All the good people would simply step into it. (Lowenstein 1973–89)

While the salvationist element of *uivvaqsaat* is uppermost here, Aviq's testimony perhaps incorporates a cosmological idea that was recorded by her father-in-law, Aġviqsiiña:

> The old Eskimos believed that there was a place all around the earth where the earth and the sky met, and there was a noise all the time... The earth stood still, and the sky moved around all the time in the same direction. They met all the way round. (Aġviqsiiña 1940:17)

While Aġviqsiiña and others from his generation spoke corroboratively of the sun, moon, stars, and aurora, of shamanistic spirit life, the fate of souls, and the creation of the Point Hope peninsula, this statement is the only surviving description of a cosmos in which earth is shown in separate and adjacent relationship to surrounding space. While the geometry of Aġviqsiiña's description is hard to visualize, the pattern is relatively clear. Surrounding a stable earth, the sky moves always "in the same direction," an orderly motion that makes a "noise all the time"—a "grinding sound," said Aġviqsiiña.

Less clear is what Aġviqsiiña means by "old Eskimos" (English from Rainey's translator, Aġviqsiiña's son, David Umigluk Frankson, a bilingual speaker and husband of Aviq). On the face of it, "old" might allude to people with a pre-contact cosmology. On the other hand, the speaker may refer not to pre-contact time but simply to the time of Aġviqsiiña's parents. Born in the 1850s, the people of this earlier generation would have been Maniiḷaq's contemporaries, who generated or first encountered *uivvaqsaat* ideas. If, as seems likely, this is the case, then the cosmology sketched by Aġviqsiiña may also have represented a nineteenth-century idiom—although Aġviqsiiña's "heaven" would be an innovative construction and not one adapted from Raven myth or moon spirit lore. Putting the statements of Aġviqsiiña and Aviq together, the spherical movement of Aġviqsiiña's upper air harmonizes with Aviq's description of *uivvaqsaat* "rotating as it arrived." And if Aġviqsiiña's sketch of a "firmament" that moved around earth indeed represented some aspect of *uivvaqsaat* cosmology, then the spiraling soul land described by Aviq might in turn be understood as part of the same cosmological machinery: the souls and/or soul land emanating from a transcendent source, as though thrown spinning onto its voyage from a revolving heaven of which it was a manifestation.

Aviq, a traditional healer and an authority on shamanistic phenomena, concluded her statement with reflections based on a lifetime of Christian practice:

> From what I understand now as a grown woman and after listening to the missionaries who came later, I read the book of Revelations in the Bible and understand it to be New Jerusalem.
>
> It would be the time when Jesus came and we would depart with some of our relatives and friends. After all, it is said that the dead would rise again and embark. They said that this land was abundant in plants and flowers and that it was a beautiful and joyful place. It was to be the land of the good people. (Lowenstein 1973–89)

These last phrases, storybookish as they may appear, radiate a belief in spiritual possibility that Aviq continued to feel. The year was 1977, and *uivvaqsaat* was almost as remote as the myths it had displaced or modified. From this elderly woman's statement, we may nonetheless grasp something of the fervor that *uivvaqsaat* generated during a period that was otherwise marked by desperation. *Uivvaqsaat* lasted until the missionaries revealed the "truth" towards which *uivvaqsaat* had been pointing; and as an expression of hope, its doctrine appears to have contributed a means of compensating for the dissolution of the traditional ideas that lay behind *uivvaqsaat* and which modern America would displace.

Notes

1 As suggested in papers by Du Bois (1939) and Hittman et al. (1973), the visions of the first Ghost Dance prophets of the late 1870s may have been influenced by Mormon doctrine. See references to origins of the *uivvaqsaat* movement at the end of the bibliography.

CHAPTER 25

Driggs's Homecoming, 1896

Driggs left Point Hope in July 1895 and returned twelve months later to take over from Edson, who went home to continue a church career in upstate New York. Shadowy as he is to us during his time on the East Coast, during his journey to New York and Delaware Driggs caused a small sensation. Driggs had left for the Arctic in 1890 as a novice. With five years of arctic experience behind him, he entered Episcopalian society and the colleges where he was invited to lecture as a pioneering hero. Driggs must have cut an impressive figure, with exotic stories of Point Hope, Jabbertown, whales, polar bears, and grizzlies—his frontier narratives justified by moral purpose. Gazing northwest across the continent from New York, Driggs must also have marveled at the scope of his activities in Point Hope. Most careers in medicine, education, or the church on the mainland were separately pursued within the context of professional groups and alongside co-workers. In Point Hope and the surrounding region, the solitary Driggs had singlehandedly dispensed medicine, education, a little Christianity, and, not least, social support to both Native and trader communities.

But praise in the east brought increased accountability in the far northwest. Driggs's first years of improvisation were no longer good enough. He had watched from a somewhat marginalized non-ecclesiastical position while Edson laid the foundations of the church; and now, back on the East Coast, he was spending unprecedented time with churchmen—very different company from his old opera colleagues, Doctor Fred in Delaware, and his Point Hope comrades. The pressure to evangelize would soon be increased by the appointment in 1895 of Alaska's first Episcopal bishop, Peter Trimble Rowe. Rowe was based in Sitka with his invalid wife, but he would travel vast distances, often by dog team, to the remotest Episcopalian communities. Driggs could therefore no longer remain a semi-autonomous pioneer with only the hurrying and irascible Sheldon Jackson to contend with; his results from now on would acquire greater visibility.

Driggs returned to his "northern home," as he called it, in July 1896, and the *Jeanie* anchored off the north side where he was rapturously greeted by Natives "climbing over the steamer's side from their *omeak*s, bidding [him] welcome amongst them once more" (Driggs, Letter of June 7, 1897). Driggs was moved by this reception and happy to be back: "Nowhere in my travels," he continued, "among the wilder tribes have I met any Natives that I like as well, or consider to have so many good traits as our Tigara people… A few days after my return the Natives and the whites came to make me a call, and for a second time bid me welcome." Driggs responded to this welcome with sensitivity: "At first they

FIG. 30. *Group portrait, summer 1895: John Driggs (second from left, back), Elijah Edson (third from left, back), with two Revenue Cutter doctors. The man on the far left (back) is "Big Mouth Jim" (O'Hare). Judging from his resemblance to descendants, the Iñupiaq man second from left (front) may be Kiḷigvak. Candace Waugamann Collection, Archives, University of Alaska Fairbanks.*

seemed rather shy, as if they felt they might be intruding; but I gave them to understand that, old and young, they were just as welcome at the mission as they had been previous to my departure." And reaffirming this hospitality, he wrote:

> I cut a hole in the wood-shed door and inserted a lifter to the latch, so that the people need not stand out in all kinds of weather awaiting the opening of the door. I also left the school-room door unlocked both night and day, in which all my supplies were temporarily stored, along with the carpenter's tools that had been borrowed to use in building my new home. (Driggs, Letter of June 7, 1897)

Driggs's trust was repaid, and he "never had the least reason to regret the confidence placed in my Eskimo friends, for not even the most trivial thing was missing." Nor is there a reference in any of Driggs's writings to Native dishonesty. Indeed, none of Point Hope's missionaries complained that their property had been interfered with or stolen.

Long before he had left on furlough, the doctor had ordered lumber for a new mission building. Driggs's letter, which would be published that November in the *Spirit of Missions,* described its layout:

> My new home, although small, is the most comfortable one in this section. It is just large enough for me and my annual supplies, in addion to the sitting-room and sleeping-room, each 9 x 10 feet, the kitchen, 9 x 11.5, and the store room, 9 x 8.5. I have a small hall, 8 x 8 feet, which is used as a consulting-room for those who are sick and require my services. There is also a good sized wood-shed… (Driggs, Letter of June 7, 1897)

The new house was larger and better insulated than the original building. Driggs's first home, with its "multiplicity of draughts" had never retained heat. A volunteer gang of Jabbertown and Point Hope men helped to build the new mission, insulating it at the base with four feet of sod and six-inch-thick cakes of ice on the windows of the roof. Heating remained a problem. Edson had left no supply of firewood, so the doctor was "caught with only a few water-soaked logs, which were frozen solid and refused to do more than smoulder." But given that the house was secure from the weather, Driggs "stood it nicely" (Driggs, Letter of June 7, 1897).

Anxious to give evidence of evangelical progress, Driggs filled the rest of his 1897 letter with missionary news. He conducted regular services until May, and Sunday attendance for church services in the schoolroom averaged more than a hundred and twenty, a huge increase on levels before Edson's arrival. A new feature of these events was that of children singing from hymnals. Driggs wrote: "The children know a number of hymns, so I distribute the Hymnals, and the singing adds to the enjoyment of our worship. I notice a tendency among the adults to try to do their share by joining with their voices, although they may not know the words" (Driggs, Letter of June 7, 1897).

Here was evidence for the Board of Missions that literate children's new skills could further advance the process of conversion, and that a policy of from-the-ground-upwards, through schooling to Christianity, was effective. Yet more conspicuous was the fact that people were enjoying themselves. The doctor's easygoing sociability and a happy life in church were parts of the same picture.

Easter 1897 came in the midst of whaling preparations, and Driggs's account of Easter Sunday service vividly illustrates the nature of Point Hope Christianity. Edson had complained that early converts used prayer for luck in the hunt. And if Driggs divined that his congregation was adapting its devotions to the hunt, he seems to have had no objection:

> On Easter-day the people assembled in full force, the benches, tables, and floor being well packed, so that it was with difficulty I managed to crowd myself in and stand while telling them about the day. Outside they crowded about the windows, and many stood in the wood-shed, all trying to hear what was said within. (Driggs, Letter of June 7, 1897)

This is Driggs at his descriptive best, and the scene brings to mind how women and children also thronged the ceremonial house skylight to glimpse *qalgi* rituals. But whereas there were six or seven *qalgit* in old Point Hope, there was only one church, and the crowd represented the largest congregation Driggs had so far reported.

Driggs's next paragraph shows another way in which increasingly versatile children may have been helping their elders:

> On my arrival last summer, I was interested in listening to those who had come on board…for the purpose of trading with the ship's people. They had learned from the school children what they considered useful words and were trying to make their wants known in English, but did not do very well. The children would have succeeded much better, but they are diffident to make the effort… In the school room the children read

> and pronounce quite nicely, and also translate Native sentences into English, but the difficulty is to get them to converse in English. (Driggs, Letter of June 7, 1897)

Thus while the children's simple English helped their elders to participate in church and thereby integrate bits of divine service into hunting ritual, Driggs suggested that children were also providing adults with a reinforced trading language, and a year or so later he offered this additional description:

> Their own language is largely made up of compound words, so they seem to experience some difficulty in absorbing the English grammar. Yet all their correspondence with distant friends, or sending word to the [Jabbertown] station of what they wish in the way of trade, is carried on through the medium of English, they not knowing how to write their own language. (Driggs, Letter of June 13, 1899)

These remarks evoke what may have been a pivotal moment in Point Hope's history of acculturation. First, we get an idea of Driggs's grasp of the Iñupiaq language. This reveals both his comprehension of linguistic structure and his insight into why Iñupiaq people found English difficult. But more significant is the doctor's account of the beginnings of literacy among adults. Presumably, the letters to which Driggs refers came mostly from the young men who attended his night school. As Driggs notes, there was no Iñupiaq writing system. Given the growing importance of communication with Jabbertown traders and the distance between the two settlements, the written word was efficient. Someone was always traveling along the beach who could act as mailman and a letter could achieve its effect even in the addressee's temporary absence. Examples of letters written in English by Iñupiaq men in the decade that followed Driggs's account of the literacy issue survive in Heinrich Koenig's papers. All of them concern trade; most express hardship:

> Feb 19 1901: My Dear Cooper
> I like flour Now.
> one tin can molasses
> we no food Now poor man Now. *Achironer ?Pat*
>
> May 1904 Dear Mr Cooper: I want use string of ? and Powder
> and wads (of paper). And molasses.
> we no have to salt. (I want that) *Nashook*
>
> December 13 1904
> I want 2 boxes cartridges 38 55 forget one tin sugar not 3 of flour
> 8 tin matches
> I have the stove 3 paper
> next spring catch whale not
>   give to you stove
> back again stove you *Ahviksinya*[1]

These are the earliest surviving examples of Point Hope people's writing. Inscribed in pencil on worn scraps of paper, the letters are remarkable both for their clarity of script and orthography; only sentence structure and punctuation are lacking. Of course, some people's spellings may have been dictated by Driggs. But if this were the case, then presumably the syntax would have also have resembled his.

The "Turning Point"

"In those [first] days," wrote Driggs, slightly contradicting what he had suggested in 1891, "they had no Supreme Being or God whom they worshipped, and they used to tell me that their Heaven and my Heaven were two different places; but now there is not an iglu, or scarcely a family on the Point, where prayer to the one and only true God is not offered" (Driggs, Letter of June 7, 1897).

Driggs's single extant letter of 1897 expresses this high moment of satisfaction, and it is worth comparing it with his despondency of 1894. Then, Driggs had admitted he had made almost no headway in his "religious instructions." Now looking back, the doctor identifies a turning point that did seem to have occurred in 1894–95. But just as he is about to proclaim his "turning point," Driggs is sidetracked into a complicated digression:

> Soon after the big storm that visited us four years ago [October 1893]...a Native "prophet" at a distance, predicted that there would be a severer [storm] later on, which would wash the Point entirely away, or as the people thought, bring an end to the world (*webuksich*).[2] They were considerably alarmed and came to me; but their anxiety soon passed away. (Driggs, Letter of June 7, 1897)

The first thing to note is Driggs's probable reference to Maniiḷaq and to a millenarian *uivvaqsaat* prophecy in relation to the 1893 storm (see Chapter 23), which coincided with famine in the Kobuk and Kivalina areas, events that may have further generated salvationist fervor. The purpose of the Kobuk shaman's prediction is hard to interpret. Perhaps first, like shaman business such as "scaring away animals: (*niġġutaiḷi*), it appears, if passive, to be hostile. On the other hand, the prediction may, like other of Maniiḷaq's pronouncements, be historically and politically neutral.

Of course, readers of Driggs's letter at New York's Episcopalian Church House had no means of comprehending the reference to *webuksich*. Why, therefore, having proclaimed a "turning point," did Driggs allude suddenly to the Iñupiaq millennium? One answer may lie in the semiconscious mechanism of word association, in that the doctor was aware that the term *uivvaqsaat* was related in meaning to the English "turning point," a phrase he had just inscribed in his letter. When Driggs returns to the nature of his own turning point, his message remains enigmatic:

> One afternoon in the winter [1894–95, before] my return to the States...I was visiting a sick boy and he requested me to teach him to pray in the Native language. I did so, writing out the prayer and giving it to him, that he might commit it memory. The boy passed away during my absence, but before his death he had made a copy of that prayer and given it to a friend. After my return a villager brought me the copy and wished me to teach it to him.
>
> So that is the history of what I consider the turning-point among the people, and I doubt if there is single city or village in the United States where the ratio to the total

> population of those who attend worship on Sundays, has been as large the past winter as it has been here on Point Hope. (Driggs, Letter of June 7, 1897)

This is a poignant scene—but perhaps too much so. Driggs had attended scores of sick people and his medical narratives had never, up to this moment, come close to the sentimental. On the contrary: while the doctor's work was motivated by compassion, his case reports were sometimes inappropriately comic in tone or gruesome in detail. Thus Driggs's description of the dying boy—with its juxtaposition of the charitable visit and the patient's request for a prayer—is a confection being served to the reader, and the episode comes across as an evangelical cliché. Like Edson's account of the woman who thought God might see her drinking, Driggs's story recalls evangelical tracts with their exemplary tales of last-minute salvation. But having proclaimed his missionizing turning point, Driggs then disguises what he may have wanted to express. And having diverted his presentation of the "turning point" with an introductory paragraph about *uivvaqsaat*, he hurries so quickly through the dying boy's story that we are left wondering which aspect of his narrative represented the climactic moment.

The facts of the story do nonetheless provide a map of an interesting series of events, and taking the narrative at face value, three things seem to happen. The young man requests a written version of a prayer in Iñupiaq. He copies it, gives it to a friend, and dies shortly afterward. A year or so later, the same friend, or a third individual who is carrying the paper, approaches Driggs and applies to have it deciphered. Which aspect of these transactions represented Driggs's turning point? We can only guess.

First: the turning point perhaps lay in the fact that the sick man may have been "saved," this perhaps constituting the first presumed entry of a Point Hope soul to the Christian heaven. Second: the turning point lay in the "interest" that at least two men, and possibly three, took in prayer. Third: the turning point consisted in the fact that they passed the script from one to the other. In this case, Driggs would be remarking on the momentum of a religious gesture that takes off through the village, thus demonstrating that Christianity had entered a Native communication system, and was no longer merely being broadcast by white men. (The prayer, in other words, *went round*: perhaps another latent *uivvaqsaat* association.) Fourth: the turning point was the prayer in the form of a written text that could then be memorized and repeated. Without literacy, which provided access to Holy Writ, there could be no Christianity. In Point Hope, the "mumbo jumbo" of "*hoodoo*" (Driggs's own locutions) has given way, or been translated, to the orthodoxy of prayer. Here then we would witness the beginnings of a literacy in the village that, paradoxically, would then revert to the oral. For, as Driggs wrote in the same report, prayer was entering the culture when spoken in "desperate situations" (this is discussed below). The turning point was perhaps a combination of some or all of these elements.

Another partial key to what Driggs meant by his turning point lies in a letter of June 1899. Significantly or not, Driggs's somewhat fragmentary reports of the previous year contain only vague generalities about religious life. Now in 1899, Driggs returns to the evangelical themes of 1897, and after opening with a description of lapses in local taboo observance, proceeds to an account of prayer in the village. First, Driggs notes a "gradual breaking down of the tribal superstitions," in particular during the whale hunt. This connects with our own earlier discussion of Ataŋauraq's use of a stove on the sea ice—the first recorded lapse in Point Hope's taboo system. Like the Ataŋauraq episode, the stories that Driggs cites also take place during whaling, but they center on taboos relating not to food preparation but to death:

> Out with the whaling canoes were two widows that had recently lost their husbands. At former times such a thing would not have been allowed, but instead, they would have been placed under the superstitious ban called "kerooktoak" (*qiruktuaq*, "is in taboo"), and isolated from the rest of the tribe. (Driggs, Letter of June 13, 1899)

Here, women who previously would have been spiritually "infected" by contact with death are allowed to participate in the whale hunt, but it is difficult to know how far these changes were influenced by the mission. But other issues may include the fact that women, who would anyway never have been in a skinboat crew, are now *required* to take part because the subsistence crisis has made their labor indispensable.[3] Point Hope people had no doubt also observed that some Jabbertown boats with women crew members caught whales. Driggs's second example (from the same letter) is more straightforward:

> Then at the latter part of April, one of my best scholars, a young man of twenty died. His remains were brought to the school room where we held the funeral service. A number of the whalesmen coming in from their canoes for the occasion. Among them being some of the older and most superstitious members of the tribe. It was very gratifying to see so many present. For by their former tribal law, if anyone that had recently come in contact with the dead should go on the ice, and join their canoes, that the whaling season would consequently be a failure.

The death here takes place at the beginning of the whale hunt and, hearing news of death in the village, hunters of an earlier period would have maintained their isolation on the sea ice. But now (spring 1899?), the opposite has happened and the rule demanding separation of land and sea ice has relaxed. Instead of repelling "superstitious" crewmen, death has drawn them to the village to comply with the demands of Christian burial. Further, the mission house on which they converge is revealed, during this highly charged ritual season, as a place of positive, or at least neutral, spiritual import.

Driggs then introduces the subject of prayer, the implication of his narrative being that people were adopting this practice because taboo rules were eroding:

> I believe one of the best things I have accomplished since my return, has been in teaching the people how to pray. They have a considerable confidence in the efficacy of prayer, and it had undoubtedly aided several in preserving their lives at times when placed in desperate situations. Their prayers giving them the confidence which has encouraged them to battle on under adverse circumstances.

Driggs had taught non-Christians to pray; and though the doctor could not bring himself to describe it earlier, this most probably is what his readers would have regarded as the evangelical "turning point." Four years later, reinforcing his view that prayer had replaced traditional incantations, Driggs goes further, claiming in a final anecdote that Christianity had spread from Point Hope to Siberia:

> This spring, only one of the whaling stations met with success, they succeeding in landing two large whales. At the second [station] they had rather an exciting time. They were twelve miles from the shore on the ice, the pack commenced crowding in threatening to crush the whale.
>
> The cry was made, What can we do? Then one of the Native helpers proposed that they should pray, at which all the Natives knelt down and prayed. Then the ice stopped crowding in and several thousand dollars worth of whale bone was successsfully landed.
>
> During the winter, one of the villagers had gone far out on the ice pack and had killed a bear. The wind had changed and he discovered that he was surrounded by water and apparently no way of escape. So he knelt down and prayed—then looking up saw a narrow strip leading from the pack to the floor ice which enabled him to reach home in safety.
>
> All the Natives now have a great faith in the efficacy of prayer and resort to it when they find themselves in danger. The results of the teaching of the mission has not been confined to these people alone, among whom we are working, but like the "Good seed," has taken root and spread to the most distant tribes so now, even over in Siberia where there are no missionaries, the Natives have their piece of paper and a pencil, each day making a mark with a longer one and a cross on Sundays so they know when the Sabbath had arrived and are beginning to observe the day. (Driggs, Letter of July 3, 1903)

Through his *qalgi* visits and hunting trips with Anaqulutuq, Driggs had acquired an understanding of Point Hope ritual and knew the place occupied by magical words and the rhythms that accompanied these. The *word*, in local religion as in Christianity, was a vehicle of power, and thus when Driggs wrote of "confidence in prayer," he understood that this involved the relationship of new verbal "amulets" to their existing, still by no means discredited, culture of shamanistic songs. And perhaps he also comprehended that it may have been the same with regard to the nature of the deity. It might not, in other words, have been important whether it was the moon spirit or a "Heavenly Father" who supported the whale hunt, so long as one or both did something useful for the community.[4]

Interaction: Tikiġaq Religion and Christianity, 1900–1902

The relationship between the old and new religions seems seldom to have been stable even by the turn of the century. Driggs encapsulated the nature of this issue in an account that both repeats and contradicts the events of 1899 concerning death rites during the whale hunt. While in a letter of June 1900 Driggs wrote that people were "gradually abandoning their superstitions," he admitted in the same paragraph that "many of the older members of the tribe still cling to some of them." To illustrate the latter point he reported how "a young man had to bring his [sick] brother ashore" from whaling camp, and the older men wouldn't let him return to the ice until a whale had been caught. Later, wrote Driggs, the same [old or young?] man "tore down one of the older elevated graves and used the wood as fuel to cook his dinner. Five years ago the same man could not be induced to touch a grave." This was no doubt true in part, although Driggs presumably understood that people had always broken taboo when circumstances compelled them to it.

As we saw in Chapter 11, when Ataŋauraq performed a shamanistic autopsy on a woman who had died during the spring whale hunt, it was profoundly inauspicious. And the metaphysically contagious nature of death was more generally expressed, as we have also seen, in the rule that death must never be allowed to occur within the home; when this did happen during the traditional period, the spiritually contaminated iglu had to be abandoned. But, wrote Driggs in 1901, a "great change has occurred during the past three years.The young men visit their sick companions, singing and praying with them, and as they pass away take their hands and bid them farewell, hoping that they may meet again in the future life" (Driggs, Letter of June 4, 1901).

Once again, there is a tractarian flavor to this anecdote that makes one suspect that Driggs was writing for a home audience; but given that it was generally Driggs's habit to recount his observations with a somewhat literal-minded bluntness, there was presumably truth in his description of this change. In the same letter of 1901, Driggs then wrote of further transformations in local behavior, suggesting that this was connected to church attendance:

> The people have shown their interest in our services by filling the school-room each Sunday, and I have the satisfaction of noticing that the work is showing good results. I believe that many are thinking seriously, and firmly believe in the instructions they receive each Sunday at the Mission.... Our Tigara people have gained the reputation of being the most orderly and peaceable tribe among the Eskimos, and there has not been any [liquor] distilling among them for the past two years.[5]

Driggs's command of Iñupiaq was by now good enough for him to converse freely, presumably on any subject. And such was his good standing in the village that not even the older and more "superstitious" were now beyond his reach: "While I have been told by others in the past that it was useless to work among the older members of the different tribes, my experience has been different. I have made the effort and found the old just as susceptible to conversion to Christianity as the younger ones" (Driggs, Letter of June 4, 1901).

Driggs next proceeded to an astonishing assertion: "I do not think that there are now any real pagans left among the Tigara people." And to illustrate this, he concluded:

> One member of a distant tribe [to the south], who was visiting one of our village families, remarked that he did not believe in the instructions that were given to their young people on Sundays at the [Society of Friends?] Mission Station in his section, and that he considered it all untrue. The father immediately ordered him out of the iglu, remarking that he did not wish his family to hear him talking that way.

In an undated interview from this period Driggs went a step further, claiming, perhaps hyperbolically, that this visitor was ostracized by the whole community for repudiating

Christianity. "Eventually," said Driggs, "I converted him" (Driggs, in Episcopal Church Archives). Given the poor man's isolation, he presumably came to a rapid decision.

Driggs's accounts of shamans who sometimes attended church when this fit with their own practice have been given already (see Chapter 22). The storyteller Asatchaq claimed that in addition to other healers and visionaries, there were ten shamans working in the village during his childhood (Lowenstein 1973–89). Taking into account their clientele, who might also have been considered as "pagans," Point Hope's traditional worldview was, in 1901, by no means a thing of the past. That said, a good many of these non-Christians may also have qualified as Driggs's "converts." And judging even by Asatchaq's own devotion, as late as 1977, to both the old and new religions, it was certainly premature of Driggs to have claimed that Point Hope's shamanistic system had been entirely superseded. As noted already in passing, Asatchaq's father was said to have been killed by a shaman (see Chapter 22), while even in 1953, old man Masiiña claimed to have flown in spirit to the Soviet Union to have dispatched Stalin (Lowenstein 1973–89).

Bishop Rowe's Visitations in 1899 and 1903

"I have not heard from Dr Driggs since last year," wrote Bishop Peter Trimble Rowe in his report on Alaska for 1898 (Rowe 1899:91). No correspondence between Driggs and the bishop during this period survives, and Rowe had been in Alaska for at least three years before; thanks to the *Bear*, he reached Point Hope after a long trip north, and wrote an account of his visit. The late 1890s saw virulent epidemics in north Alaska, and in summer 1898 influenza was rampant along the coast. Rowe witnessed terrible scenes of disease when he landed at Port Clarence, but further north found Cape Prince of Wales in good health. He sailed north anxiously. Iñupiat on the Bering Strait had told him that Driggs was sick, and when he reached Point Hope, he raced to the mission. "But to my great relief," wrote the bishop, "I found him in his lonely post quite well, though he had been ill with the grippe (flu) and delighted to see me" (Rowe 1900:87).

Point Hope had not, as described later, escaped the flu, and Rowe praised Driggs for opening his own supplies to the village. Interestingly, in the light of Driggs's future history, the bishop wrote: "I was also glad to hear men, as far north as Point Barrow, say that Dr Driggs was the only [true] missionary in the country, because he attended strictly to what he was sent to do, minded his own business and did not combine trading with missionary work" (Rowe 1900:88). Given Driggs's anxiety about trading when he arrived in 1890, and his dismissal by Rowe in 1908 for precisely that activity, the bishop's remarks must have been reassuring to Driggs, while in retrospect they have a poignantly ironic ring.

Perhaps Rowe did not expect much of the school, and he skates tactfully around the fact that people seemed to speak little English. The bishop did however take note of the subsistence crisis, and perceptively made the connection between the absence of game animals and low resistance to disease:

> The children whom I met understood English fairly well. There is a need of large pictorial scenes from Holy Scriptures; also a supply of good woolen clothing for summer use so that their fur winter garments may be saved for winter. It is becoming more and more

> difficult for the Natives to get the furs necessary for winter clothing. This partly accounts for the epidemic which has decimated them so fearfully this year... (Rowe 1900:88)

And merging human and natural history in a paragraph that might have come from the pen of John Muir, Rowe recorded these impressions:

> This must have been a large village from many, many years ago. Here is one of the oldest and most interesting graveyards in all the north. It extends for two miles or more, and everywhere stand scaffolds, six feet high, made of the jaw bones of whales, upon which the bodies of the dead are laid. Some were falling to pieces and beneath, amid the fresh forget-me-nots, lay the skulls and bones of a people long since dead. (Rowe 1900:88)

As though some evangelical forget-me-not had germinated in the pagan humus, Rowe also notes the fragile growth of missionary endeavor: "now interspersed amid these scaffolds one sees the cross at the head of a grave, an evidence of the presence and influence of the faith of our Lord Jesus Christ. In other places the dead are hidden amind the cliffs and rocks" (Rowe 1900:88). It was a sensitive appreciation of the Native cemetery that Driggs's successor (perhaps with Rowe's agreement?) would destroy nine years later.

This first Episcopal visitation amounted to a fact-finding exercise. But Driggs was encouraged and supported, while the bishop left satisfied that the mission was in reasonable order—although, as he noted, it was "something of a refrigerator." Rowe did not, however, go to great lengths to praise his Point Hope missionary, and since most people were absent during his visit, he had only secondhand reports to go on: "The doctor reported some progress, a good attendance at school and much improvement among the Natives. On all hands I heard good accounts of these Natives" (Rowe 1900:88).

Most significantly, Rowe gave Driggs the authority to conduct services of marriage and baptism. These Driggs started in the winter, writing a sketch about a baptism that went some way beyond news of missionary activity:

> [I have married one couple & have baptized four.] Among the baptized was an infant that the parents had named after me. Early one morning last August the father who is a pupil at the mission, came to inform me that his wife had just presented him with a son and that they had decided to name him Doctor. The Natives pronounce it *Docte*. He was sure that the little fellow must be very hungry so, he begged for a can of condensed milk and then went skipping back over the trail to the village whistling "There's a New Coon in Town," a tune that he learned from some whalesman.[6] [Driggs, Letter of Sept. 14, 1900)

Aside from the cheerful whistling, the naming event represents a convergence of Native custom with baptismal practice. Two other people—"Electuna Driggs," one of the first Point Hope reindeer herders and another man on the Kobuk River—took the name Driggs during this period. But the story of the baby, whose name would have been pronounced *doctuh*, provides the first instance of Driggs being drawn, at the point of birth, into Point Hope's naming system (see Chapter 13). The significance of taking "doctor," the potentially medicinal aspect of Driggs's name, can be well understood. And in the context of the disease crisis, Driggs no doubt comprehended this well.

Before long, there were more Point Hope and Jabbertown people applying for baptism:

> On Easter Sunday the school room was packed to its utmost capacity. After services I went up to and baptized five of the children belonging to the [Jabbertown] whaling station. I have baptized twenty-seven the past season, twenty-two of whom were Eskimos. The other five had white fathers and Native mothers. In one Native family I baptized the grandfather, the father and mother and the two grandchildren...I have also solemnized the marriage ceremony twice. (Driggs, Letter of Sept. 14, 1900)

Twenty-seven baptisms represented a significant proportion of the combined population, and it is interesting to see that Driggs's new powers extended to both the Point Hope and Jabbertown communities. This set a seal of legitimacy on an "alliance" between the two villages which since 1887 had been forged by both trade and common-law marriages: an association that had produced a new generation of mixed-race children, who now emerged as the first baptized Point Hope people. Some of these (the children of Koenig and Hachmann, for example) would soon leave the region. But more significant for Point Hope is the final detail in Driggs's record, which shows the baptismal spread going vertically along three generations of one Native family. Here was extra evidence for Driggs's earlier claim that his missionary effort was to some extent reaching the older generation.

Rowe's visit had left its mark, but, wrote the bishop four years later, "To reach Point Hope oftener, is not an easy matter," and his next visit was not until 1903. "Several official acts," he continued, "make this year's visit...monumental and worthy of note" (Rowe 1903:50). The first two of these would be to ordain Driggs as a deacon and then to catechize and confirm the men and women Driggs had prepared to be Point Hope's first communicants.[7] Mindful perhaps of the bishop's remarks on the subject of proficiency in English, Driggs had been anxious to prepare his candidates thoroughly. About one month before the bishop arrived, Driggs wrote that he was "instructing those whom I wish to present to the Bishop when he visits this summer. But the candidates are finding it quite difficult to memorize English, although I believe they understand very well in the Native, and are living up to their instructions" (Driggs, Letter of July 3, 1903). Driggs follows this, echoing Edson's comment in 1895, with some uncharacteristically unfeeling remarks about language and intelligence:

> These are a primitive people, having existed here, cut off from the outside world from time immemorial, consequently, they have never undergone the brain development that those who are living in the more favored latitudes have undergone, and memorizing seems to be especially difficult.
>
> Probably one great drawback is, that they are no sooner through the school room, than they are back to the village, where the Eskimo is the medium of communication.

Still, Driggs observes:

> there is scarcely any of the younger members of the community between fifteen and thirty, but what understand English when spoken to and can reply.
>
> It is a condition that time will rectify, especially when the older inhabitants die off, which they are doing quite rapidly. I look forward to the day when our language will take the place of the Eskimo.

In contrast to what Driggs had just written, Rowe was overcome with admiration for the people he met, and it is gratifying to see Driggs so warmed and lit up by the Bishop's admiration: "This honor [of the diacanate] was due him after so many years of true and faithful service—of lonely life and unselfish toil, the fruits of which are now appearing; besides the needs of the people and the work demand it" (Rowe 1903:50).

The bishop's visit was without doubt a high point in Driggs's career, and he made sure that there would be a good turnout. In an ordinary week of August there would have been almost no one on the Point except the old, the sick, some children, and a few couples working on their iglus. No doubt, to guarantee a crowd, Driggs had trailed the event colorfully, insisting that his students were in residence so as to impress the bishop and lead the congregation. The impact of this on summer subsistence activities must have been considerable, but at least Driggs was not disappointed:

> Thursday, August 6th… was chosen for the service. It was an ideal Arctic summer day. At the appointed hour, the bell rang out its clear call and soon the spit was alive with the Eskimo, issuing from their iglus as rabbits out of the ground, a picturesque group in their varied skin and [cotton] drill "berugas" [parka covers] as they came toward the school house. There were 200 or more. I think they understood the meaning of the service, for I had explained it to them the day before through a Native interpreter [Uyaġaq Sam Rock]. (Rowe 1903:50)

It is important not to devalue the significance of the event that Rowe describes here. A community that had been so ravaged by disease and decimated by outward migration was clearly receptive to and grateful for this degree of ritual intensity:

> They filled the school-house, transformed into a chapel as far as our means permitted. The service was held. Many of the younger Natives seemed to be able to repeat intelligently the Creed, the Lord's Prayer and the Ten Commandments, while all sang the several hymns. They were intent, reverent and interested. The doctor received the Holy Communion—"the first time in seven years," as he said to me with tears in his eyes. What a deprivation to this lonely banner bearer of the Cross so far off! After service the Natives pressed round and shook our hands, so touched were they with this blessing conferred on their friend and leader of thirteen years. (Rowe 1903:50)[8]

Three days later the ceremonies resumed with yet more momentous scenes. This time it was the turn of Driggs's most advanced students to take center stage. This, surely, was the climax of the work that Driggs had been sent to accomplish. And while within the next eight weeks the *qalgi* ceremonies would start and the shamans and *umialit* would be drumming up their spirits, the experience represented a historical moment—an initiation that no doubt included shamans—of unprecedented splendor:

> Sunday August 9th was appointed for the confirmation of such candidates as Dr. Driggs felt he could present. The days preceding were spent in examining them. I was happily surprised at their knowledge, brightness, intelligence and ability to answer questions of the Catechism. Their understanding of the truth, of their duty as confirmed Christians, was most satisfactory to me. Again the bell called them to service—the same congregation. First came some to be baptised, others to be married; then seven young, reverent, noble-looking men and women knelt to receive the blessings of the laying-on of hands. The first Natives of so far north—the first Eskimo—to be confirmed and become communicants of the Church! (Rowe 1903:50)

Overwhelmed by this display of piety, the Bishop concluded: "It is an event happy to relate—encouraging and blessed to me! These fresh recruits received their Communion as reverently, as intelligently, as one could wish to see."

If Point Hope people impressed the bishop with their dignity and intelligence, the event, in its length and solemnity, was also marked by an ecstatic fervor that might have appeared too "enthusiastic" for the generality of Episcopalians. To the Point Hope congregation, the sheer physicality of the occasion no doubt resembled the long, drawn-out autumn rituals in the *qalgi*: "The services lasted for hours. The building was filled. The heat was intense, for the building is made for winter use, and we all dripped and dripped with perspiration" (Rowe 1903:50).

The bishop never doubted the sincerity of his congregation, and his impression was presumably justified. Yet the doctor must have known what Rowe could not see. As September approached, Driggs would open school and, fired up by the bishop's visit and Driggs's ordination, Sunday services would start again. At exactly the same time, the ceremonial houses would reopen. Masked dancers would invoke the souls of animals, and the moon would re-enter its season of authority. The Christian God must, for now, share heaven with the lunar spirit. Despite Driggs's tolerance, however, the last two standing *qalgit* would soon lie derelict, and by 1909 all the houses would be abandoned (see Appendix B). Many *qalgi* rituals, dances, songs, and games would nonetheless survive the ruins and be transferred to new frame buildings. And affiliation to the two last *qalgit*, Uŋasiksikqaaq and Qaġmaqtuuq, would survive as the organizing parties that to this day pursue the whale hunt and celebrate on the sites of the old houses. But that very un-Episcopalian religious enthusiasm that flowed free in the *qalgi* and which had flooded the schoolroom in the days of Rowe's visit would soon transfer, if not entirely, to a Christianized environment.

Notes

1 Probably Aġviqsiiña, whose information on *uivvaqsaat* was quoted in the previous chapter.

2 *Webuksich* is *uivvaqsaat* in Point Hope dialect. The final *t* used in Kobuk and Barrow plural nouns becomes a sibilant, as transcribed here by Driggs.

3 See also Brower's account of women at the Point Barrow whale hunt in Chapter 11.

4 Stefansson (1913:81–82) gives this account of the use of prayer on the Colville River in 1908: "when it turned out that my Eskimo had been to Herschel Island, where there was known to be a missionary, the local people enquired eagerly whether we had brought any new prayers

with us… What people especially wanted, they told us, was a new prayer for caribou from Kotzebue Sound. It had worked so well for the first two years that they had secured plenty of caribou through the use of it, not only during the summer season when the skins are good for clothing, but also (so efficient was the prayer) during the winter, when under ordinary circumstances they would not have been able to get any. But this year the prayer did not seem to be working so well. They supposed that white men's prayers, like their rifles and other things, no doubt deteriorated with age, and now they were anxious to secure a new and more efficient prayer. Akpek told them that he had a very good one, and he proceeded to teach it to them."

5 As suggested by Mary A. Larson (2004:66), this change in behavior may have been occasioned by visitations of typhoid, scarlet fever, whooping cough, and chicken pox.

6 "New Coon in Town," by J. S. Putnam (1883), from the "Coon Song" repertoire of black-face "Ethiopian Minstrels" of the 1880s. While the whistling father may have learned the tune in Jabbertown, Driggs no doubt recognized it from his theater days in the 1880s.

7 A deacon is an assistant priest, third in order of seniority to bishop and priest.

8 In fact, twelve years and one month.

CHAPTER 26

Driggs: The Final Years and Deposition

Relations with Authority

Bishop Rowe had affirmed some vital relationships during his stay, and arriving for his second visit on the naval ship *Thetis*, he also connected the village with a benign church paternalism. Adding his voice to those of Episcopalian "Friends of the Indian" in the 1880s (see Appendix C), Rowe laid down a demanding code of behavior and spoke out resolutely against the exploitation of Alaska Natives. In a letter of 1909, he cried:

> I am despair in regard to the condition of the Natives of Alaska, especially in the interior... A perfect carnival of lust, debauchery and drunkenness prevail among them. They are demoralised, diseased and pauperized in consequence...there seems to be no check to it anywhere... The government owes it to these people, its wards, to defend and protect them from the evils which white men ply among them... (Rowe, Letter of June 5, 1909)

In Point Hope, the main issue remained sexual assaults on women. These came mostly from visiting whalers, but Jabbertown men also contributed to the violence. Driggs, unlike Edson, had hung back from confrontation. Following Rowe's visit of 1899, however, Driggs started to speak more forcibly. In January 1902 he wrote:

> I have written to the U.S. Commissioner at Kotzebue and am in hopes that he may come here next spring and hold court. There are two cases that should be tried: both [involving] Portuguese [Cape Verdean] whalers that were landed from a whale ship last October. One brought liquor with him and has been trading it to Natives. The other one has been guilty of committing rape on a girl of tender age... (Driggs, Letter of Jan. 10, 1902]

Following a second rape, Driggs reported:

> [The Portuguese] had been complained of by one of the [Jabbertown] whalemen and had received a letter of warning from [Healy] telling him not to repeat the offence... this element has no respect for American laws and do not fear them. (Driggs, Letter of May 23, 1902)

Ruminating on his earlier passivity, Driggs continued: "I have always tried to avoid these whalesmen's troubles and have overlooked a considerable [number] of their wrong doings."

And perhaps echoing Rowe's comment that he "minded his own business," he went on, "it being my aim to attend strictly to my own affairs. But I can not overlook the raping of the two little girls by the Portuguese, and shall do all in my power to bring him to justice."

Driggs understood, however, that a single missionary could hardly regulate the behavior of Point Hope and Jabbertown—about two hundred people. The only viable answer lay in the legal system, but for the present, the courts in Kotzebue, Sitka, Juneau, and Nome were too far off.

Driggs was also anxious to exonerate the Jabbertown majority—after all, it had been Jabbertown men who had reported the rapes—and in a somewhat xenophobic attack on the Cape Verdeans, Driggs declared:

> It seems to be a difficult matter to get the law enforced in this section. A portion of the foreign element apparently consider that our laws are not worth paying any attention to, at least so far north… There are only a few of these undesirable non-citizens. The rest of the whales men are as a rule a peaceable set, and most of them seem to be well disposed towards the Natives and are somewhat a benefit to the place.

But the nearest court involved a five-day sea journey, a route which was open only in the summer: "The courts at Nome are not near enough to be available. If the *Bear* should arrest the criminals and take them Nome, the question would arise: 'How would complainants and witnesses get home for the winter?'" To which question Driggs offered a solution whose fulfillment would take another two decades: "[I]f the authorities in Washington could send one of their many commissioners and a marshal here, twice a year, that would show these people that the United States has laws and is capable of enforcing them."

These thoughts recall the legal vacuum of the 1880s. Troublemakers like Bayne, Ataŋauraq, and now these whalers represented a perennial difficulty. In 1889 Commander Stockton had appointed a headman—an illusory position—and pleaded for a missionary presence. Driggs at last saw himself as an embodiment of Stockton's ideal missionary worker. After the Ataŋauraq period, and in contrast to Jabbertown, where the ruffian minority did much as it pleased, there was peace in Point Hope. Such tranquility, argued Driggs, followed the introduction of Christianity, for which, as of 1902, he took credit: "I have been the means of introducing Christianity to these people. They are now a well behaved peaceable tribe, so I feel it my duty to do all in my power to protect them" (Driggs, Letter of May 23, 1902).

Perhaps somewhat unrealistically, Driggs saw himself as the embodiment of the new order. It was, he argued, the individual and not just his teaching that created the milieu through which transformation happened: "Missionaries," Driggs wrote in 1901, "are accomplishing a great deal. Each doing a share in uplifting our fellow men for the advancement of His kingdom" (Driggs, Letter of Aug. 18, 1901). Continuing in this vein, the doctor also claimed a more humanistic position: one that placed himself and his work socially among people with whom he had established both individual and collective friendships: "I myself am a great believer in the brotherhood of man [and] derive a great pleasure in my work."

Was this a stray thought that occurred to Driggs in 1901 or the expression of a long-held conviction that slipped out in a letter dominated otherwise by anxiety about the loss of his supplies that summer? As with much of what the doctor did and thought, an intriguing complexity lies within this somewhat abrupt and unprecedented proclamation of philanthropy. The complete statement, written as a single utterance without the conjunction "and," comes as near to the truth of Driggs's motivation as we are likely to get. Literally, given his phrase "great believer," this sentence represents Driggs's credo. The altruism and the pleasure—the latter may be understood also to contain the happiness in a free life beyond conventional American society—are complementary. And no other Euro-American in the territory qualified better, in both the social and the humanistic sense, as a brother to Native Alaskans than Driggs: an aspect of his fraternal feelings lying in his wholesale commitment (without, in the received sense, "going native") to life in the Arctic. His impact on Point Hope's social order represents, of course, another question.

Missionary and Baleen Trader

Thanks to Rowe, Driggs had consolidated his place in the church. But the new century saw a development of new interests. Driven by loathing for the abusive Jabbertown minority, sympathy for Point Hope, and a personal interest in whaling, Driggs now entered the period that would lead to his deposition. The first hint of this convergence of philanthropy and self-interest comes in a letter of 1902:

> This spring, the Portuguese and one or two other boats came flocking down to the village and did their best to crowd the Natives out, thereby destroying their chance of doing anything for themselves and enforcing a year of poverty on them. The Natives came and complained to me but I had no authority to do anything. (Driggs, Letter of May 23, 1902)

Driggs did however have one means of helping the village. He would become a whale-boat owner. To launch a whaling boat, a Native *umialik* needed two kinds of resources: hunting equipment and enough provisions to feed a crew for up to six weeks. Given the crisis in local food sources, this meant obtaining flour, rice, and other imported foods from Jabbertown or from passing ships. Few Point Hope men had sufficient surpluses of skins and ivory for such trade. Driggs, on the other hand, could easily afford the goods with which to pay a whaling crew. And so in the heart of Point Hope, a whaling enterprise that approximated that of Jabbertown evolved, and there were now three categories of hunt on the peninsula:

- The traditional, but modified, Native hunt, some of which was now partially subsidized by trading;
- The Jabbertown hunt, which employed both Euro-Americans and Natives;
- Driggs's hunt using Point Hope crews. Baleen went to Driggs, meat and blubber to crewmen.

Driggs's crews differed from Native outfits because the former were entirely underwritten by a white man. But his boats also differed from those in Jabbertown because they were manned by Point Hope men and the edible whale products went to the community. The thrust of Driggs's operation was thus both commercial and philanthropic—though in what proportion is hard to determine.

The next passage in Driggs's letter becomes somewhat over-confiding but indicates to later readers how closely he followed Native whaling:

> Last year, the Natives were unsuccessful in their whaling with the exception of Anaqulutuq, who had been fortunate in landing one, with seven foot bone.
>
> This year the villagers were all very poor and it looked as if Anaqulutuq would be the only one that could put out a crew for the whaling season. (Driggs, Letter of May 23, 1902)

Driggs provides evidence here that commerce was now built into Point Hope's subsistence effort, for Anaqulutuq's haul of baleen in 1901 provided him the capital to subsidize the 1902 hunt. The next paragraphs show Driggs moving from the position of observer to a participant in the whaling economy: "So some of them applied to me and I loaned them a few supplies which enabled them to put out two other crews making three crews for the villagers only." Driggs saw two practical reasons for this intervention. One was to help supply the village with food. The other was to make it possible for Natives to enter the baleen trade. But large, mature whales with marketable baleen had become scarce:

> One of the crews I aided succeeded in landing a small calf, so I have the satisfaction of knowing that, although they have not captured any marketable bone to trade with the ships for their winters supplies, yet they have succeeded in laying by a considerable amount of fresh meat and blubber in their caches for next winter's use.

But even if the Natives had landed a mature whale, they would still have been at a disadvantage:

> The whites and Portuguese receive a dollar a pound for their bone from the whale ships and pay two dollars a sack for their flour... The Natives do not receive anywhere near that price and have to pay the highest rates for every thing. It places them in a very great disadvantage and to a considerable extent at the mercy of the foreign element.

There is confusion in the final detail. Unwilling to disparage his Jabbertown friends, Driggs conflates the "foreign element" with the Euro-American majority—although most of them offered Natives more or less the same terms. But Driggs's point is clear. Natives were routinely treated as commercial subordinates in a trading system that was biased in favor of Euro-American profit. This imbalance lay at the heart of white-Iñupiat trade relations, and Natives, however practiced they were in barter, had no choice but to accept the conditions offered.

Driggs's intervention was timely. By provisioning some boat crews against an anticipated return for himself of baleen and perhaps ivory—the term was to "grubstake"—he both supported the Native hunt and helped to redress the imbalance between white man's commerce and Native subsistence. But Driggs's philanthropy was also informed by self-interest. We saw how, back in 1890, the doctor had been tempted (and tormented) by the prospect of trade. He had, so far, scotched that snake. Over the past decade his only recorded acts of trade had involved friendly goods exchange with Jabbertown people and payment to Natives for services such as fuel and ice gathering. Driggs's status in the village had been endorsed by Bishop Rowe. But the doctor had also assimilated into both Point Hope and Jabbertown societies. There was more to his life than the professions that he had brought with him and continued to pursue. His interests were nonetheless divided. In Point Hope he was *Taaktaiḷaq* (that puzzling term meaning "lacks a doctor"), teacher, and missionary. To his Jabbertown friends he was both a man of the professions and also one of commerce. And what, after all, did most men, Native and white, do in the nineteenth-century Arctic? They hunted and traded. And without apparently compromising professional obligations, the doctor had slipped quietly into these activities. Since 1891, he had, after all, always closed school when the whale hunt started. His new "*umialik*" activity was in the nature of holiday work, not unlike that of many an impoverished nineteenth-century village schoolmaster.

Driggs was in his fiftieth year in 1902, and while he seemed happy to continue living in the Arctic, presumably his thoughts must have turned to what old age might bring and the means he would need during his retirement. It is thus not surprising that along with his interest in whaling, statistics about gold prospecting run though Driggs's journal from 1899 to 1901, and one isolated entry in July 1895 shows Driggs had long been active among Jabbertown treasure hunters:

> Start home in boat with Hackman. We visit the beach on north side and locate slate and black quartz which has a good amount of gold, silver and perhaps pyrites in it. We have agreed to keep it a seacrit (sic), and next year to return bringing the proper tools and chemicals and work it for gold and silver. We have located this as our claim

In addition to scientific knowledge of his own, Driggs may have read John Muir's *Cruise of the Corwin*, where he notes of the Seward Peninsula: "The rock containing mica, slate, and a good deal of quartz, seems favorable for gold" (Muir 1881:130). Driggs's interest in prospecting was also heightened in 1900 by the Nome gold rush, which had excited several other Alaskan missionaries, some even abandoning their stations for Nome's golden beaches. Thomas Lopp, Alaska's reindeer superintendent and formerly one of the teachers at Cape Prince of Wales, was among them. Driggs, who appears to have been sufficiently well connected to maintain an interest by proxy, stood his ground in Point Hope from where he planned his involvement with Jabbertown colleagues:

> Kelly has arrived back and brings me a letter from Lopp. I sent Lopp my power of attorney to locate me a claim. Dan [?] starts for the gold diggins [sic] tomorrow...
>
> John [Hachmann] and Mr Johnson the miner called and brought wood. The miner thinks [Point Hope] a good prospecting country. Marlin and John start for Nome 15 March...
>
> I start with Marlin and return to Cooper's and spend the night there. I sign my name to notices for coal and gold claims. Kelly writes that he arrived home last evening. He has steaked [sic] me out a placing and also a coal claim... (Driggs, Journal entry for January–March 17, 1900)

None of this activity of course ever reached Driggs's employers—though an interest in local gold prospecting might have been regarded as harmless pottering. Then in 1899,

Driggs first records his interest in commercial whaling. The multiple journal entries that follow show Driggs's keen observation of both the Jabbertown and Native whale and beluga hunt:

> April 21–22 1899
> First Native canoes go on ice… Bain [sic] is said to have arrived and is at Nelson's [whaling station at Piñuatchiaq]. Eight Native canoes go out today.
>
> April 26 1899
> [?] shot the first whale but it got away
>
> Saturday 29th/99
> s-everal white fish were shot yesterday but they sunk [sic]. 9 rifles and a musket were lost overboard.
>
> Sunday 30th\99
> Nelson's boats came down. Nelson spends the night at the house.
>
> Thursday 4th May
> Two whales were killed to day. One by Mr Liebs man & one by Nelson's man a small one
>
> Monday 8th/99 The whale Oayugyah [Uyaġaq] killed for Stine was found under the ice yesterday (a small one)
>
> Wednesday 17th /99 Two whales taken. One by Kileegewuk & one by Kinukto.
>
> Thursday 18thJoe [Tuckfield] is said to have taken a small whale at Cape Lisburne. bone about four feet long.
>
> Sunday 21st/99 Anakaloota [Anaqulutuq] killed a very small whale yesterday. No bone.
>
> Wednesday 24th/99 Cooper's [Koenig] man kills a small whale last night.
>
> Saturday 27th Mr Lieb called yesterday & says Joe's bone is a little over 5 feet long…
>
> Sunday 28th Annokkok takes a small whale (ingotu)
>
> Wednesday 31st Chuchailah [Sisailaq] kills a big whale for Jim [O'Hare].
> (Driggs, Journal 1894–1901)

Thus while Native and white crews competed for whales, Driggs monitored their progress, noted the names of individuals who made each strike, and not least recorded the baleen harvest. His journal for the same spring also provides a picture of social, medical, and subsistence activity that suggests frequent and more or less equal association with Native and Jabbertown figures:

> Marlin, Cooper and Jim called… I pulled Cooper's eye tooth. I went to village this afternoon to set Nacowra's shoulder. I saw Ayagaiyaq who is very sick. Loaded shotgun cartridges. Ducks flying thick. Bought my first ducks. Ayagaiyaq died this morning. The funeral services were held in the school room. I told Ogglago to come and be cook if she wished. Niruwuna calls. Ogglatia calls and presents me with a pair of boots. (Driggs, Journal selected from entries for April and May 1899)

But some entries for June 1900 suggest a new dimension to Driggs's involvement. On June 8, Driggs notes that "First Native boats haul in," signifying the conclusion of the whale hunt. During the whaling feast, which started the next day, Koenig and Joe Tuckfield spent a couple of nights with Driggs. Unusually, the whale hunt started up again. Suddenly Driggs records an artifact in his possession that had so far floated innocently in the background: "Cooper and four men bring my boat to the mission" (Driggs, Journal entry for June 12, 1900).

Driggs's first mention of any boat had come two years earlier. Coal had been discovered at Cape Dyer in summer 1897, and because Driggs had relayed this information to his Jabbertown friends, Hachmann repaid the courtesy by building him a boat. Driggs's mention of his boat in June 1900 suggests that it had now been deployed for another purpose. Given the time of year and the fact that all the whaling crews had just come in, presumably Driggs's boat had been in the whale hunt. That assumption is supported by his next entry on the subject: "Painted boat" (Driggs, Journal entry for June 19, 1900), which suggests that the boat has been in use and its sides scraped by sea ice—although this could merely have represented weathering (Bockstoce, personal communication). But Driggs seems to have had at least two dinghies. Later in June, as the seal and walrus hunt got under way, Koenig and Tuckfield called to "take away the seal boat." And on the same day, referring perhaps to a third boat, Driggs noted, "Put skiff in water to soak."

A picture emerges of Driggs both grubstaking Natives in support of the Point Hope hunt and lending boats for Koenig to manage in a whaling enterprise. In partnership with Koenig, Driggs would have taken a boat owner's share of baleen, sealskins, and ivory from the bowhead, seals, and walrus that had been taken. From the Native-manned boat, Driggs would also have acquired baleen shares. The philanthropist who aimed to boost the village food supply by simply equipping a Native crew in the interest of their subsistence had thus also assumed the roles of commercial whaler and baleen trader. The young doctor, who in 1890 had written with such anxiety about having accepted a trade outfit from his San Francisco supplier, was now confident and powerful enough within both the local Native and Euro-American communities to far outreach that first potential breach of Episcopalian propriety. As discussed in Chapter 16 and in Appendix C, the Missionary Society's expectations, albeit unwritten, were clear and Driggs must have comprehended very well that he was exposing himself to danger.

E. J. Knapp at Point Hope and Events Leading Up to Driggs's Deposition

In summer 1904, Driggs was due for a holiday and a replacement was found in Edward J. Knapp, a New York lawyer and lay missionary. Knapp had first gone to Alaska in 1899, serving with John Chapman in Anvik and at Rampart mission on the Yukon until 1903. Knapp arrived in Point Hope in July 1904 and his twelve months of service appear to have been uneventful. But it was through his friendships in both Point Hope and Jabbertown that Knapp learned of Driggs's commercial activities; it was this information that led to Driggs's downfall.

Knapp's first letter to the Board of Missions depicts a busy scene somewhat in the colonial genre: "Dr. Driggs had seen the steamer rounding the Point, came down with several Eskimo men and boys to meet and welcome me… The Eskimo men and boys seized my luggage and carried it up to the mission house" (Knapp, Letter of July 29, 1904). Knapp found Driggs in good spirits and gave an enthusiastic account of the doctor's achievement:

> Dr. Driggs I found very well. The winter had proved a long and lonely one… It would be difficult for me to express in adequate terms my admiration for the work that Dr. Driggs has done here. His patient devotion can not have been excelled in Alaska or anywhere else in the mission field. He entertains a sincere affection for the Eskimo people and has won them by his gentleness. They on their part are devoted to him. The relationship existing between them is something akin to the patriarchal.

Driggs the patriarch with his Native porters is matched by Knapp's suggestion that the doctor had (somewhat paradoxically) "lifted" Point Hope to a condition of Christian humility:

> Dr. Driggs has done much to uplift these Native people in every way. He has improved their morals, enlarged their mental vision, and taught them spiritual truth. Last Sunday it was touching to see members of the Native congregation stand forth at the close of the service and with bowed head offer their petition to God, praying Him to teach them what He would have them do...

Knapp also commended Driggs's success both in the teaching of prayer and in the schoolroom: "Dr. Driggs talks the vernacular like a Native, and is able to teach in it… During the past winter he has begun to teach the grammar of their own tongue, instructing them how to conjugate verbs etc... The younger people speak English and write it a little."

Driggs, for his part, reviewed the year's work with "considerable satisfaction" and wrote of "a band of sincere Christians, growing up around the mission" (Driggs, Letter of June 20, 1904). Those whom Bishop Rowe had confirmed in 1903 "remained faithful to the vows they made." Building on this, the doctor had prepared thirty more candidates against the bishop's next visit. Driggs continued to encourage weddings, and by summer 1904 there were twenty-six married couples in the village. But rather than marry the population en masse, as many Point Hope people urged him to do, Driggs believed it "best to observe caution, and marry only those that will prove a shining example to the others" (Driggs, Letter of June 20, 1904).

Driggs notes also how the bishop's visit of the previous summer "proved quite a stimulation to the work." Rowe had presented small crosses to the people he had confirmed, and

FIG. 31. *Driggs's "iglu church." Driggs received materials for this building in 1906 and completed it in 1907. Two years later Bishop Rowe and Reverend Hoare demolished and replaced it. The three Europeans have not been identified. Note the bell tower and water barrel on a trolley. Episcopal Church Archives, Austin, Texas.*

just as prayer had joined a repertoire of traditional incantation, so the cross was added to the amulets that all Point Hope people owned. Driggs wrote:

> During the winter, Lily Kilman's infant was very ill. I found the father and mother wearing the crosses, while they watched their little one struggle for life… Sam Rock came and enquired, Should the cross be buried with him, when he died? Later in the season, he was taken ill, on going to his iglu, I found him wearing the cross.

In another of his teaching experiments, recounted in the same letter, Driggs hit on yet another method: "I am having them conjugate their long compound words, then turn around and conjugate the whole English sentence, that such words represent." But this method had poor results: "The scholars find it difficult to master English pronunciation, but by daily drills they are doing better." Fascinating as it is to witness Driggs demonstrating to his students the phonological structures of their own language, it is hard to imagine that such demanding exercises can have contributed to their fluency in English.

As in the 1890s, Driggs's work was hampered by continued emigration. In a postscript to this letter, Driggs noted that "Fifty of our Natives have gone North [for shore whaling at Barrow and Herschel Island] so the village population will probably be small next winter." Still, taking the longer view, Driggs identified changes that had taken place in the village. As so often, the doctor expresses himself with tantalizing vagueness: "If I take the people as a mass today and compare them with what they were a few years back, I find that it has been a gradual improvement for the better, all through."

Only slightly more tellingly, he imagined "some old villager, that has been taking his long sleep for many a year…brought back to life. He would find a great change among the people, from what they formerly were." One of these changes is evoked in Knapp's first letter, which includes a vivid meditation on Point Hope's burial ground:

> a short distance to the southward [of the village] is the weird Eskimo graveyard two miles and more in length by about a third of a mile in width… Exposed to the weather the jawbones have bleached so that they resemble trunks of blasted trees, and the bodies have dissolved. Many of these ancient so-called graves have fallen into utter ruin and the bones and clothes that shrouded the dead lie scattered on the ground. (Knapp, Letter of July 29, 1904)

Like Rowe before him, Knapp sketches a picture of strange coloration: the old, metaphysically ambiguous bones of the Point Hope dead with flowers and driftwood crosses rising among them:

> It is a curious sight… There are human crania and other bones lying above the surface of the soil, but it is pleasant and hopeful to see growing up among them delicate wild flowers of the most beautiful forms and colors—the daisy, the yellow poppy, the forget-me-not, both blue emblems of the resurrection at the last day when these dry bones shall live. And more hopeful still it is to see the rude wooden cross that marks the more recent graves where rest the bodies of the Christian dead, sometimes buried under the ground, but as often placed above it…

Here was on-the-ground evidence that at least some Point Hope souls had been saved and that Driggs's effort had taken root. But Knapp's sensitivity to ambience (he detected "weirdness" and an "atmosphere that hangs about the place") did not inhibit a critical appraisal: "The combined school house and chapel is not a suitable building for the purpose. It is a high building with a high peaked roof, is built of lumber unsodded, and moreover is old and in the winter time is said to be very cold."

Driggs himself had designed the building, and as early as 1902, acknowledged:

> I feel that we are in need of a new more appropriate building for our Sunday services. The present one is too high. The many high gales have shaken open the seems [sic] so, that it is no longer possible to warm it in severe weather even with two stoves. It is not an infrequent thing for me to be breaking icicles from my moustache while speaking to the people Sundays, to have the audience stamping their feet to keep them warm. Then on school days, the children are much annoyed by the freezing of the ink on their pens. (Driggs, Letter of May 23, 1902)

Still, Knapp ended his letter of 1904 with a comfortable, if alarming, picture of the living quarters he would inherit: "The Mission's house in which we live is a rather low structure with a low peaked roof built of lumber with double walls, so low that gales sweep over it, and so tight that sound of the surf thundering on the beach near by does not penetrate" (Knapp, Letter of July 29, 1904).

Knapp ran the school and mission until summer 1905. His was a safe pair of hands and to his knowledge of guns, tools, and boats he added an impressive commitment to community health issues. Finding agreeable companionship in Jabbertown, his written notes to Koenig cover a wide range of subjects, both domestic and social:

> The shot gun that you offered to return to me I shall be very glad of. I will either borrow it from you, or buy it back, paying you what you paid the Doctor… Thank you for the papers which I will return to you soon. Can you lend me a nail drawer? I need one for taking down the partition wall… Tuzroyluk's baby has the mouth disease that seems prevalent among the young children. I wish you would look in the child's mouth if possible tomorrow morning… There is a well developed case of diptheria here on the Point… Detain your people from coming to church tomorrow. I want to take counsel with you about what I believe is going to be a serious epidemic… (From Knapp's letters to Koenig, December 1904–March 1905, in Koenig 1880–1914)

Knapp ran the school efficiently, but the huge cultural distance dividing Point Hope from mainland America is illustrated in a description of how he tried to teach the children about the 1904 presidential election, which returned Theodore Roosevelt for a second term (made popular by his "Square Deal" slogan, a phrase that Bishop Rowe would apply to Driggs's deposition in 1909). Closing school on November 8, Knapp traveled to Jabbertown and enjoyed his afternoon disputing the precise timing of the election, noting humorously that he would not hear the result until the summer.

Summer arrived, Driggs returned, and Knapp traveled south, stopping in Seattle where he published an article in the *Seattle Post-Intelligencer* on September 24, 1905. This describes the plight of coastal Iñupiat, focusing largely on the liquor trade and the abuse of Point Hope women by whaling captains. Knapp's assertion that he would raise these issues in Washington came to nothing. But on the rural outskirts of Seattle he also made a social call on Hachmann and his Point Hope family, who had settled on a farm. Driggs had also visited Hachmann, and by the end of both men's visits, the former Jabbertown whaler must have known that Driggs was in trouble. When Driggs returned to Alaska in July 1905, his place at Point Hope was already in the balance. Next May, en route back to Alaska with Archdeacon Hudson Stuck, Knapp hints at Driggs's insecurity, as leaked when he traveled to Nome with Stuck, where it was reported that "Dr Driggs would have to relinquish his work at Point Hope" (*Alaska Crossbearer* 2, no. 4 [July 1906]:4). In view of the doctor's possible departure for the south, Knapp ordered a year's worth of supplies for himself and waited for developments. But when Driggs's own order arrived at Foster & Co., Knapp continued his journey with Stuck, while Driggs stayed put in the village.

Driggs's one letter of 1906 is filled with mission building plans, for which lumber arrived that summer. "When the snow has melted and the ground dry" he would start constructing a new church and schoolroom (Driggs, Letter of May 25, 1906). Driggs was expecting a visit from Stuck that summer, but to his dismay it was Knapp who arrived, and by the time Knapp had spent a few days at the mission, the two men were not on speaking terms. Knapp moved to Jabbertown where, in anticipation of the Koenigs' departure for Washington, he tutored the Koenig children in English. It was this, perhaps, that convinced Driggs that he should be teaching English more energetically, and he "pretty much discarded Native tongue in the school" (Driggs, Letter of May 25, 1906). How far his use of Iñupiaq in the classroom had been motivated by his own educational ambitions for his students, and how far it represented a continuing gesture of courtesy to the home community, is impossible to know. Clearly Driggs enjoyed his mastery of the language, and presumably while this took him deep into village life, it enhanced his status in the classroom if not the thrust of what he had been employed to do. Again, the truth is impossible to determine.

Driggs spent the summer of 1907 working on new mission premises, managed to improve his old quarters, completed the first stages of a new church, and erected a bell tower. "The bell," he wrote, "now rings forth each Sunday from its new home, calling the people to worship" (Driggs, Letter of Aug. 23, 1907). But a new isolation had descended. Hachmann and Koenig had gone, the white man's society on which he had depended was breaking up, and he was being ignored by his employers in New York. Stuck had disappointed him the previous summer. Now Rowe, who had promised to visit, failed to arrive. Finally, Driggs had fallen out with Knapp, who was, for his part, contemptuous of Driggs's improvisational methods.

Worse was to come. That autumn Driggs wrote:

> No supplies for the Mission have arrived. If the vessel delays...it will be doubtful if anything can be landed; the fall gales having already begun...
>
> At present I do not know what consequences will be. Fuel and oil are about exhausted and my larder growing slim. (Driggs, Letter of Sept. 28, 1907)

For the first time, Driggs faced an autumn in circumstances almost equivalent to those of the Native community:

> The Natives were unsuccessful in their whaling last spring, so they have no meat laid by in the ice houses. Neither have they bone [baleen] to trade with the ships for supplies. It looks as if we [sic] may experience a hard time next winter. Tomorrow, worship will be held in a cold church, there being no fuel to warm it.

In his poverty and distress, Driggs abruptly, and for the first time, suggests the employment of female colleagues. "I would like to see," he writes, "one or two ladies connected with this mission as workers.... I hope to make room for such an increase of the working force by building onto the mission home. At present the quarters are only large enough for one man." If the transitions in this letter point to disorganization and perhaps, for the first time, a hint of social deprivation or emotional stress, the visual appearance of these pages is also striking. The doctor's handwriting has become huge and loosely formed, while new letters of the alphabet join his habitually hyphenated *s*, all of which give the orthographic appearance of mental unraveling.

Jumping from his call for women helpers, Driggs suggests building a new mission station at Candle, a mining camp about two hundred and fifty miles inland. Then he wants to see if "a whale boat...could be procured... Should the supply schooner be wrecked any year, it would make the mission much safer than at present." Driggs now lurches perilously close to the taboo subject of commercial whaling. Innocent that Knapp had already compromised him, Driggs writes:

> For a number of whaling seasons I have tried to introduce prosperity among the Natives by lending aid to the young men telling them if they were successful in capturing a large whale to bring me one half of the bone. When enough had been accumulated to be sent down, their supplies would be ordered up for a store, they to run the store themselves for their own benefit.

Driggs's idea for a Native store might have been a helpful innovation. But by now the church authorities were more interested in terminating Driggs's existing commercial activities than in introducing new ones. Anticipating the silence that was closing in on him, Driggs concludes this distressing letter: "I am afraid the length of my letter will weary you, but it is so seldom there is a chance for me to converse in English that I am liable to allow the pen to run away with me." Driggs's final phrase is, perhaps semi-consciously, insightful, for it describes precisely what, both in content and in form, these pages reveal about his psychological condition.

Slim Living at the Mission

Driggs's supplies never did come: they were lost when the schooner *Vine* was wrecked at Deering in mid-September (Driggs, Letters from Foster & Co., Sept. 30, 1907), and with resources at his lowest ever, the doctor wrote that for the first time since he had arrived, he had closed the school that autumn: "School," he wrote, "will not be opened till the return of the sun as there is no oil to light the room with. I expect lumber up next summer for the new building and am anxious to complete the work; then if there is money left over, I wish to enlarge the Mission house" (Driggs, Letter of Nov. 8, 1907).

Nor could Driggs do anything to alleviate Point Hope's continuing subsistence crisis in which game animals were scarce and even an abundance of polar bears contributed few commercial opportunities: "The Natives brought in fifteen [polar] bears this week. The people are all hard up, so this would be a good winter to help them out by purchasing their bear skins for my friends, but unfortunately I am as hard up as they are."

But in contrast to his anxious message of the previous summer, Driggs's letters of November 1907 are imperturbable: "It is slim living at the Mission, but I will hold the work together and I hope to pull through the winter all right." Silence then falls. In January 1908 Driggs dispatched a letter with a group of Natives who were traveling down the coast. This was Driggs's first letter written in the winter, and he sent it in response to one from New York that had been dispatched in mid-September and arrived in Point Hope in December—the product of a vastly improved communications system. Isolated communities on the arctic coast were now being served by dog-team carriers, with a northern postal station at Kotzebue. Driggs outlined first how the "S-eattle s-chooner" had run aground, and then in the absence of local driftwood how he remembered

> an old ship-wreck twelve miles from the mission...that had laid submerged fifteen years, I offered the Natives five dollars a sled-load if they would visit the wreck with axe and crow bar, dig down in the ice and pry off any pieces they could get hold of. It was like a dog gnawing an old bone. Nevertheless, I procured twenty loads of wood. It is water soaked and not good, the old room is not comfortable. On opening school,

> the temperature was more degree below zero than there were scholars. (Driggs, Letter of Jan. 26, 1908)

Thus Driggs kept both school and church going. Behind the scenes, however, the authorities were moving to depose the doctor. Twelve months previously, Bishop Rowe had written requesting him not to continue building the new mission until he arrived. Driggs did not know that the bishop himself wanted to supervise the work, with a view to installing a new resident as soon as possible. Poor Driggs had looked forward to architectural improvements and an augmented mission presence, including "one or two ladies to take charge of the school and conduct the music Sundays" and his letter ends with these beautifully evocative paragraphs. First: "During the fall, a rare phenomenon occurred. There was a shower of meteoric dust which contained magnetic iron, at Kivalinya and further south. At one place it was an inch deep on the snow" (Driggs, Letter of Jan. 26, 1908). This is followed by a vivid description of polar bears. After a hunt in which Natives had taken forty-five bears, wrote Driggs, "one was shot on the roof of an iglu; two have been near the mission at night and one entered the village of an evening to watch the children play football by the light of the new moon." These events contain, for the first time in Driggs's letters, something of the magic of traditional Point Hope stories. It is as though, having participated in but now partially let go of the effortful factuality of nineteenth-century life, Driggs had allowed himself to enter the realm of local poetic experience which Rowe and Knapp had sensed in their visions of Point Hope's "weirdness."[1] No doubt Driggs had many such experiences, and if his relations with employers had been less formally constrained, he might have evoked more of old Tikiġaq.

In December, Driggs had an unexpected visitor. This was the Icelandic explorer Ejnar Mikkelsen of the Anglo-American Polar Expedition. His own schooner had been wrecked in the ice, and Mikkelsen was on his way to catch a steamer for Seattle. Mikkelsen's account of his journey contained this brief but appreciative description of Driggs's mission:

> His work had brought its own reward; he is beloved in the village and the young men and young women look upon him as a father, wiser and better than their own, who does all he can to make the people for whom he has sacrified his life a useful and self-dependent race. (Mikkelsen 1909:373)

A difficult winter may have suited Driggs better than a summer of lassitude. The doctor's one letter of summer 1908 finds him writing in the disintegrated hand of the previous season. The recipient was a Miss Day, Driggs's one extant female correspondent. The letter opens with a sympathetic reference to the death of Bishop Hare. Then the doctor runs on blindly: "Recently I had a young child to treat, the upper lip and part of the nose were eaten off in four days by cancrum oris" (Driggs, Letter of Aug. 12, 1908). This is another of Driggs's gruesome wilderness-life close-ups that genteel readers must have recoiled from. (The first of these, in which dogs had eaten a dead man's head, appeared in Driggs's very first letter of September 1890.) Driggs had intended to visit Miss Day on his next furlough. But he never traveled home again. His pen, from this moment, with all its hesitations, glitches, digressions, and, not least, sympathetic insights, falls silent. There are no more Driggs letters and the rest of his story is narrated by others.

Deposition and Exile

The question of Driggs's mental health clearly exercised the church authorities. Rowe wrote understandingly that the conditions endured by Driggs would have put "ninety-nine men out of a hundred in an asylum, or worse, long ago" (Rowe, Letter of May 29, 1909). And a decade after Driggs's death, based on what he heard from colleagues, Archdeacon Stuck wrote that Driggs had become "eccentric and absent"—a euphemism, perhaps, for a mental disorder (Stuck 1920:39). Stuck also claimed that Driggs had resorted to drugs. There is no documentary evidence for this—and if perhaps the church was anxious about Driggs's consumption of tobacco and alcohol, the Episcopal archives appear to have nothing to say. The storyteller Asatchaq, who attended Driggs's school from about 1898, spoke of Driggs's tobacco habit, and described how the doctor periodically spat into a can as he worked his way through the wad in his cheek. Driggs, as Asatchaq also noted, also drank (Lowenstein 1973–89). But judging from his supplies list from Foster & Co. in July 1901, relatively late in Driggs's career when he was perhaps overconfident of his autonomy, this was moderate. Eight bottles each of Madeira and whiskey may have been excessive in the missionary society's view, but it was scarcely enough to intoxicate a man for the twelve months between deliveries. Perhaps Driggs drank more towards the end, but there is no evidence.

Just before his last message to Miss Day, Driggs's removal was, from a distance, being cautiously and somewhat callously managed. First, Driggs was to be cut off, if not starved, if he continued at the mission. In April 1908, a telegram to the Missionary Board reads "Order supplies. One man two years. Substitute found for Driggs. Cancel Driggs order" (Rowe, Letter of April 4–5, 1908).

But Rowe still had to produce evidence to justify the doctor's deposition. Driggs was lying low, and by February 1909 he was not responding to letters. Rowe continued nonetheless to assemble incriminating evidence, most of which came from Knapp: "I have received all the correspondence so far as I know on this matter. While the Committee has acted rightly [in deciding against Driggs], it throws a responsibility on me [to confront him] which I do not hesitate to bear" (Rowe, Letter of Feb. 28, 1909).

Determined to act judiciously, Rowe continued: "I shall not take any decided action until I have visited Point Hope and satisfied myself in reference to Mr Knapp's charges." Acknowledging Driggs's extraordinary service, and echoing Roosevelt's election slogan, Rowe admitted: "against the years he has spent at Point Hope, [Driggs] is entitled not only to a 'square deal,' but to a charity and sympathy."

Driggs's whereabouts had become obscure, though presumably he was in or around Point Hope that winter. When Driggs reappears in February 1909, he is in Seattle. Rowe finds him there, perhaps by arrangement, though Driggs may have been purchasing supplies:

> I met Dr. Driggs in Seattle getting ready to go north. While unable to pass judgement on him…in the face of his denial of all charges, I felt it but right to oppose his return

> to Point Hope...and that if he were set on going north again this year he would have to [settle] at some other point other than Point Hope. (Rowe, Letter of Feb. 29, 1909)

Driggs may already have decided to decamp, and oral history suggests that he settled initially on the Kuukpak River where he had an iglu or cabin (Lowenstein 1973–89). Rowe nonetheless refused to act finally until he got to the village. Then, en route to Point Hope in May, Rowe gave a second account of his meeting with Driggs:

> I met him in Seattle. He was preparing to return to Point Hope. That would be inadvisable. He and Mr Hoare [Driggs's replacement] can no more live together than he and Knapp. So I told him that if he went north he must [settle for] the winter at Cape Lisburne where there is a small Native village [while I] investigate. That should I find any foundation for, or truth in, the charges preferred, I would expect him quietly to retire. This it seems to me but a fair and just dealing with a man who has served so long in conditions that would have put ninety-nine men out of a hundred in an asylum, or worse, long ago. Again we should be slow to punish in this case and long in mercy. (Rowe, Letter of May 29, 1909)

Rowe's even-handedness is seriously compromised just once during this process. Apparently oblivious of the work he had praised so sincerely on his first two visits, Rowe asserts: "he has never been a satisfaction" (Rowe, Letter of Feb. 29, 1909). Certainly Driggs had never been an insider. On the contrary, his pioneering service had defined a self-imposed exile. Nonetheless, he had, in his necessarily improvised manner, fulfilled the Missionary Society's demands, and if in the process he had also gone his own way, that was his manner of surviving both as an individual and as a member of the Point Hope and Jabbertown communities. In almost all respects he had succeeded in accomplishing the difficult task he had been employed to undertake, and Rowe's negative assessment could surely relate only to Driggs's most recent activities. Generated perhaps by his own fatigue, exasperation, and the necessity of making a clean start with someone more straightforward, Rowe's verdict nonetheless expressed the Board's view. And while clearly the doctor was facing a penalty, Rowe was anxious to temper justice with mercy: "the fact of his long service, such as it was, constrains one to extend mercy rather than summary punishment; and therefore should I find any warrant for the charge, Dr. Driggs will be quietly left to withdraw within the year" (Rowe, Letter of Feb. 29, 1909).

Rowe arrived in Point Hope in August and spent about a month there. Reverend Hoare had traveled from Anvik and had taken over the previous summer. There is no record of the two missionaries' interaction, but judging from Rowe's earlier comments, Driggs had not given Hoare a friendly reception.[2] With Hoare now in residence, the bishop presumably summoned Driggs to the village, where their decisive confrontation took place. By early September, Rowe was in Nome and wrote as follows:

> I have made my visit to the mission at Point Hope. The matters affecting Dr. Driggs were investigated as carefully as possible. A just and full opportunity was given the doctor in his own defence against the things of which he had been charged. Neither defence nor explanation were enough to refute the same. I am satisfied that the charges made by Mr

FIG. 32. *Joe Tuckfield (left) and Bishop Peter Trimble Rowe seated smoking together. They sit facing south, perhaps outside the mission building or Tuckfield's cabin. Anonymous photo taken at Point Hope ca. 1909–1915.*

> Knapp were justified and are sustained. The Rev A.R. Hoare had come to this conclusion from things that had been told him and from evidence of neglect of work which he could see for himself. He said that it might be difficult to obtain the proofs for every charge, only because of conditions not because they did not exist, and I agreed with him.
>
> Therefore I beg to recommend that the name of Dr. Driggs be dropped from the list of missionary workers in Alaska. On August 7th I notified Dr. Driggs that I would make the above recommendation to the Board of Missions through you. I further advised him that it would be only his duty to apply to me for deposition. (Rowe, Letter of Sept. 10, 1909)

Only one surviving document makes Driggs's offense explicit. This comes in an undated letter from Rowe:

> Dr. Driggs returned to Alaska, in spite of my advice to the contrary until final action was taken. Persisting as he did, I prevented him from returning to Point Hope. He went to Cape Lisburne, is settled there, and will engage in whaling, which beginning to engage in about 1898 under the desire to help the Natives as stated by him in 1898, developed finally into an enterprise for his own profit. It is unnecessary to go further into the details of this unhappy affair...

This ended the matter, and whether Driggs received a pension is not recorded. Rowe spent the rest of his weeks in Point Hope demolishing all traces of Driggs's tenure: "Hoare and I did swift work at Point Hope in constructing a church in three weeks. This one, apology for one, built by Dr Driggs was a terror. The new one will be all right, I think" (Rowe, Letter of Sept. 1909).

And in a second account he wrote:

> The church we built at Point Hope from materials sent in by Dr. Driggs fortunately was on hand. He had built what he called a church, but it was a thing fearfully made. It was like an iglu, so low, narrow, inadequate for the congregation. When the latter assembled the air became so dense, the lamps went out, it condensed on their heads in a rain and this froze under their feet. We pulled this down and erected a building that is fine and will be used for school as well as church... (Rowe, Letter of Oct. 6, 1909)

Rowe praised Hoare and Knapp for present and past work at Point Hope. But Hoare had been at the mission for four months at the most, and Knapp's tenure had been brief. Sadly, what the bishop was commending—as he had done in 1903—was Driggs's now-discredited effort.

Driggs in a Vision: A Ghostly Story

Driggs retired to Cape Lisburne, where he lived for five years with a small community of Point Hope people before succumbing to a stroke in September 1914. If, as his old employers suggested, Driggs had become distracted, he also spent his last years piecing together roughly the same routines he had pursued in the main village. He ran a small school, held church services, operated a whaling boat, and did some gold prospecting. Hunting at Cape Lisburne is difficult at the best of times, and gold has never been found there. No letters from this period survive, though presumably Hoare let Driggs use the Point Hope mail. Still, Driggs did not entirely abandon his old idealism, and an insight into his activities is provided by Maligian, a Point Hope woman who spent her childhood as Driggs's neighbor. Her narrative contains this remarkable passage:

> My parents' house at Uivvaq [Cape Lisburne] was bigger than the others. Since Dr. Driggs was living up there at the time, he would sometimes hold a service in our house, though most of the services took place in his. The other houses at Uivvaq belonged to Aniqsuayaaq, Auvaq, Putqaq, Iñuuġvik, Auvaŋauraq... Since there was a mission there, Driggs held school for the children in the afternoon. He taught us the white man's way of saying things. He would teach us to pronounce the names of everything in his house, like "door, ventilator, stove, wall, floor." Everything there was in the house, we had to name it. We learned numbers too, and the ABC...
>
> Once at Christmas we received presents from Dr. Driggs at the mission. White people's toys: dolls, clothes, crayons, pencils, books and marbles. All these presents were distributed by Dr. Driggs, and I didn't get anything until almost all the gifts were gone, and everybody had their presents. I don't know why, but at the end, he gave me two big bags of toys, dresses, paper, more than anybody else. Maybe Dr. Driggs was pleased with what I did at school...All the children went to school in one class. The youngest must have been five or six.
>
> Sukaŋana cooked for Dr. Driggs. She used to come out of the house with any food that Dr. Driggs left, and the children came running to her because they knew

> she would have pancakes and bacon for them. When one of the children saw her, he would shout, "Taaktaiḷuich! ('From the Doctor!')." As soon as the other children heard the cry, they would rush out, and Sukaŋana would divide the food up equally. That was lots of fun. That's how we lived at Uivvaq. (Lowenstein 1973–89: Told by Beatrice Maligian Vincent, 1904–79)

The end came in September 1914: a brief account Driggs's death being provided by Augustus Hoare, who sent this letter of condolence to Mary Elizabeth Turner, one of Driggs's two older sisters who was still resident in Smyrna, Delaware:

> My dear Mrs Turner:
>
> You have no doubt heard in recent letters from Dr Driggs of his illness and increasing debility and are somewhat prepared to hear of the fatal termination. I am very sorry to have to inform you that Dr. Driggs passed away on September 21st last. He appears to have been seized with a second paralytic stroke and was paralyzed from the waist down—the doctor on the revenue cutter 'Bear' saw him and the captain promised to take him down to Seattle but has not yet returned. We buried him in a beautiful spot on the side of a hill at Cape Lisbourne [sic]. I am sending you his watch chain (his watch had been lost sometime), Bible with inscription, etc. Although he fully realised 24 days before his death that he was dying, he left no letter or message. No will has been found… All letters, etc. that appear to be personal or private [I] am burning. The house with what small supplies (no outfit was sent in this year) was turned over to Harry Kilbear [sic; see Appendix A] who has lived with him all these years and to whom Dr. Driggs had promised them…I will erect a suitable stone & fence around the grave and will (all being well) send you a photo next summer. There was no one present at the last but two or three natives living at the cape for although I left the moment I [heard] the news of his serious illness, was too late. (Hoare, Letter of Sept. 27, 1914)

A final and more disturbing view is provided by Hudson Stuck, Archdeacon of the Yukon and the Arctic Coast. Stuck wrote briefly in his survey of the arctic missions:

> The isolation and solitude of his life at last told on Dr. Driggs's mind. He began to grow eccentric and absent, and it was thought wise that he return to the United States. But Dr. Driggs was wedded to the Arctic and to the Eskimos; long absence had made him strange to civilization… (Stuck 1920b:39)

Stuck recorded a more personal meditation on Driggs in a book describing a journey through north Alaska in 1918. Accompanied by his Athabascan friend Walter Harper, the archdeacon spent several months at Point Hope assisting the missionary William A. Thomas.[3] One evening, Thomas and Harper went out to hunt, leaving Stuck alone at the mission:

> The night before we had talked much of Dr. Driggs, his long work here and his miserable end. There is no doubt that his solitary residence had told upon him and that he had become mentally unbalanced, and little doubt that towards the last he had addicted himself to the use of drugs. I cannot see any good in hushing up such matters. To acclaim a man for years a hero in the high-flown manner of missionary publica-

> tions, and then suddenly drop him and mention him no more at all, is likely to rouse a suspicious bewilderment that is worse that the commiseration that would follow a knowledge of the facts.
>
> That he was mentally unbalanced his eccentric doings and sayings establish, and that he fell latterly into the use of stimulants, I think very likely. Anyone who has spent eighteen years alone in the Arctic regions and has retained his full faculties and self-control, is entitled to throw the first stone at his memory, I think, and no one else. It became necessary to remove him, there is no question about that; and there can be no question in the minds of those who know the Bishop of Alaska that it was done with all gentleness and tenderness and consideration. I warrant he had rather have cut his hand off than do it, but, as we say in the north, "he had it to do."
>
> But Dr. Driggs took it ill; refused to accept his passage out and returning in dudgeon some twenty miles further up the coast made his residence with an Eskimo family; venturing a little income of his own in a Native whaling enterprise. It is said that whenever the weather permitted he would continually walk the beach, looking towards the sandspit which had been his home so long, muttering and gesticulating. Here, some years later, he fell very ill. Word of his plight came to his successor at Point Hope on the wings of a gale that denied return against it for some days, and when it was possible to travel he was found already dead. (Stuck 1920a:128–29)

In contrast to Rowe's unequivocal condemnation, Stuck then defends the doctor's achievement: "The change in Point Hope from the conditions described by Lieutenant Commander Stockton (in 1889) to those which now prevail, is largely the result of Dr. Driggs's labours." There follows a short account of the "drunken, despotic, polygamous chief Ataŋauraq, who ruled this community with terror in those early days." These comments are relevant to Driggs, for Stuck continues with a point that had been overlooked in Episcopalian writings: namely, that Driggs arrived in the village only eighteen months after the troubles that followed Ataŋauraq's murder. "It was into such scenes that Dr. Driggs entered when he landed at Point Hope and started a school. How very slowly and gradually he made an impression upon the people and, little by little, won their confidence and respect; how many times his own life was in danger" (Stuck 1920a:130). Stuck continues with a disquieting tale in the genre of ghost stories of the period.

> My mind [that night] was full of these things, and especially full of Driggs, his faithful labour and his miserable end, when the two young men went polar bear hunting and left me alone in the house. I read awhile in a desultory way and then went to bed. Meanwhile the wind had risen again and whistled and whined about the house, and a loose dog, I think, had crept for shelter between the floor and the ground and made strange noises. Again and again after I had put out my light I started up in bed thinking that I heard footsteps below. Most stairs creak when they are being trodden upon, but some have the miserable habit of creaking without being trodden upon, and the mission house stairs were of that kind. Frequently I was sure I heard someone coming upstairs and entering the little room across the hall from mine. I listened and listened—and lay down again, already creepy and afraid. But my mind instead of composing itself to sleep brought up visions of the old doctor, in ragged and dishevelled Arctic attire,

FIG. 33. *Ataŋauraq's grave at Point Hope, 1889. The remains of the wooden platform are surrounded by whale's jawbones; the large bone stands as the* umialik's *personal marker. The ridge on the horizon marks the bluffs of the north beach. Photograph by the author, 1973.*

> pacing the beach near Cape Lisburne, raising his clenched hand against Point Hope and those who had dispossessed him.
>
> I was taken with the notion that he would not lie quiet until his bones had been translated to the place where his life work was done. Presently I dozed off and dreamed, and the same haggard figure rose before me, grew gigantic and ghastly, gnashing its teeth and slavering, and I started awake with the feeling that someone was entering my room. Looking at the door in the faint light that filtered from the moon through double sashes obscured by encrusted snow, I was certain that it was moving, that very slowly it was opening, and then that someone, something was in the room with me. The wailing of the wind took a tone of human despair that pierced my excited brain and for a while I lay in an agony of fright, utterly unnerved and abject… I never saw Dr. Driggs in life, but the unshaven, dishevelled, minatory figure in greasy ragged furs of my dream, is stamped indelibly on my mind. Presently I recovered myself, but with a resolution that I would never be left alone at night in that house again. I should really like to know that Dr. Driggs's body had been translated. (Stuck 1920a:130–31)

E. J. Knapp had spoken of the "weirdness" of Point Hope's atmosphere. The archdeacon's experience suggests that he, too, was sensitive to the place and to the unearthlier aspects of its historical echoes. Stuck's is an expressive tale, but with that combination of the gothic and the psychological that he may have derived from ghost stories of the period.[4] Still, colored as it may be by literary convention, Stuck's story has value. It shows that Driggs was not forgotten. At the very least, his presence had been reasserted in that mysterious

FIG. 34. *Point Hope cemetery with* umialik's *grave in the foreground marked by the jawbone of a large whale. Behind this lies the gravestone of Reverend Hoare, Point Hope's missionary 1908–1920, who created this cemetery. Point Hope village is visible on the eastern horizon. Photograph by the author, 1973.*

old Point Hope to which Stuck had become quickly attuned. And the doctor's case also presented Stuck with meditational possibilities. Where, for example, lay the value of the solitary and idiosyncratic struggle in the work of a man such as Driggs, if the very stoicism and initiative that had generated his earlier successes led him to transgress an unwritten contract? How could it be that a man who had devoted so many years to an altruistic cause should end his career in such wretched and perhaps disillusioned solitude?

Having rebuilt the mission in summer 1909, Hoare proceeded, as described in more detail in the next chapter, to uproot virtually all the grave markers in Point Hope's pre-contact burial ground. Hoare and his helpers drove these whale jawbones into the tundra just west of the village to create a fence for a newly consecrated cemetery. Driggs's remains were never brought to this culturally hybrid space, which used old Point Hope monuments to enclose a burial ground for Christians. As Stuck notes, Ataŋauraq's grave markers were left *in situ*. "The bloody, defiant, old heathen's body was not admitted within the consecrated precincts, and lies outside, marked by two jawbones that tower over the rest" (Stuck 1920a:129).[5]

FIG. 35. *Driggs's grave at Uivvaq, Cape Lisburne. This stone on the site of Driggs's burial is thought to have replaced an earlier wooden marker sometime in the 1960s. There is no record of a previous stone. Photograph by the author, 1977.*

Nor did Driggs gain access to this resting place. His grave at Cape Lisburne was marked with a wooden cross, and after this disintegrated, a stone was erected in the 1960s. That both Driggs and Ataŋauraq still lie in semi-exile remains one of paradoxes of this history.

Notes

1 It is possible and perhaps even likely that Knapp had used the word "weird" in one of its older senses: "of a mysterious or unearthly character" (*Shorter Oxford English Dictionary*).

2 Reverend Hoare continued as Point Hope's missionary priest until he was murdered in a premeditated gun attack on April 27, 1920, by his young teaching assistant, James Maguire,. Maguire and his Native lover escaped to Cape Lisburne, where they were apprehended by a group of Point Hope men and brought back to the village. From Point Hope, Maguire was taken to court in Candle, Alaska, where he was sentenced after a full confession. Hoare was buried in the new cemetery he had created at Point Hope in 1909 (see Chapter 27).

3 William A. Thomas relieved Hoare for the year 1918–19. After Hoare's death in 1920, he worked in Point Hope from 1920 to 1925.

4 Examples of the genre include Sheridan le Fanu, *In a Glass Darkly* (1872); R. L. Stevenson, *Island Nights' Entertainments*(1894), and M. R. James, *Ghost Stories of an Antiquary* (1904).

Comparable fiction from this period did reach Point Hope in the early twentieth century. In July 1977 I witnessed a mission clear-out operation and noted some fine editions of early twentieth-century authors; whether any of those books survived, I don't know.

5 In 1976 there were seven jawbones to Ataŋauraq's grave. In addition to the two mentioned by Stuck, there were the smaller bones that held the burial platform. Two other prominent early-twentieth-century graves remain outside that burial ground: the grave of the shaman Ayagiiyaq lies southeast of the village. Sisailaq, an early-twentieth-century shaman, is buried on the edge of Ayagutaq lagoon.

CHAPTER 27
Reburying the Ancestors

> A man and woman had a son. They lived alone in Tikiġaq. One day the boy fell sick. His parents could do nothing for him. At that time there were trees in Tikiġaq. Try as they did to keep their son alive he died. His parents wept much for their boy. Then one day the man said to his wife, "Make me a pair of caribou skin gloves for me to work in." When the gloves were ready, the man left the iglu. But once outside, he started weeping for his son. He removed one glove and swinging it by its cord, he threw it at a tree. When the glove hit the tree, the tree fell over. As the man continued weeping, he felled all the trees. Otherwise, there would be trees in Tikiġaq.
>
> —Traditional Point Hope story told by Asatchaq, June 1977

Once Rowe had gone, Hoare conducted a census of the Point Hope region, which, as though to introduce a new era, he recorded in a new parish register. The census took place in September and October 1908, and Hoare included "Tigara" village, the Jabbertown stations, and the recently arrived reindeer herders whose work kept them moving through Point Hope's interior. To illustrate the census, Hoare drew sketch maps that showed the houses on the peninsula and listed names of iglu residents.

According to Driggs, Point Hope's 1906 population had stood at 121. By 1908, the Great Sickness had passed, and the population had risen. Point Hope village contained at least 181 inhabitants, while the Jabbertown population of Iñupiat stood at forty-four, about thirty of whom had been or would become Point Hope residents; an unknown proportion of these were recent migrants from other Eskimo societies. At reindeer camp, there were a further sixteen Natives, bringing the total to about 245. This number marks a defining moment. While tuberculosis continued to take its depressing toll until the 1960s, the epidemics were over. Even the flu pandemic of 1918–19, which struck elsewhere in Alaska, spared Point Hope peninsula its visitation.

While population growth confirmed Point Hope's future, Hoare's new church, and not least the transformation of the Native cemetery, added extra symbolic weight to a process of modernization that was developing alongside the institutionalization of Christianity. The new religion, which had developed via Driggs, Edson, and the Iñupiat who had brought it from Barrow in the mid-1890s, was becoming less a hybrid shamanistic-Christian or salvationist cult (*uivvaqsaat*) and more a well-comprehended doctrine, in which, as Bishop Rowe is quoted as remarking, "the responses...Canticles, Psalter and...hymns" (*Spirit of*

FIG. 36. *Monument with skulls. No single burial would have contained such a burden, and the rifle poised against the rack suggests an artificially constructed composition. Perhaps a comment on the turmoil of the period. Alaska State Library, Historical Collections.*

Missions 1909:946) were now known—despite the fact that church language was the only English that many of the congregation could repeat. Many Point Hope people thus enjoyed a relationship to Christianity not unlike that of pre-literate European congregations whose faith was partially embedded in an incomprehensible Latin.

Besides the gradual conversion of shamans, a turning away from them by the laity, and a decline in ceremonial house ritual that marked the demise of indigenous belief, it was the Christianization of the cemetery that represented the most dramatic token of the new figuration: the "forest of stakes"—the whale-bone uprights that Beechey observed in 1826—which had appeared to Europeans as a wilderness of barbarism and disorder. To Charles Brower in 1884 it "looked like a forest of small trees with the tops cut off. There were thousands of these whale jaw bones" (Brower nd: part 1, 43). Similarly, to E. J. Knapp, this "weird Eskimo graveyard two miles and more in length resembled trunks of blasted trees…the bodies…dissolved. Many of these graves," continued Knapp, "have fallen into utter ruin and the bones and clothes that shrouded the dead lie scattered on the ground" (Knapp, Letter of July 29, 1904). All these images harmonize somewhat uncannily with that of Tikiġaq's primordial forest—both in its upright and felled character—as described in the legend given at the beginning of this chapter.

Now with the obliteration, or the *translation*, of this intimately lived-in burial environment with its palpable reminders of the ancestors—many of whom could be identified by location, bone, and relics—the landscape of the Point was, in the space of twenty-four hours, suddenly changed forever. The extent of the transformation, with the apparent cooperation of the Native community, was described in an anonymous article titled "Cemetery and Sanctuary" in *Spirit of Missions*:

> One of the first duties, not of Christianity only, but of common humanity, is the seemly and safe burial of the dead. Yet to guard effectively the graves of the departed is not an easy thing in some parts of Alaska. The fence which may be erected so promptly in [our temperate] part of the world becomes a problem, in a land where practically all the wood is—or is *not*—brought in by the tide. To this problem [Rev. Hoare] addressed himself with the following result:
>
> "I am sending you a picture of our new graveyard, which is probably the only one in the world enclosed by a fence composed of the jawbones of whales. Each bone is from six to ten feet long, and curved like a rib; the butt is buried in the ground and the bone curves outward, forming a picket fence; as in old times no one was allowed to bring a jawbone on shore until he had killed four whales, and then only one of the jawbones, and as there are over 2,000 bones on the point, you can see how many whales have been killed in times past.
>
> "The graveyard was completed in one day. Every living person of working age on the Point assisted. While some were building, others were patrolling up and down, collecting the skulls and bones of those who, in time past, have been laid on the surface of the ground to await decay. In one common grave we buried over 1,200 skulls, and about three cartloads of bones; and since then a heap of a hundred or more skulls awaits burial. This marks the passing of a superstition connected with the exposure of bodies. The people have accepted Christianity, and are doing their best to carry out its teachings." (*Spirit of Missions* 1909:945)

The assumption that the new enclosure with its subsurface burials represented spiritual protection for the dead was no doubt as well intentioned as it was uninformed. If traditional Tikiġaq people had wanted to inter the deceased they would, over the previous centuries, have contrived it. In fact, that interment had preceded rack burial in a remote past was acknowledged in oral lore. And the proto-Eskimo Ipiutak people (ca. 0–500 AD) had also dug graves into the peninsula that lasted until the 1940 excavations (Larsen and Rainey 1948).[1] But if the reburial of human remains was an expression of Christian generosity, the removal of the old *umialit*'s grave markers also subordinated ancestral autonomy to the "shelter" of the new dispensation. Here too was an enactment of the doctrine that Sheldon Jackson had learned from Alexander Duff, whose rhetorical idealism ("While we throw down, we also rebuild") was given concrete reality by Hoare.

This act of translation was perhaps the most revolutionary communal event in Point Hope since Jabbertown was established or Driggs opened his school. And the same enthusiasm that drove Hoare to gather the bones (both of humans and the whales' jaws that accompanied each *umialik*) went into the destruction of Driggs's church, which Rowe had dismissed as "like an iglu." Of this same church, Rowe had written, again echoing Duff, "We pulled this down and erected a building that is fine" (Rowe, Letter of Oct. 6, 1909). With those two acts of clearance, both Point Hope's ancestral population and Driggs's "iglu Christianity" were demolished.

Only recently Driggs had imagined "some old villager brought back to life who would find a great change among the people" (Driggs, Letter of June 20, 1904). The speed of this change had been astounding, most of it having taken place since 1850 and most especially between 1880 and 1900. In fifty years, Point Hope had thus reached a stage of acculturation that many other Native societies in mainland America had known for two or three centuries. One reason for this lay in nineteenth-century America's own accelerated advances. Earlier contact had occurred between Natives and less technologically developed Euro-American cultures. Few as the non-Natives in the Alaskan Arctic were, they brought with them technologies that were capable of implanting virtually instantaneous representations of modernity on the small societies they touched.

But a *layer* this remained for at least another fifty years, and a visitor to Point Hope in the first half of the twentieth century would find a vigorous and self-confident Native people living in a territory that was their own, bringing up their families in subterranean iglus, dressed in skin clothes, pursuing the animals they knew, and speaking the Iñupiaq language. And while hunting technology was dominated by rifles, shotguns, and, for the whale hunt, exploding harpoons, much of the traditional assemblage of tools remained intact. When Froelich Rainey visited Niġuvana's iglu in 1940, she named for him in Iñupiaq over one hundred and eighty items of men's and women's tools and other cultural artifacts still in traditional or modified use (Nirowana 1940:17–31). Similarly, Ququk Dives identified some forty traditional and transitional terms connected to seal hunting alone (Dives 1940:6–16). An analysis of this material would require a separate study.

The traditional cultures of singing and dancing also survived. And while by the early twentieth century there were no more *qalgi* houses, the old autumn and spring festivities were still celebrated in frame buildings such as the school, the Native store, and Browning Hall, a spacious building donated to the village by Mrs. J. Hull Browning of New York in 1930 (Goodman 1932). In addition to these, ceremonial grounds for *Uŋasiksikaaq* and *Qaġmaqtuuq*, the two surviving *qalgi* groups, remained near the sites of the old ceremonial house buildings. These ritual areas, marked by tripods of whale ribs and whale jaw uprights, remain, to this day, central to the community's annual whaling feast in June. In the early twentieth century, storytelling sessions, displaced from the ceremonial houses, took place in the household or at camp. Some elders nonetheless understood that this tradition was in jeopardy, and this, most significantly, was expressed by the storyteller Samaruna (the young man who had defied Ataŋauraq), who began systematically to transmit his repertoire to his nephew Asatchaq in about 1920. Others, such as Christopher Tingook, who had arrived in Point Hope from the Kobuk as a child at the turn of the century, began recording stories on reel-to-reel tape machines in the early 1960s. The Point Hope generation born in the 1950s still spoke Iñupiaq, hunted with dogs, and retained detailed knowledge of local geography. With language skills that enabled them to communicate with the elders, they had access to local history and tradition. Those born in the 1960s belonged to a world that was changing, once again, with unprecedented speed; this would be the first generation for which the Iñupiaq language must, on the whole, be sought after and acquired, and this distanced them further from the world that missionaries and schoolteachers, from Hoare onward, had taught even their parents and grandparents to distrust and forsake.

But socially and culturally, the years 1900 to 1960 represented a period of successful reintegration and reinvention. The experience of contact had been intense and often

FIG. 37. *Point Hope Council, ca. 1920. This council of elders and* umialiks *was effective in establishing and maintaining social harmony in the early twentieth century. While two separate* qalgi *groupings remained, the council represented a cross-*qalgi *authority that remains to this day. Back row, right: Uyaġaq Sam Rock. Front row, left: Peter Kunuyaq and Samaruna.*

FIG. 38. *Old Point Hope in the modern era: a traditional iglu, skinboat frame, and sled on the iglu owner's equipment rack. Photograph ca. 1915.*

traumatic. But it had largely been contained within a fifty-year period. The decline of commercial whaling, the departure of Jabbertown, and the deposition of John Driggs all happened to converge on the first decade of the new century. By 1910 the period of worst affliction had come to an end, and a new era, whose evolution had of course started much earlier, took progressive shape. Serious poverty remained. Tuberculosis and pulmonary and ear infections continued to be endemic. But the distillation of liquor went out with the whalers, and dangerous levels of drinking subsided. The elders, women, and the elected village councils of the 1920s forbade the sale and public consumption of alcohol.

Thus the survivors of contact created a harmonious amalgam of the traditional and contemporary, which in the view of later-twentieth-century elders represented something of a golden age in Point Hope's long history. Arctic people had always, when necessity demanded, improvised and mended their subsistence tools with whatever materials came to hand. A broken harpoon shaft that was spliced with sinew or baleen in 1850 might, in a similar way, be mended in 1930 or 1999 with the very same materials or with a strip of metal that had been cut from a food tin. While a seal shot by rifle has a tendency to sink, Iñupiat of the late nineteenth century, no longer entirely dependent on harpoon line and float, improvised the one-man umiaq and wooden float, making use of bent nails for grapple hooks. Neither traditional nor modern technology was ever completely foolproof. Nor could the improvisations of early-twentieth-century subsistence life (dangerous, difficult, and demanding as it continued to be) become perfect. But just as technologies evolved to patch the fault lines that must always occur, a new Iñupiaq world that made the best use of available materials, both physical and intellectual, was put together by the people themselves.

Thus Point Hope survived. But as we have seen, the demography of the village had also changed, and the families that emerged into the twentieth century were not genealogically the same as those who had met Beechey in 1826. While the pre-contact population had always, albeit occasionally, been enriched by incomers from other Eskimo societies, the influx in the late nineteenth century from the Kobuk, Noatak, Kivalina, Seward Peninsula, and Diomede Islands took place over a mere three decades and was immeasurably greater in numbers and in impact on a reduced Point Hope population. Jabbertown men had likewise made their contribution. But forged by the place and its subsistence demands, all these newcomers became Point Hope people. This ancient, far-flung continental outpost thus outlived the culture that had threatened to submerge it. Like the whale that the Raven harpooned and which transformed to become the Tikiġaq peninsula, it had changed. But it had also lasted, as it does to the present, creatively and with spirit.

Notes

1 Archaeological work on Ipiutak houses started late in 1939, and excavation of graves was carried out the following year.

APPENDIX A

The Point Hope and Jabbertown Census of 1908

Introduction

Augustus Hoare conducted a census about six weeks after his arrival in 1908, and included all the households at Point Hope, Jabbertown, and Beacon Hill (or "Coopertown," as the latter was also named after "Cooper" Koenig's station). Hoare took the Jabbertown census on September 24 and Point Hope's ("Tigara") census on October 17. While Hoare wrote that the Tigara population stood at 168, the actual number seems to have been 181.[1] The reindeer herder census was conducted roughly in the same period. The population of the three communities was as follows:

Point Hope	181	[Hoare's figure: 168]
Jabbertown area	48	
Reindeer camp	16	
Total	245	

Hoare headed his record "Families at Tigara"; it is dated October 17, 1908, and is in two parts:

1. A "Key plan of houses with families" showing Point Hope's twenty-five houses, each represented by a small rectangle.
2. A list of the individuals who lived in each of the numbered houses.

1. The Key Plan shows uniformly clustered houses numbered 2 to 24 in two parallel rows running east to west with one single house and one group of houses lying outside the main body of the village. The single House #25 inhabited by Peter Kunuyaq's family lies a little to the north of the main grouping. House #1 lies furthest from the main community to the east of the village and was perhaps built next to the mission during Driggs's residence. House #1 is represented by four rectangles, one of which stands for the mission, while the others presumably represent three households in interconnected, possibly modified, iglu dwellings.

2. The list of individuals includes the names of every resident in Tigara village, reindeer camp, and Jabbertown on October 17, 1908. Each entry records names (Iñupiaq and/or English), sex, and age. The letters B, C, and C indicate baptism, confirmation, and status

as active communicant, respectively. All the resident Point Hope population, with the exception of sixteen individuals, had been baptized by that October. Of those sixteen, at least two were "old people" in their fifties or sixties; one came from Icy Cape; two were babies; seven are registered with question marks; five in House #2 have their names penciled in, perhaps hastily. Two of these latter are marked as having been baptized; only one's age is given. A proportion of these sixteen, two with English names, had presumably been baptized.

A breakdown of the Point Hope village population as of October 1908 reads as follows in tables constructed from Hoare's register by Elizabeth Lowenstein:

Point Hope

Age Group	Total Number	Male:Female Ratio	Years Born
0–11	39	19:20	1897–1908
12–20	30	16:14	1888–1896
21–30	81	40:14	1878–1887
31–40	9	5:4	1868–1877
41–60	12	5:7	1748–1867
61–80	1	0:1	1847
Total	172 + 9 of unknown age or gender 181 inhabitants in total		

Jabbertown and Coopertown

Euro-Americans	3
Mixed Race, Euro-Eskimo	5
Eskimos with Euro-American Names	12
Total	48

At Coopertown there were three Iñupiaq households:

- House #1 contained a married couple and three sons (no ages or baptismal record given).
- House #2 contained eleven individuals, including at least three couples, two fifty-five-year-old women, with two baptisms noted.
- House #3 contained two couples and an "orphan adopted" (who was a sixteen-year-old boy), all baptized.

At Jabbertown there were three non-Iñupiat:

- Joe Ferreira, Portuguese (50)
- Arthur James Allen, white American (35)
- William John Seymour, white American (45)

Their households and/or stations consisting of more than one dwelling were as follows:

- Joe Ferreira, his wife Konyanna (34) and their children Tillie (13), Maggie (11), Nellie (5), and Mary (1 month).
- Arthur James Allen, his wife Eleanor (32), and Billy (born Oct. 5).
- Seymour was resident with the Allen family.

Next door to Allen lived ten Iñupiat, including at least three couples and one twelve-year-old boy called Walton. Against Walton's name is written "R.I.P. May 24 '31" in Archdeacon Frederick Goodman's later hand. Against the name of Rachel, aged sixteen, is written "R.I.P. 1937."

One further household contained six individuals consisting of two married couples, a one-year-old girl, and Nyakok, an unmarried man described as "Ida's brother," an entry probably referring to one of the married Iñupiaq women of the household whose name is recorded in Iñupiaq only.

Baptism is not recorded against the names of thirteen individuals, including Ferreira and Seymour.

Iñupiaq and Euro-American Names

In the population of 181, eighty-nine individuals have Euro-American names. Hoare listed thirty-two of these by English name only. This did not of course signify that they had no Iñupiaq name. He listed, for example, two young people as Nanny and Jimmie whose Iñupiaq names were Uyataun and Asatchaq, both born in the 1890s. Hoare's grasp of Iñupiaq names was good. Nanny and Jimmie's parents, neither of whom took Euro-American names, are noted as Killilwuk, amended in parenthesis to Killigiwuk, and Nukka.wonna.

Below the written surface is evidence, in both cases, of erasure and emendation. Both transcriptions are good versions of the names Kiḷigvak and Niġuvana.[2] Hoare's attention to such detail is evident in several later emendations. One example is in his addition of an Iñupiaq name to Dick, "reindeer boy."

The Point Hope Census: Discussion

A point of interest in the register lies in the distribution of English names. Entries for House #1 provide the best information about this:

M	Harry Killbear (Iñuuġvik)	B.C.C	25
F	Lily (Samaruna)	B.C.C	22
M	Tommy	B	5
M	Ben Bolt (Ah-kayvruk)	B	27
F	Weet-ayuk	B	24
M	Dick Ahgaiuyak	B	18
M	Ikey	B	14
M	Hank	B	21

Here, one of eight people is without an English name and three males are known to Hoare by English names only—a higher ratio of English to Iñupiaq names than in any other household.

Of the people in House #1, we know that prior to Hoare's arrival, Harry Killbear had been one of Driggs's confidantes, and it may be assumed that the whole group was associated with him and/or mission business. According to Hoare (Letter of Sept. 27, 1914), Killbear had "lived with [Driggs] all these years." What Hoare might have meant is uncertain, but presumably he was alluding to contiguous but separate iglu habitation, first in Point Hope and later at Cape Lisburne. Driggs's further association with people in House #1 is suggested by Hoare's note, dated January 1909, that Dick (Ahgaiuyak) became a "reindeer boy": that is, he had gone to the Seward Peninsula to train as a herder and returned to help manage the Point Hope reindeer, which were mission property at this time.

The names Harry Killbear and Lily (his wife) are particularly interesting. Iñuuġvik means "one who kills people," and Driggs had changed this name first to Killman and then to Killbear: the transformations from Iñuuġvik to Killman to Killbear representing a progressively "civilizing" movement, and the name "Harry Killbear" also proclaiming a satisfactory combination of hunter and Christian. Iñuuġvik's wife Samaruna is given the name Lily, which suggests the flower associated with moral purity.

Not all the English names carry this level of significance. Out of 181 individuals at Point Hope, about eighteen were not known to have been baptized (Hoare either omits information or writes a question mark in the "B" column). Ninety-three of the village population have English names, forty-four of which are biblical or quasi-biblical (i.e., Ikey for Isaac, Jim for James). Thus while a large proportion of Point Hopers had been baptized with English names, baptism did not necessitate a name associated with religion.

Of the seventy-five people to whom Hoare ascribed only an Iñupiaq name (of whom some perhaps did have an English name), only sixteen had not been baptized. But Hoare signifies the uncertainty of his figures with four question marks in the "B" column.

Finally, the associations attributed to names such as Harry Killbear and Lily were no doubt attached to many Euro-American names in the village, and it is therefore possible—despite the argument above—that biblical names such as Eva, Mary, James, or David had spiritual name power (*atiq*) that supplemented the ownership of crosses (see Chapter 13).

Conversely, it is possible that a number of individuals to whom no Euro-American names were given specifically held out against acquiring new names. There is no information about this nor evidence of an anti-American traditionalism in turn-of-the-century Point Hope. It is nonetheless possible that certain people or groups were rooted more strongly in traditional patterns of thought and subsistence than others, and that these people were immovably identified with their Iñupiaq names. With the luxury of hindsight, it is inconceivable that Samaruna, who passed his repertoire of stories and subsistence lore to his nephew Asatchaq, should have taken a surname. The same seems true of Samaruna's sister Niġuvana and her husband, Kiḷigvak, who were similarly versed in narrative, subsistence, and healing traditions. And while on the one hand it may have been mere chance that none of this closely associated group took Euro-American names, on the other, they might all have been so strongly identified with their Iñupiaq names and the namesakes who preceded them that "Christian" names and surnames were unthinkable—to themselves, to the community, and even to the missionary. But this is speculation, and there is likewise no evidence that knowledgeable figures such as "Frank" Aġviqsiiña and the biblically named "Dives" Ququk were any less Iñupiaq in their affiliations than was, for example, Kuwana, who himself took no Euro-American name but who gave his sons Siġliuna and Uqpik the names Walter and Augustus (after the missionary Hoare).

Nothing of significance about Point Hope life survived in Hoare's correspondence, and fieldwork from Rainey onward in 1940 has focused principally on the traditional and late-nineteenth-century periods. With the departure of Driggs, the death of Ataŋauraq, and the disappearance of the Jabbertown stations, day-to-day information about life in the village and contact with whalemen and traders recedes.

Notes

1 At the end of his "Families at Tigara" list, Hoare entered the total for Point Hope's population as 168. There are, however, 181 named individuals in his Point Hope village census. Hoare may have named the individuals normally in residence but, in order to arrive at the figure of 168, subtracted those who were on the Kuukpak River. He also gave names of three people recently deceased. I have not included these in my own tally of 181.

2 These are the parents of Asatchaq or "Jimmie," all of whom figure in this volume.

APPENDIX B

The Point Hope *Qalgi* and Its Changes

According to oral testimony there were six or seven *qalgit* in traditional Point Hope, and as outlined already in Chapter 1, these were centers where shamans and *umialit* organized whaling rituals and conducted divination events, gift exchange, masquerades, competitive games, and shamanistic séances and where the elders told stories (Rainey 1947:244ff; Lowenstein 1993:81ff).[1] The *qalgit* were also kin-based social organizations that, in the pre-contact period, provided the basis for skinboat crew affiliation (Burch 1975:22, 246). But from the 1870s onward, as suggested in recent work by Mary Larson, other factors also determined which *qalgi* an individual might belong to (Larson 2004:88ff). Kinship groups surrounding particular *umialit* still accounted for part of individual *qalgi* identity, but following the epidemics and population shifts in the late nineteenth century, people might also join a *qalgi* for other reasons. Marriage ties, namesake alliances, and place of origin were among these extra factors; given the subsistence crisis and thus frequent and sudden manpower emergencies, people changed skinboat crew and thus their *qalgi* affiliation more readily than they had in the pre-contact period.

In parallel to the depletion of Point Hope's population came a reduction in the number of *qalgi* buildings that the community could support (Larson 2004:88). By the late 1870s there were five *qalgit* in the village, and according to Kuwana (born ca. 1876) there were, by the mid-1880s, just three left: Qalgiiraq, Uŋasiksikaaq, and Qaġmaqtuuq (Kuwana 1940:1). At the end of the 1880s, Qalgiiraq had been eroded by the sea—a force, as in the 1893 flood, that may also have destroyed other ceremonial houses (see Chapter 8 and Larson 2004:86). By 1890, there were just two *qalgi* buildings left: Uŋasiksikaaq and Qaġmaqtuuq. These nonetheless remained the major sites for traditional activities such as equipment making, ceremonials, shamanistic events, and inter-*qalgi* trading (Kuwana 1940:38). And these two *qalgit*, with their great whale-bone markers standing on the probable location of the original buildings, are the ceremonial groups to which all Point Hope people still belong.

While the missionary Driggs had not preached against *qalgi* activity ("he had even visited the Uŋasiksikaaq *qalgi* building on at least one occasion for a dance"),[2] it remains possible that his successor Augustus Hoare accompanied his reorganization of Point Hope's cemetery with teaching that discouraged *qalgi* life, and that the disappearance of the two remaining buildings took place within the first year of Hoare's ministry. On this Larson writes:

> The two surviving *qalgi* structures were abandoned sometime between 1908 and 1910. What precisely happened is unknown. It is possible that they also washed out and that the structures were no longer necessary because of the larger gathering places createdby the mission buildings. It is equally possible that the coming of the new missionary in 1908, the Reverend Augustus Hoare, was a factor in their disappearance. Driggs, being a medical doctor, had been a dedicated missionary but seems to have seen the value in many of the local cultural traditions. Hoare, however, was less tolerant, and the demise of the remaining *qalgi* buildings within a year and a half of his arrival probably is not coincidental. The truth may never be known, however. (Larson 2004:86–87)

As both Larson and I discovered in the course of our separate researches, Hoare left no correspondence that touched on Point Hope daily life or culture.

A remarkable story told by Asatchaq (Jimmie Killigivuk) and recorded by Don Foote (Asatchaq 1961) contains what is probably the last record of any single event in a Point Hope *qalgi*. Part of the story is given below and some discussion of it may help to give a picture of changing early-twentieth-century *qalgi* life. The event described was a Messenger Feast (*kivgaq*): an inter-regional tradition in which runners from a host village visited people of a separate community and invited them to feasting, gift exchange, and dances. *Kivgaq* was an occasional midwinter event and part of its object was to stimulate goods exchange and to enhance the prestige of the host *umialit*, who would thereby receive a return invitation. Point Hope's guests in this case were a group of people from the Noatak River, and in a brief appendage to the story, Asatchaq summarizes a return visit that probably took place the following year. This latter fragment is not given here, since Asatchaq disclaimed knowledge of what happened.

Asatchaq, who was born in 1891, says that he was *nukatpiaruq* (about twelve years old) at the time of the feast, and this dates it to 1902 or 1903, when one or two of the Point Hope *qalgi* buildings were still standing. As was the case with most *qalgi* events, children were fascinated and mystified by ceremonial house business, which they (and often women) could watch only from the skylight. As described already, traditional *qalgi* life was in its final days, and Asatchaq's sense of childhood distance from what he could glimpse is paralleled by elements in the story—Jabbertown men and white trade goods—which evoke a changing tradition. Before arriving at his description of the *qalgi* event itself, Asatchaq sketches the preliminaries, and these, in paraphrase, are as follows:

When it was decided to hold the feast in Point Hope, two runners were sent, with accompanying guides, to the Noatak River, where they put on a dance display. One aim of this was to encourage the hosts to signify acceptance of their invitation with reciprocating dances. It was difficult on this occasion to get the hosts to dance, and only after repeated Point Hope performances did the Noatak people agree to come. The reason for this is unclear. While the storyteller described the hosts as "always lazy at that time," it would seem possible that ill health, hunger, and depopulation might have been responsible—acceptance, after all, would commit them to a long journey. It is not clear from Asatchaq's narrative how many Noatak people came, but we may assume from the three or four names he gives that the numbers were more than that, but not great. The month was February, and the journey on foot and by dogsled having largely been achieved, the travelers were met at Piŋutchiaq, a bluff on the south shore about fifteen miles from Point Hope village.

In spite of the cold weather, the Point Hope men who went to meet the visitors were wearing "thin trousers made of calico," because "they didn't want to wear anything heavy." These were probably gifts from a principal *qalgi* host, but also perhaps a property display that also made running easier. Once met, both parties ran on towards Point Hope, passing Jabbertown en route for the *qalgi* where gift exchange and competitive dancing would begin.

When the runners arrived at Jabbertown, contact details in the story multiply. One of the runners encountered a non-Eskimo dog from a whaling ship, which belonged to an Iñupiaq woman who was married to a Jabbertown whaler called William, and both William and the dog joined the running. On arrival, there was an interlude for sleep, as some of the Point Hope runners were coughing and/or suffering from minor food poisoning. Apparently they then returned to the east of the village to sleep, and the next day Asatchaq and about four other boys were chosen to go and meet the visitors beyond the burial ground. Asatchaq had been told not to eat that morning and was given a small piece of *maktak* to present to one of the visitors, but "the Noatak people told me not to let anybody take that *maktak* away from me, so I held it," until a girl of about his age took the *maktak* away from him. Which party she belonged to is not stated. Asatchaq and a boy called Ataŋauraq then ran together along one of the beaches (apparently a mistaken route) and they arrived late. This scene in the *qalgi* followed:

> When the guests from Noatak arrived, the Tikiġaġmiut brought each one of them a dish of food that they had requested. After they had eaten, they slept, and the next day started to exchange gifts (*yugaq*). The Tikiġaq people made (or encouraged) the Noataġmiut to *yugaq*. I couldn't get into the *qalgi* because I was just a boy. I tried to look inside, but I couldn't because people [surrounding the skylight] would trample on me. Then they started dancing (*uamit*). And by dancing they tried to win the *qalgi*. I don't know how to explain this because I was just a boy at the time: but the Noatak people are trying to "get the *qalgi*."[3]
>
> Then the Noatak people stood up and started to sing and drum (*qiḷausiq-*), thinking to get the *qalgi* away from the Tikiġaq people. Nauġiaq started to entertain them by looking at them and singing a long time so that people would get tired and leave, and they would win the *qalgi* easily. Then Nauġiaq put on his caribou skin mittens with the thumbs turned backwards. He didn't really dance, he just swung his arms, and they just smiled at him because this was the first time they had seen anyone try to take the *qalgi* by dancing like that.
>
> Then they let them out [this detail is unclear], and the next day the Noatak people took off the *qalgi* skylight. I didn't know what this meant, because I was a boy. They brought red fox skins, caribou meat and sinew [through the skylight]. And the Tikiġaq people gave each of them a dish of food they had asked for. Then the Noatak people requested certain other foods. There were no white man's goods in the Noatak area; it was hard for them to get. And the Noatak people asked for onions [*aiñan*—a loan word] so some Tikiġaġmiut went to fetch onions from Jabbertown. On return, they cut the onions up and made the Noatak people stand in a row and starting at one end, fed them raw onions. They wouldn't let them touch the onions themselves. Nasugluk

> was feeding them, and because they respected him, they went on chewing. But they couldn't swallow, and they didn't spit them out. Nasugluk's older brother then whispered to my uncle Kunak, "How come he's so high and mighty that we have to obey him?"[4] And they finished the ceremony.

The first thing to say about this story is that it was narrated by the one of the last surviving Point Hope witnesses of the event. Told in 1961, when Asatchaq was seventy, the tape was translated in 1979, the summer before the storyteller's death, but Asatchaq was by this point too disabled to respond to questions.

Second, there is the question of the *qalgi*'s identity. Given that the storyteller Asatchaq was a member of Uŋasiksikaaq, it is likely, although not certain, that this *qalgi* staged the feast. Since, furthermore, no *qalgi* name is given, it is possible that Uŋasiksikaaq was the one *qalgi* still standing—albeit Qaġmaqtuuq remained an active ceremonial grouping. This is speculative, for the names of Pauyuuraq and Ataŋauraq (son and a namesake of the late "chief") who were probable members of Qaġmaqtuuq, are given among those who met the runners at Jabbertown. Even if two *qalgi* houses were still in use in 1902–03, it was possible that the feast represented a combined *qalgi*, all-village, event for which just one building was required.

Space forbids a longer description of the traditional *kivgaq* and details of the northern Iñupiaq feast are given in Spencer (1959:210–28), while the Yupik event is described by Nelson (1899:361–63). New, perhaps improvised, aspects of the feast following culture change should however be noted here. First, as a result of nineteenth-century population movement, there are Noatak residents in Point Hope who guide the hosts' runners south. Second, some of Point Hope's second group of runners wear calico trousers—cotton drill having routinely been put to use as parka covers in the late nineteenth century. Third, the returning runners used Jabbertown as a staging post and are joined by a white man and his dog. Fourth, Nasugluk (one of the hosts, who had spent part of his childhood on the Noatak Rver) feeds onions to the guests when they have asked for white man's food. An autobiographical narrative that Nasugluk recorded for Froelich Rainey may help to shed light on this detail. What follows is paraphrased from Rainey:

Born in Kivalina of a Point Hope mother in ca. 1876, Nasugluk was adopted by Noatak people after the death of his mother in Kivalina. His mother's death put the ten-year-old boy in a condition of contaminating taboo, and he was several times both abandoned and occasionally succored as the people with whom he had sought refuge moved along the river. In his old age, Nasugluk described this survival story in painful detail (Nasugluk 1940:1–7). On his return to Point Hope, Nasugluk, still a boy, was adopted by Ataŋauraq (see Chapter 12), who abused him. Finally, in about 1888, Nasugluk also witnessed the "chief" humiliating some of his own Noatak partners who had come to trade with him (see Chapter 9).

The thirty-year-old Nasugluk's role in the Messenger Feast, so evidently puzzling to his own brother, could perhaps be interpreted with reference to this past history: as an act of power towards representatives of a community he resented, and in imitation of Ataŋauraq's treatment of his Noatak partners. Whatever partial explanation one may suggest, the onion incident occurs within an inter-regional relationship between communities whose interest in each other had traditionally expressed both generosity and hostility. Nasugluk, a victim and representative of displacement following regional epidemics and famine, and who had lived in both communities, mediates this tension with recourse to a third party—the white man—whose influence lay within the interchange without itself belonging to either. The strangers in the *qalgi*, who chew raw onion without wanting to swallow, stand dramatically within this complex situation. It was a painful moment, but it does not appear to have destroyed positive relations between the two communities, since a return visit took place the following year.

The storyteller Asatchaq concluded his recitation with the following song, which is probably derogatory:

Qikiqtaġruŋmiuni
Taiguġniaġaatin
Aiñaniglu
Manniŋiglu imma
Taaŋaġniglu
Kotzebue people who live here [or in Kotzebue]
they call you
onions
eggs
and whiskey[5]

Which individual, from what precise community, composed the song is unknown. And if there is a cheerful insult here, no doubt this was part of the give and take contained in inter-regional trade, while perhaps Nasugluk's action, too, was an expression of similarly routine teasing. Whatever the explanation, this final expression of *qalgi* life is characteristically intense and filled with robust Iñupiaq imagination.

Notes

1 The *qalgi* names were Qalgiiraq (the oldest), Agriqtuaq, Suaġvik, Qaŋiḷikpak, Qaġmaqtuuq, and Uŋasiksikaaq.

2 Larson (2004:87), quoting Rainey (Aġviqsiiña 1940:87).

3 *Qalgiiqsaaqtigraŋi taapkua*, "they wanted to get the Point Hope *qalgi*" (i.e. take it away from others in a contest to see who would open the *qalgi* ceremony). Translation of this detail by Lawrence D. Kaplan (personal communication, 2007).

4 Nasugluk was born ca. 1875. In famine conditions of the mid-1880s he walked to Point Hope, where he was adopted by Ataŋauraq. See Chapter 14.

5 Transcription and translation revised by Lawrence D. Kaplan (personal communication, 2007).

APPENDIX C
Driggs and the Issue of Missionary Trading

Both Driggs's dilemma of 1890 as described in Chapter 16 ("Anger at Jackson. Anxious About Trade") and his eventual deposition on account of having engaged in baleen trading can be seen against a milieu of post–Civil War reformist ideas in the aftermath of the Indian wars in the 1860s and 1870s. Episcopal Church thinking about the Alaskan missions to Iñupiaq communities in relation to whalers and traders was however never separately or explicitly formulated.

The discord between churchmen and traders represented a split that had long been built into American discourse and nowhere more acutely than on the subject of the "Indian question." In acknowledgment of broken treaties, white land appropriations, and Indian agency corruption of the past decades, Congress set up a committee in 1865 to investigate the state of Indian tribes. In response, ex–Civil War general President Ulysses Grant sought "The proper treatment of the original occupants of the land" and welcomed "any course towards them which tends to [their] civilization and ultimate citizenship" (Prucha 1984:158). To accomplish this, Grant's so-called "Peace Commission" of 1867 determined that the War Department, which was "spotted with fraud" and "tainted with corruption," should withdraw from the Indian Service.

Grant proceeded to dismiss existing reservation agents and opened the Indian Service to Missionary Society management, the aim of which was, through Christian organizations, to "provide churches and schools that would lead Indians to understand the benefits of a Christian civilization" (Prucha 1984:153). Leading participants in this initiative were the Episcopalians and Quakers.

Washington politicians were not enthusiastic about the enterprise, and they exploited disagreements between Protestants and Catholics to undermine it. The churches themselves, criticized for not having done more for dispossessed Indians in previous decades, now had the task of rehabilitating the service.

It was an almost impossible brief. The previous generation of agents had lived by trade (liquor included), embezzling annuities intended for schools and agricultural projects, and by handing reservation jobs to cronies. But while the U.S. economy enjoyed a postwar boom, the churches had difficulty scraping together salaries for the evangelicals whom missionary societies were sending to the last "wild" Americans.

Unlike their predecessors, the new missionary agents were strongly discouraged from trading with either Natives or the white people within reach of their stations. On salaries of $1,500 or less, and caught between "pagan" Indians and an often hostile settler culture, missionary life was one of crushingly hard work, arduous traveling, isolation, and microscopic influence.

Abuses against Native societies continued, and the next decade saw the emergence of new voices, including that of Helen Hunt Jackson, whose *A Century of Dishonor* (1881), which decried the forced removal of the Ponca tribe from Nebraska in 1877, became a cause célèbre. Another influential voice was that of H. W. Whipple, Episcopal Bishop of Minnesota, who used church status and sheer force of personality to lambast Washington with demands for reforming Christian action.

Like many of his progressive contemporaries, Whipple was painfully aware of what a "dominant, greedy, avaricious Saxon race" had offered Native American people, and he became a leading member of a society inaugurated in 1883 called the Friends of the Indian. The Friends, one of several groups demanding reform, was an ecumenical association that met at the mansion of Quaker philanthropist Albert K. Smiley at Lake Mohonk in upstate New York, and whose pressure on Congress eventually led to the Dawes Act of 1887, which—idealistic in intent but ultimately unworkable—granted parcels of reservation land to Indian men who could claim to head a European-style family.

The moral ambience that Driggs entered in 1890 had been created by individuals such as Bishop Whipple, whose many writings contained denunciations of traders' activities. "Missionaries," wrote Whipple, "should confine themselves as much as possible to their own calling and their own sphere of work" (Whipple 1899:341). Too many of the agencies and trading posts on which many Indians had become dependent were still "nurseries of fraud" engendered by a "vicious system of trade." This was a theme picked up in other missionary society publications, most prominently by William H. Hare, Episcopal Bishop of Niobrara (Nebraska; see for example Hare 1873), and in a series of articles by Susan Fenimore Cooper, serialized in *The Living Church* in 1886. This latter focused on the "evil-minded among the white traders and speculators" among Oneida Indians and looked to the guidance of Whipple and Hare in redressing traders' abuses (Cooper 1886:753). To these voices may be added that of Hudson Stuck, Archdeacon of the Yukon and the Arctic, who wrote about trade in Alaska in much the same idiom: "The trader, the beach-comber and the squaw-man have always been hard upon the heels of the explorer" (Stuck 1920b:Chapter 2). But as the reader may have noted, it was Stuck who, on a visit to Point Hope long after the doctor's demise, partially rehabilitated Driggs's standing, and who evoked Driggs as a victim of circumstance whose isolation had led him to a relatively minor transgression against what remained an unwritten church rule.

BIBLIOGRAPHY

Aġviqsiiña, Frank. 1940. Frank's story, as told to Froelich G. Rainey. Unpublished field notes from Point Hope. Rainey Collection, Box 2, folders 8–10, Archives, Elmer E. Rasmuson Library, University of Alaska Fairbanks.

Aldrich, Herbert L. 1889. *Arctic Alaska and Siberia, or Eight Months with the Arctic Whalemen*. Chicago: Rand McNally and Company.

Allen, Arthur James. 1978. *A Whaler and Trader in the Arctic*. Anchorage: Alaska Northwest Publishing Company.

Annual Report on Domestic Missions of the Board of Managers. 1891–1910. Protestant Episcopal Church, New York. In Episcopal Church Archives, Austin, Texas.

Asatchaq (Jimmie Killigivuk). 1961. Reel-to-reel tape 9A recorded January 19, 1961 by Don Charles Foote. Translated by Ada Ward. Archives, Elmer E. Rasmuson Library, University of Alaska Fairbanks.

———. ca. 1970. Transcription from an anonymously recorded and translated tape in the author's possession.

Asatchaq and Tom Lowenstein. 1992. *The Things That Were Said of Them: Shaman Stories from Tikiġaq, Alaska*. Told by Asatchaq, translated by Tukummiq and Tom Lowenstein. Berkeley: University of California Press.

Asatchaq and Edna Ahgeak MacLean. ca. 1978. Undated and unpaginated transcription from tape in manuscript translated by Edna Ahgeak MacLean.

Axtell, James. 1981. *The European and the Indian: Essays in the Ethnohistory of Colonial North America*. New York: Oxford University Press.

———. 1988. *After Columbus: Essays in the Ethnohistory of Colonial North America*. Oxford: Oxford University Press.

Bailey, George W. 1880a. Letter to Hon. John Sherman, Secretary to the Treasury, U.S. Senate, 46th Congress, 2d Session, Ex. Doc. No.132.

———. 1880b. *Report upon Alaska and Its People Giving Statistics as to the Numbers, Location, Pursuits, and Social Condition of the Inhabitants; The Climate, Productions, and General Resources of the Country; and of the Commerce, Ocean Currents, etc.* Washington, DC: Government Printing Office.

Bancroft, Hubert Howe. 1886. *History of Alaska, 1730–1885*. Darien, CT: Hafner Publishing Co.

Beaglehole, J. C. 1955–74. *The Journals of Captain James Cook on his voyages of Discovery*. 4 vols. London: Hakluyt Society.

Beaver, R. Pierce. 1966. *Church, State and the American Indians*. St. Louis: Concordia Publishing House.

Beechey, Frederick William. 1831. *Narrative of a Voyage to the Pacific and Beering's Strait to Co-operate with the Polar Expeditions; performed in His Majesty's Ship "Blossom"...in the years 1825, 26, 27, 28*. 2 vols. London: Colburn and Bently,

Belcher, Edward. nd. Private Journal. HM Ship *Blossom* on Discovery During the Years 1825, 6, 7... Typescript copy in University of British Columbia Library, Special Collections Division, Vancouver.

———. 1861. On the Manufacture of Works of Art by the Esquimaux. *Transactions of the Ethnological Society of London* n.s.(1).

Benjamin, Marcus. 1925. *Charles Herbert Stockton*. Protestant Episcopal Church, Diocese of Washington.

Berkhofer, Robert F., Jr. 1965. *Salvation and the Savage*. Lexington: University Press of Kentucky.

Bertholf, E. P. 1899. Report of Second Lieut. E.P. Bertholf, R.C.S., Point Hope, Alaska, July 15, 1898. In *U.S. Treasury Department, Report of the U.S. Revenue Cutter "Bear," and the Overland Expedition for the Relief of the Whalers in the Arctic Ocean, from November 27, 1897 to September 13, 1898*, pp. 18–27. Washington, DC: Government Printing Office.

Bingham, Charles. 1998. Barrow Church Plans Yearlong Centennial Celebration. *Arctic Sounder* (September).

Bockstoce, John R. 1975–2005. Personal communications on whaling history, technology, archaeology, sea-ice hunting, nineteenth-century culture change.

——. 1977a. *Eskimos of Northwest Alaska in the Early Nineteenth Century*. Oxford: Pitt Rivers Museum, Monograph Series, no. 1.

——. 1977b. *Steam Whaling in the Western Arctic*. New Bedford, MA: Old Dartmouth Historical Society.

——. 1980. A Preliminary Estimate of the Reduction of the Western Arctic Bowhead Whale Population by the Pelagic Whaling Industry: 1848–1915. *Marine Fisheries Review* 42(9–10).

——. 1983. The Historical Status and Reduction of the Western Arctic Bowhead Whale Population by the Pelagic Whaling Industry, 1848–1914. *Scientific Reports of the International Whaling Commission Special Issue* 5:107–41.

——. 1986. *Whales, Ice, and Men: The History of Whaling in the Western Arctic*. Seattle: University of Washington Press.

Bockstoce, John R., ed. 1988. *The Journal of Rochfort Maguire, 1852–1854. Two Years at Point Barrow, Alaska, aboard H.M.S Plover in the search for Sir John Franklin*. 2 vols. London: The Hakluyt Society (2nd series, nos. 169 & 170).

Bodfish, Hartson H. 1936. *Chasing the Bowhead*. Cambridge, MA: Harvard University Press.

Bogoras, Waldemar. 1918. Tales of Eastern Siberia. *Journal of the American Museum of Natural History* XX(1):73–74.

Brevig, Tollef Larson. 1944. *Apaurak in Alaska*. Philadelphia: Dorrance & Co.

Brooks, Alfred Hulse. 1973. *Blazing Alaska's Trails*. Fairbanks: University of Alaska Press.

Brower, Charles D. nd. *The Northernmost American: An Autobiography*. Unpublished typescript. Arctic Collections, Elmer E. Rasmuson Library, University of Alaska Fairbanks.

——. 1942. *Fifty Years Below Zero*. New York: Dodd, Mead and Company.

Burch, Ernest S., Jr. 1974. Eskimo Warfare in Northwest Alaska. *Anthropological Papers of the University of Alaska* 16(2):1–14.

——. 1975. *Eskimo Kinsmen: Changing Family Relationships in Northwest Alaska*. St. Paul, MN: West Publishing Co.

——. 1978. Traditional Eskimo Societies in Northwest Alaska. In Yoshinobu Kotani and William B. Workman, eds. *Alaska Native Culture and History*, pp. 253–304. Suita, Osaka: National Museum of Ethnology.

——. 1979. The Nonempirical Environment of the Arctic Alaskan Eskimos. *Southwestern Journal of Anthropology* 27(2):148–65.

——. 1998. *The Iñupiaq Eskimo Nations of Northwest Alaska*. Fairbanks: University of Alaska Press.

——. 2003. *The Cultural and Natural Heritage of Northwest Alaska*. Vol. VI. *The Organization of National Life*. NANA Museum of the Arctic, Kotzebue, Alaska. U.S. National Park Service, Anchorage.

——. 2005. *Alliance and Conflict: The World System of the Iñupiaq Eskimos*. Lincoln: University of Nebraska Press.

Call, S. J. 1899. Report of the Surgeon S. J. Call. In *Report of the Cruise of the U.S. Revenue Cutter "Bear" and the Overland Expedition for the Relief of the Whalers in the Arctic Ocean from November 27, 1897 to September 13, 1898*. Washington, DC: Government Printing Office.

Cantwell, John C. 1899. A Narrative account of the exploration of the Kowak River, Alaska. In M.A. Healy, *Report of the Cruise of the Revenue Marine Steamer* Corwin *in the Arctic Ocean, 1884*. Washington, DC: Government Printing Office.

Carlton, Rosemary. 1992. Sheldon Jackson: The Collector. Unpublished master's thesis, University of Oklahoma, Norman.

Cathell, D. W. 1883. *The Physician Himself and What He Should Add to His Scientific Acquirements*. Baltimore: Cushings & Bailey.

Chance, Neville. 2007. The 'Lost' Reindeer of Arctic Alaska. http://arcticcircle.uconn.edu/NatResources/reindeer.html (accessed Feb. 20, 2008).

Chapman, Henry W. 1947. Untitled article from the Archives of the American Episcopal Church, Austin, Texas.

Chapman, John W. 1891–92. *Annual Report on Domestic Missions of the Board of Managers*. Protestant Episcopal Church, New York. In Episcopal Church Archives, Austin, Texas.

——. 1914. *Ten'a Texts and Tales from Anvik*. Publications of the American Ethnological Society, vol. 6.

Collier, Arthur J. 1906. Geology and Coal Resources of the Cape Lisburne Region, Alaska. *U.S. Geological Survey Bulletin* No. 278.

Collinson, Richard. 1889. *Journal of the HMS "Enterprise" on the Expedition in Search of Sir John Franklin's Ships by Behring Strait, 1850–1855*. London: Sampson Low.

Collinson, Richard, ed. 1876. *The Three Voyages of Martin Frobisher in Search of a Passage to Cathaia and India By the North-west A.D. 1576–78*. Reprinted from the first edition of Hakluyt's *Voyages*. London: The Hakluyt Society.

Cook, John A. 1926. *Pursuing the Whale*. Boston: Houghton Mifflin.

Cook, Sherburne F., and Woodrow Borah. 1971. *Essays in Population History*. Berkeley: University of California Press.

Cooper, James Fenimore. 1986. *The Pioneers*. Harmondsworth: Penguin Books. (Orig. pub. 1823.)

Cooper, Susan Fenimore. 1886. Missions to the Oneidas. Serialized in *The Living Church* II (April 18, 1885):26–27. Available online at http://anglicanhistory.org/indigenous/oneida/cooper/02.html (accessed Feb. 20, 2008).

Cox, James L. 1991. *The Impact of Christian Missions on Indigenous Cultures*. Lewiston, NY: Edwin Mellon Press.

Crosby, Alfred W. 1972. *The Columbian Exchange: Biological and Cultural Consequences of 1492*. Wesport, CT: Greenwood Press.

Dall, William H. 1870. *Alaska and Its Resources*. Boston: Lee and Shepard Company.

Damas, David, ed. 1984. *Handbook of the North American Indians*.Vol. 5: Arctic. Washington, DC: Smithsonian Institution.

Daurkin, Nikolai I. 1765. *Sentral'nyi Gosudarstvennyi Voenno-Istoricheskii Arkhiv* [Central State Military-Historical Archive], Moscow. Fond VUA map No. 23435.

Dives (Qukuq). 1940. Dives's story, as told to Froelich G. Rainey. Unpublished field notes from Point Hope. Rainey Collection, Box 2, folders 8–10, Archives, Elmer E. Rasmuson Library, University of Alaska Fairbanks.

Dobyns, Henry F. 1966. Estimating Aboriginal American Population. *Current Anthropology* 7:395–416.

——. 1992. Native American Trade Centers as Contagious Disease Foci. In John W. Verano and Douglas H. Ubelaker, eds., *Disease and Demography in the Americas*. Seattle: University of Washington Press.

Driggs, John B. nd. Undated interview in Archives of the Protestant Episcopal Church, Austin, Texas.

———. 1887–91. Private letters in the possession of John B. Driggs's collateral descendants.

———. 1889–1908. Letters from Point Hope, Alaska, to officials of the Domestic and Foreign Missionary Society, New York. Archives of the Protestant Episcopal Church of America, Austin, Texas.

———. 1891. Leaflet on Alaska, Point Hope, April 21, 1891. This, with deletions and emendations, was published as *Dr Driggs' Work at Port (sic) Hope, Alaska, Spirit of Missions.* New York: Domestic and Foreign Missionary Society of the Protestant Episcopal Church of America.

———. 1894–1901. Journal of John B. Driggs in Point Hope. In the possession of collateral descendants of Dr Driggs.

———. 1897. Report of flu epidemic of summer 1896, dated June 1897. *Spirit of Missions* (November). New York: Domestic and Foreign Missionary Society of the Protestant Episcopal Church of America.

———. 1898. Missionary Intelligence. *Spirit of Missions* (May). New York: Domestic and Foreign Missionary Society of the Protestant Episcopal Church of America.

———. 1899. Alaska. *Spirit of Missions* (December). New York: Domestic and Foreign Missionary Society of the Protestant Episcopal Church of America.

———. 1903. A Busy Year at Point Hope. *Spirit of Missions* (November). New York: Domestic and Foreign Missionary Society of the Protestant Episcopal Church of America

———. 1906. *Mission Life at Point Hope.* Hartford, CT: Church Mission Publishing Co.

———. 1908. *Short Sketches from Oldest America.* Philadelphia: George W. Jacobs and Co.

Driggs, John B., Elijah Edson, and Augustus Hoare. 1893–1909. Point Hope Mission School Record Book. In possession of St. Thomas's Mission, Point Hope.

Edson, Elijah. 1894–96. Letters to Domestic and Foreign Missionary Society and Heinrich Koenig. Archives of the Protestant Episcopal Church of America, Austin, Texas.

———. 1895–96. Report to Domestic and Foreign Missionary Society. *First Annual Report of the Bishop of the Missionary District of Alaska for Year 1895–96.* Protestant Episcopal Church, New York. In Episcopal Church Archives, Austin, Texas.

Episcopal Church. 2006. Constitution and Canons of the Episcopal Church, 1789–2006. Available online at http://www.episcopalarchives.org/e-archives/canons/CandC_FINAL_11.29.2006.pdf.

Fienup-Riordan, Ann. 1991. *The Real People and the Children of Thunder: The Yup'ik Eskimo Encounter with Moravian Missionaries John and Edith Kilbuck.* Norman: University of Oklahoma Press.

Fitzhugh, William W. 1985. Early Contacts North of Newfoundland before A.D. 1600. In William Fitzhugh, ed., *Cultures in Contact.* Washington, DC: Smithsonian Institution Press.

Fitzhugh, William F., and Aron Crowell, eds. 1988. *Crossroads of Continents: Cultures of Siberia and Alaska.* Washington, DC: Smithsonian Institution Press.

Foote, Don Charles. 1959–61. Unpublished field notes from Point Hope and Noatak. Foote Collection, Archives, Elmer E. Rasmuson Library, University of Alaska Fairbanks.

———. 1960. The Eskimo Hunter at Point Hope, Alaska. Report for the Bio-Environmental Studies of Project Chariot, Atomic Energy Commission.

———. 1964. American Whalemen in Northwestern Arctic Alaska. *Arctic Anthropology* 2(2):16–20.

———. 1965. Exploration and Resource Utilization in Northwestern Arctic Alaska before 1855. Unpublished PhD dissertation, Department of Geography, McGill University, Montreal.

Foote, Don Charles, and H. Anthony Williamson. 1996. A Human Geographical Study. In Norman J. Wilimovsky and John N. Wolfe, eds., *Environment of the Cape Thompson Region, Alaska,* pp. 1041–1107. Oak Ridge, IL: U.S. Atomic Energy Commission.

Fortuine, Robert. 1971. The Health of the Eskimos, as Portrayed in the Earliest Written Accounts. *Bulletin of the History of Medicine* 45(2):97–114.

——. 1985. Lancets of Stone: Traditional Methods of Surgery among the Alaska Natives. *Arctic Anthropology* 22(1):23–45.

——. 1992. *Chills and Fever: Health and Disease in the Early History of Alaska.* Fairbanks: University of Alaska Press.

Friis, Herman R., ed. 1967. *The Pacific Basin: A History of Its Geographical Exploration.* New York: The American Geographical Society.

Gambell, Vene C. 1910. *The Schoolhouse Farthest West: St. Lawrence Island, Alaska.* New York: Woman's Board of Home Missions of the Presbyterian Church.

Giddings, J. Louis. 1961. *Kobuk River People.* College: University of Alaska Press.

——. 1973. *Ancient Men of the Arctic.* New York: Alfred A. Knopf.

Goodman, Frederick W. 1925–32. Diary of Rev. Fred Goodman, St. Thomas's Mission, Point Hope. Archives, Elmer E. Rasmuson Library, University of Alaska Fairbanks.

——. 1932. St Thomas Mission, Point Hope, Arctic Alaska. Typescript in Archives, Elmer E. Rasmuson Library, University of Alaska Fairbanks.

Goodykoontz, Colin B. 1971. *Home Missions on the American Frontier.* New York: Octagon Books.

Gough, Barry M. 1973. *To the Pacific and Arctic with Beechey: The Journal of Lieutenant George Peard of H.M.S. "Blossom" 1825–1828.* London: Hakluyt Society (2nd series, no. 143).

Gregg, David W. 2000. Technology, Culture Change, and the Introduction of Firearms to Northwest Alaska, 1791–1930. Unpublished PhD thesis, Brown University.

Hakluyt, Richard. 1913. *The Principal Navigations.* Glasgow: University Press

Hare, William H. 1873. First Annual Report of The Missionary Bishop of Niobrara to the Board of Missions of the Protestant Episcopal Church. Available online at http://anglicanhistory.org/indigenous/hare1873.html.

Healy, Michael A. 1887. *Report of the Cruise of the Marine Steamer* Corwin *in the Arctic Ocean in the Year 1885.* Washington, DC: Government Printing Office.

——. 1889. *Report of the Cruise of the Marine Steamer* Corwin *in the Arctic Ocean in the Year 1884.* Washington, DC: Government Printing Office.

——. 1893a. Letter to Captain Shepard, 22 January. Michael A. Healy Papers, 1881–1900. San Marino, CA: Huntington Library microfilm MS20.

——. 1893b. Letter to Captain Shepard, 5 March. Michael A. Healy Papers, 1881–1900. San Marino, CA: Huntington Library microfilm MS20.

Hinckley, Ted C. 1965. Sheldon Jackson, Presbyterian Lobbyist for the Great Land of Alaska. *Pacific Northwest Quarterly* 40.

——. 1972. *The Americanization of Alaska, 1867–1897.* Palo Alto, CA: Pacific Books.

Hoare, Augustus R. 1908. Census of Tigara, Coopertown, Jabbertown and Reindeer Camp, September 24, 1908. Mission record book viewed at St. Thomas's Mission, Point Hope.

Hooper, Calvin L. 1881. *Report of the Cruise of the U.S. Revenue Steamer* Corwin *in the Arctic Ocean: November 1, 1880.* Washington, DC: Government Printing Office.

——. 1884. *Report of the Cruise of the U.S. Revenue Steamer Thomas* Corwin *in the Arctic Ocean, 1881.* Washington, DC: Government Printing Office.

Hopkins, David M., ed. 1967. *The Bering Land Bridge.* Stanford, CA: Stanford University Press.

Howe, M.A. Dewolfe 1911. *The Life and Labours of Bishop Hare, Apostle to the Sioux.* New York: Sturgis & Walton Co.

Hunt, William R. 1975. *Arctic Passage.* New York: Charles Scribners Sons.

Irving, Washington. 1836. *Astoria, or, Anecdotes of an Enterprise beyond the Rocky Mountains.* Philadelphia: Carey, Lea, & Blanchard.

Ivrulik. 1940–41. Ivrulik's story, as told to Froelich G. Rainey. Unpublished field notes from Point Hope. Rainey Collection, Box 2, folders 8–10, Archives, Elmer E. Rasmuson Library, University of Alaska Fairbanks.

Jabbertown documents. Letters, memoranda, trading notes and other miscellaneous materials written by Jabbertown whaler-traders, including: Heinrich Koenig, Joe Tuckfield, Nuvuk Koenig, John Kelly, Peter Bayne, and anonymous Native and Euro-American hands. Holograph documents in Koenig Collection, Archives, Elmer E. Rasmuson Library, University of Alaska Fairbanks.

Jackson, Sheldon. 1880. *Alaska, and Missions on the North Pacific Coast.* New York: Dodd, Mead & Co.

——. 1890a. Article in *The Evangelist.* Microfilm held in Archives of the library of Sheldon Jackson College, Sitka.

——. 1890b. Article in *The Occident—A Presbyterian Family Newspaper* (San Francisco), June 18.

——. 1890–91. *Annual Report on Education in Alaska.* U.S. Bureau of Education, Washington, DC: Government Printing Office.

——. 1890–1900. *Sheldon Jackson Travel Diaries 1890–1900.* Microfilm held in Archives of the Alaska State Library, Juneau.

——. 1891. Article in *The Evangelist,* March 3.

——. 1891–92. *Annual Report on Education in Alaska.* U.S. Bureau of Education, Washington, DC: Government Printing Office.

——. 1894–95. *Annual Report on Education in Alaska.* U.S. Bureau of Education, Washington, DC: Government Printing Office.

——. 1896a. The Arctic Cruise of the U.S. Revenue Cutter "Bear." *National Geographic Magazine* 7(1).

——. 1896b. *Report on Introduction of Domestic Reindeer into Alaska, 1896.* Washington, DC: Government Printing Office.

——. 1900. *Ninth Annual Report on Introduction of Domestic Reindeer into Alaska, 1899.* Washington, DC: Government Printing Office.

Jarvis, D.H. 1899. Report of First Lieutenant D.H. Jarvis. In U.S. Treasury Department, *Report of the U.S. Revenue Cutter "Bear," and the Overland Expedition for the Relief of the Whalers in the Arctic Ocean, from November 27, 1897 to September 13, 1898*, pp. 28–114.

Jenkins, J.T. 1921. *A History of the Whale Fisheries.* London: H.F. and G. Witherby.

Jenkins, Thomas. 1943. *The Man of Alaska: Peter Trimble Rowe.* New York: Morehouse-Gorham Co.

Jennings, Francis. 1974. *The Founders of America: How Indians Discovered the Land.* New York: W.W. Norton.

——. 1975. *The Invasion of America: Indians, Colonialism and the Cant of Conquest.* Chapel Hill: University of North Carolina Press.

Josephy, Alvin M., ed. 1992. *America in 1492.* New York: Alfred A. Knopf.

Kaplan, Lawrence D. 1975–2005. Personal communications on Iñupiaq language and translation.

Kelly, John W. 1890. Ethnographical Memoranda Concerning the Eskimos of Arctic Alaska and Siberia. *Society of Alaskan Natural History and Ethnology, Bulletin No. 3,* Sitka.

——. 1904. Point Hope Happenings. II. Looking Forward. *Spirit of Missions* (October):726–28.

Kiple, Kenneth F. 1993. *The Cambridge World History of Human Disease.* Cambridge: Cambridge University Press.

Knapp, Edward J. 1904–05. Letters from Point Hope, Alaska, to officials of the Domestic and Foreign Missionary Society, New York. Archives of the Protestant Episcopal Church of America, Austin, Texas.

Koenig, Heinrich. 1880–1914. Trade journals, journals, letters, memoranda: holograph documents. In Koenig Collection, Archives, Elmer E. Rasmuson Library, University of Alaska Fairbanks.

Koenig, William F. 1962. Letter to Don Charles Foote, May 19. Don Charles Foote Collection, Archives, Elmer E. Rasmuson Library, University of Alaska Fairbanks.

Kotzebue, Otto von. 1821. *A Voyage of Discovery, into the South Sea and Beering Straits, for the Purpose of Exploring a north-east Passage...in 1815–18.* Trans. from the Russian by H. E. Lloyd. London: Longman, Hurst, Rees, Orme & Brown.

Kunuyaq, Peter. 1940–41. Peter's story, as told to Froelich G. Rainey. Unpublished field notes from Point Hope. Rainey Collection, Box 2, folders 8–10, Archives, Elmer E. Rasmuson Library, University of Alaska Fairbanks.

Kuwana. 1940. Kuwana's story, as told to Froelich G. Rainey. Unpublished field notes from Point Hope. Rainey Collection, Box 2, folders 8–10, Archives, Elmer E. Rasmuson Library, University of Alaska Fairbanks.

Krauss, Michael E. 2005. Eskimo Languages in Asia, 1791 on, and the Wrangel Island–Point Hope Connection. *Etudes/Inuit/Studies* 29(1–2).

Larsen, Helge, and Froelich Rainey. 1948. Ipiutak and the Arctic Whale Hunting Culture. *Anthropological Papers of the American Museum of Natural History* 42(1).

Larson, Mary. 2004. The evolution of the *qalgi* in post-epidemic Point Hope, Alaska. Unpublished PhD thesis, Brown University.

Lowenstein, Tom. 1973–89. Field notes from Point Hope, Kotzebue, Shungnak, Kobuk, Kivalina, and Barrow.

———. 1975. *Stories from Point Hope.* Translated from recordings by David Umigluk Frankson, Dinah Aviq Frankson, Patrick Attungana, and Bob Qimiuraq Tuckfield. Alaska Multimedia Education Project, Alaska State Museum, Juneau.

———. 1980. *Some Aspects of Sea-Ice Hunting in Point Hope, Alaska.* Barrow: North Slope Borough, Commission on Iñupiaq History Language & Culture.

———. 1992. *The Things That Were Said of Them: Shaman Stories from Tikiġaq, Alaska.* Translated by Tukummiq and Tom Lowenstein; told by Asatchaq. Berkeley: University of California Press.

———. 1993. *Ancient Land, Sacred Whale: The Inuit Hunt and Its Rituals.* New York: Farrar, Straus and Giroux.

MacInnes, Tom. 1932. *Klengenberg of the Arctic.* London: Jonathan Cape.

MacLean, Edna Ahgeak. nd. Holograph transcript of historical and ethnographic recitation by Asatchaq/Jimmie Killigivuk. Tape recording, ca. 1970. Fairbanks: Library of the Alaska Native Language Center, University of Alaska Fairbanks.

———. 1980. *Abridged Iñupiaq and English Dictionary.* Fairbanks: Alaska Native Language Center, University of Alaska Fairbanks, and Barrow: Iñupiaq Language Commission, North Slope Borough.

Maguire, Rochfort. 1988. *The Journal of Rochfort Maguire 1852–1854. Two Years at Point Barrow, Alaska, aboard HMS Plover in the search for Sir John Frankli*n. Ed J.R. Bockstoce, 2 vols.

Marquette, Willman M., and John R. Bockstoce. 1980. Historical Shore-based Catch of Bowhead Whales in the Bering, Chukchi and Beaufort Seas. *Marine Fisheries Review* 42(9–10).

Melville, Herman. 1851. *Moby-Dick, or The Whale.* New York: Random House, 1930.

Mikkelsen, Ejnar. 1909. *Conquering the Arctic Ice.* London: William Heinemann.

Milner, Clyde A., II, and Floyd A. O'Neil. 1985. *Churchmen and the Western Indians 1820–1920.* Norman: University of Oklahoma Press.

Mitchell, Donald Craig. 1997. *Sold American: The Story of Alaska Natives and Their Land 1867–1959.* Hanover, NH: University Press of New England.

Moore, George W. 1966. Arctic Beach Sedimentation. In Norman J. Wilimovsky and John N. Wolfe, eds., *Environment of the Cape Thompson Region, Alaska,* pp. 587–608. Oak Ridge: U.S. Atomic Energy Commission.

Morison, Samuel Eliot. 1921. *The Maritime History of Massachussets, 1783–1860.* Boston: Houghton Mifflin Co.

Morris, William Gouverneur. 1879. Report to Department of the Treasury, July 23, 1877, in 45th Cong., 3rd session, 1879, S. Exec Doc. 59:28.

Muir, John. 1917. *The Cruise of the Corwin.* Boston: Houghton Mifflin and Co.

Murdoch, John. 1892. *Ethnological Results of the Point Barrow Expedition.* Ninth Annual Report of the Bureau of Ethnology. Washington, DC: Government Printing Office.

Murray, Andrew E. 1985. *The Devil and Mr Duncan.* Victoria, BC: Sono Nis Press.

Nasugluk. 1940. Nasugluk's story, as told to Froelich G. Rainey. Unpublished field notes from Point Hope. Rainey Collection, Box 2, folders 8–10, Archives, Elmer E. Rasmuson Library, University of Alaska Fairbanks.

Nasugruk. 1940. Nasugruk's story, as told to Froelich G. Rainey. Unpublished field notes from Point Hope. Rainey Collection, Box 2, folders 8–10, Archives, Elmer E. Rasmuson Library, University of Alaska Fairbanks.

Nelson, Edward William. nd. *Journal,* typescript. Ed. William W. Fitzhugh.

——. 1899. *The Eskimo About Bering Strait.* Eighteenth Annual Report of the Bureau of Ethnology. Washington, DC: Government Printing Office.

Newlin, Angeline, Hannah Loon, and Ruth Ramoth-Sampson. nd. Maniiḷaq stories translated from recordings by Beatrice Anausuk Mouse, Charlie Aqpaliq Sheldon, Nora Paaniikaaluk Norton, Frieda Anniviaq Goodwin, Dinah Aviq Frankson, Robert Nasruk Cleveland. Published on *Alaskool: Online Materials about Alaska Native History, Education, Languages, and Cultures* (www.alaskool.org). Institute of Social and Economic Research, University of Alaska, Anchorage.

Newman, Marshall. 1976. Aboriginal New World Epidemiology and Medical Care, and the Impact of Old World Disease Imports. *American Journal of Physical Anthropology* 45:668.

Nirowana (Niġuvana). 1940. Nirowana's story, as told to Froelich G. Rainey. Unpublished field notes from Point Hope. Rainey Collection, Box 2, folders 8–10, Archives, Elmer E. Rasmuson Library, University of Alaska Fairbanks.

Nome Nugget (Alaska), July 5, 1905.

Orth, Donald J. 1971. *Dictionary of Alaska Place Names.* Geological Survey, Professional Paper 567. Washington, DC: Government Printing Office.

Oswalt, Wendell H. 1999. *Eskimos and Explorers.* Lincoln: University of Nebraska Press.

Paton, William. 1923. *Alexander Duff, Pioneer of Missionary Education.* New York: George H. Doran Company.

Peake, Frank A. 1966. *The Bishop Who Ate His Boots: A Biography of Isaac O. Stringer.* Ottawa: The Anglican Church of Canada.

Pearce, Roy Harvey. 1988. *Savagism and Civilization.* Baltimore: Johns Hopkins University Press. (Orig. pub. 1953.)

Petroff, Ivan. 1884. Report on the Population, Industries, and Resources of Alaska. In *Tenth Census of the United States.* Washington, DC: Department of the Interior, Census Office.

Porter, Roy. 1997. *The Greatest Benefit to Mankind: A Medical History of Humanity from Antiquity to the Present.* London: Harper Collins.

Price, A. Grenfell, ed. 1971. *The Explorations of Captain James Cook in the Pacific As told by Selections of his own Journals 1868–1779.* New York: Dover Publications.

Prucha, Francis Paul. 1976. *American Indian Policy in Crisis: Christian Reformers and the Indian, 1865–1900.* Norman: University of Oklahoma Press.

——. 1984. *The Great Father: The United States Government and the American Indians* (abridged edition). Lincoln: University of Nebraska Press.

Puiguitkaat. 1981. The 1978 Elder's Conference. North Slope Borough Commission on History and Culture, Barrow.

Rainey, Froelich G. 1940–41. Unpublished field notes from Point Hope. Rainey Collection, Box 2, folders 8–10, Archives, Elmer E. Rasmuson Library, University of Alaska Fairbanks.

———. 1947. The Whale Hunters of Tigara. *Anthropological Papers of the American Museum of Natural History* 41(2).

———. 1992. *Reflections of a Digger*. Philadelphia: University Museum of Archaeology and Anthropology, University of Pennsylvania.

Rasmussen, Knud Johan Victor. 1927. *Across Arctic America: Narrative of the Fifth Thule Expedition.* London: G.P. Putnam & Sons.

———. 1952. *The Alaskan Eskimos. Report of the Fifth Thule Expedition, 1921–24.* Edited by H. Ostermann. Copenhagen: Gylendalske Boghandel, Nordisk Forlag.

Ray, Arthur J. 1974. *Indians in the Fur Trade 1660–1870.* Toronto: University of Toronto Press.

Ray, Dorothy Jean. 1967. *Eskimo Masks, Art, and Ceremony.* Seattle: University of Washington Press.

———. 1975. *The Eskimos of the Bering Strait, 1650–1898.* Seattle: University of Washington Press.

———. 1977. *Eskimo Art: Tradition and Innovation in North Alaska.* Seattle: University of Washington Press.

Ray, Patrick Henry. 1885. *Report of the International Polar Expedition to Point Barrow, Alaska.* Washington, DC: Government Printing Office.

Rock, Sam. 1940. Sam's story, as told to Froelich G. Rainey. Unpublished field notes from Point Hope. Rainey Collection, Box 2, folders 8–10, Archives, Elmer E. Rasmuson Library, University of Alaska Fairbanks.

Rosse, Irving C. 1883. Notes and Memoranda: Medical and Anthropological. In *Cruise of the U.S. Revenue Steamer* Corwin *in Alaska and the N.W. Arctic Ocean in 1881.* Washington, DC: Government Printing Office.

Rothstein, William G. 1992. *American Physicians in the Nineteenth Century.* Baltimore: Johns Hopkins University Press.

Rowe, Peter Trimble. 1899. *Annual Report of the Bishop of Alaska.* Protestant Episcopal Church, New York. In Episcopal Church Archives, Austin, Texas.

———. 1900. *Annual Report of the Bishop of Alaska.* Protestant Episcopal Church, New York. In Episcopal Church Archives, Austin, Texas.

———. 1903. *Annual Report of the Bishop of Alaska.* Protestant Episcopal Church, New York. In Episcopal Church Archives, Austin, Texas.

———. 1906–1909. Letters to officials of the Domestic and Foreign Missionary Society, New York. Archives of the Protestant Episcopal Church of America, Austin, Texas.

Sahlins, Marshall. 1985. *Islands of History*. Chicago: University of Chicago Press.

Sauer, Martin. 1802. *An Account of a Geographical and Astronomical Expedition to the Northern Parts of Russia, 1785–94.* London: Cadell.

Seattle Post-Intelligencer (Washington), September 24, 1905.

Sheldon, Kim E. W., and David J. Rugh. 1995. The Bowhead Whale: Balena mysticetus, Its Current and Historical Status. *Marine Fisheries Review* 57(3–4):1–20.

Shepard, Isabel S. 1889. *The Cruise of the U.S. Steamer* "Rush" *in Behring Sea Summer of 1889.* San Francisco: The Bancroft Company.

Simpson, Dr. John. 1988. Observations on Western Esquimaux and the Country they Inhabit. In Bockstoce ed. 1988, p. 501.

Spencer, Robert F. 1959. The North Alaskan Eskimo: A Study in Ecology and Society. *Bureau of American Ethnology, Bulletin 171.* Washington, DC: Government Printing Office.

Spirit of Missions. 1891–1910. Protestant Episcopal Church.

Stefansson, Vilhjalmur. 1909. The Eskimo Trade Jargon of Herschel Island. *American Anthropologist* n.s., 2.

——. 1912. *My Life with the Eskimo*. New York: Harper and Brothers.

——. 1913. Christianizing the Eskimo. *Harpers Magazine* 127.

Steller, Georg Wilhelm. 1988. *Journal of a Voyage with Bering 1741–1742*. Stanford: Stanford University Press.

Stevenson, Charles H. nd. Whalebone: Its Production and Utilization. Washington, DC: U.S. Bureau of Fisheries Doc. No. 626.

Stevenson, Robert Louis. 1987. *The Beach of Falesa* (orig. pub. 1893). In Jenni Calder, ed., *Island Landfalls*. Edinburgh: Canongate.

Stewart, Robert Laird. 1908. *Sheldon Jackson, Pathfinder and Prospector of the Missionary Vanguard in the Rocky Mountains and Alaska*. New York: Fleming H. Revell Co.

Stockton, Charles, H. 1890. The Arctic Cruise of the U.S.S. *Thetis* in the Summer and Autumn of 1889. *National Geographic Magazine* 2.

Stone, Thomas. 1981. Whalers and Missionaries at Herschel Island. *Ethnohistory* 28:2.

Stuck, Hudson. 1920a. *A Winter Circuit of Our Arctic Coast*. New York: Charles Scribner and Sons.

——. 1920b. *The Alaskan Missions of the Episcopal Church*. New York: Domestic and Foreign Missionary Society.

Thornton, Harrison R. 1931. *Among the Eskimos of Wales, Alaska*. Baltimore: Johns Hopkins University Press.

Thornton, Russell. 1987. *American Indian Holocaust and Survival: A Population History since 1492*. Norman: University of Oklahoma Press.

Trigger, Bruce G. 1996. *The Cambridge History of the Native Peoples of the Americas*. Vol. 1. *North America*. Cambridge: Cambridge University Press.

Turner, Frederick Jackson. 1948. *The Frontier in American History*. New York: Henry Holt & Co.

Utley, Robert M. 1963. *The Last Days of the Sioux Nation*. New Haven: Yale University Press.

Vancouver, George. 1984. *Voyage of Discovery to the North Pacific and round the World, 1791–95*. Vol. 2. Edited by W. Kaye Lamb. London: Hakluyt Society.

Vanstone, James W. 1958. Commercial Whaling in the Arctic Ocean. *Pacific Northwest Quarterly* 49(1):1–10.

——. 1962. *Point Hope: An Eskimo Village in Transition*. Seattle: University of Washington Press.

——. 1964. Some Aspects of Religious Change among Native Inhabitants in West Alaska and the Northwest Territories. *Arctic Anthropology* 2(2):21–24.

——. 1977. A.F. Kaseverov's Coastal Explorations in Northwest Alaska, 1838. *Fieldiana Anthropology* 69.

Washburne, Wilcomb E. 1988. *Handbook of the North American Indians*. Vol. 4. *History of Indian-White Relations*.Washington, DC: Smithsonian Institution.

Whipple, Henry Benjamin. 1868. Report on the Moral and Temporal Condition of the Indian Tribes on our Western Border. Letter to the Board of Missions, Protestant Episcopal Church, New York. In Episcopal Church Archives, Austin, Texas.

——. 1899. *Lights and Shadows of a Long Episcopate*. New York: The Macmillan Company.

White, James Taylor. 1889. Diary of Cruise of the Revenue Steamer *Bear* in 1889. Typed unpaginated transcript. Seattle: University of Washington Archives.

White, Richard. 1994. Animals and Enterprise. In Clyde A. Milner II et al., eds., *The Oxford History of the American West*. New York: Oxford University Press.

Wilimovsky, Norman J., and John N. Wolfe, eds. 1966. *Environment of the Cape Thompson Region, Alaska.* Oak Ridge: U.S. Atomic Energy Commission.

Wolfe, Robert. 1982. *Alaska's Great Sickness.* Philadelphia: American Philosophical Society Publications.

Woolfe, Henry D. 1893. The Seventh District. In *Report on Population and Resources of Alaska at the Eleventh Census: 1890,* House of Representatives, 52nd Congress, 1st Session, Resources Doc. No. 340, part 7. Washington, DC: Government Printing Office.

Wrangell, Ferdinand P. 1980. *Russian America: Statistical and Ethnographic Information.* Kingston, ON: Limestone Press.

Wyatt, Victoria. 1994. Alaska and Hawai'i. In Clyde A. Milner II et al., eds., *The Oxford History of the American West.* New York: Oxford University Press.

References to possible origins of the *uivvaqsaat* movement

Du Bois, Cora. 1939. The 1870 Ghost Dance. *Anthropological Records* 3(1), University of California, Berkeley.

Hittman, Michael. 1973. The 1870 Ghost Dance at the Walker River Reservation: A Reconstruction. *Ethnohistory* 20(3):264ff.

Kehoe, Alice. 1989. *The Ghost Dance, Ethnohistory and Revitalization.* New York: Holt, Rinehart and Winston.

Mooney, James. 1892–93. *The Ghost-Dance Religion and the Sioux Outbreak of 1890,* pp. 663, 692, 703, 708, 716. 14th Annual Report of the Bureau of American Ethnology, Washington, DC.

Powers, Stephen. 1877. *Tribes of California, Contributions to North American Ethnology: 3. U.S. Geographical and Geological Survey of the Rocky Mountain Region.* Washington, DC, pp. 42–43.

Ray, Verne. 1937. The Bluejay Character in the Plateau Spirit Dance. *American Anthropologist* n.s. 39(4):593–601.

Spier, Leslie. 1935. "The Prophet Dance of the Northwest and Its Derivatives": The Source of the Ghost Dance. *General Series in Anthropology,* No. 1. University of Wisconsin, Menasha.

Underwood, Grant. 1993. *The Millenarian World of Early Mormonism.* Urbana: University of Illinois.

Walker, Deward E. 1969. New Light on the Prophet Dance Controversy. *Ethnohistory* 16:245–53.

Wallace, Anthony F. C. 1949. *King of the Delawares.* Philadelphia: University of Pennsylvania Press.

INDEX

H

I

J

K

L

M

N

O

P

Q

R

S